AF352811

# Prospects Challenges and Sustainability of the Agri-Food Supply Chain in the New Global Economy II

# Prospects Challenges and Sustainability of the Agri-Food Supply Chain in the New Global Economy II

Editor

**Dimitris Skalkos**

Basel • Beijing • Wuhan • Barcelona • Belgrade • Novi Sad • Cluj • Manchester

*Editor*
Dimitris Skalkos
University of Ioannina
Ioannina, Greece

*Editorial Office*
MDPI
St. Alban-Anlage 66
4052 Basel, Switzerland

This is a reprint of articles from the Special Issue published online in the open access journal *Sustainability* (ISSN 2071-1050) (available at: https://www.mdpi.com/journal/sustainability/special_issues/3SY43ENQBQ).

For citation purposes, cite each article independently as indicated on the article page online and as indicated below:

Lastname, A.A.; Lastname, B.B. Article Title. *Journal Name* **Year**, *Volume Number*, Page Range.

**ISBN 978-3-0365-8995-4 (Hbk)**
**ISBN 978-3-0365-8994-7 (PDF)**
**doi.org/10.3390/books978-3-0365-8994-7**

# Contents

# About the Editor

**Dimitris Skalkos**

Dr. Dimitris Skalkos is currently an associate professor in Food Business Management of Innovation at the Laboratory of Food Chemistry, Department of Chemistry, University of Ioannina, Greece. He has been an assistant and associate professor at the University of the Aegean, Greece (Department of Food Science and Nutrition) and a visiting assistant professor at the University of Toledo (Department of Chemistry), Ohio, USA. Dr. Skalkos' career has expanded beyond academia, covering a period of more than 30 years, since he has worked as: (1) the founder and first director of the Business Innovation Center (BIC) of Epirus—Greece, (2) the owner of Synthesis Consulting Ltd, and (3) the co-owner of Paskal Herbal extracts S.A., and he has worked as a business consultant for more than 20 years with 50 national and multi-national companies of various sectors of activity, mainly in the food sector. He has also worked as a project manager or scientific director on more than 20 EU funding programs involved in the promotion of innovation, innovative pilot actions, the establishment of academic spin-off companies, food product development, and food business management. His research interests currently focus on the development of all innovative aspects within agri-food supply chain businesses, including production, processing, organization, marketing, consumers' motives, and knowledge transfer. He has published 60 research papers and more than 20 conference proceedings and has authored about 40 oral communications at international scientific conferences.

# Preface

A year ago, when we announced the first edition of this Special Issue, the world was changing rapidly due to the effects of the coronavirus pandemic. The first edition was entitled "Innovative Agrifood Supply Chain in the Post-COVID-19 Era". This Special Issue was focused on innovative scientific insights and technological advances in natural resources, organic pollutant identification, new food product development, traceability, and packaging, chain management, consumer attitudes, and eating motivations, with the aim of tackling the foreseen changes to the global economy and society. A year later, however, with the completion of this second Special Issue on the agri-food supply chain (AFSC), the world is still changing in an unpredictable and unprecedented way, with unforeseen consequences. The (AFSC) is already at the center of these changes and worldwide studies due to such global economic change.

The AFSC will have to change drastically to adjust and to cope with the new conditions. The process of "from farm to fork" will be a key factor in sustainability and the progress of the food produced at the end of the process for consumers worldwide. Innovation will also play a vital role in modernizing the AFSC. In addition, scientific developments in areas such as artificial intelligence (AI), the circular economy (CE), harvest and production planning for food crops, blockchain technology (BKCT), Industry 4.0 (I 4.0), and eco-design concepts will also be critical for AFSC growth and development in the new era.

In this Special Issue, selected papers on the prospects, challenges, and sustainability of the AFSC in the new global economy which are currently emerging are presented. The driving force of the chain is no doubt the end users of the food, namely the consumers. The topics cover a range of areas, from food choice motives to consumers' cheese preferences, from insect-based food to alternative protein food's acceptance by farmers and the public. Contract farming analysis, beekeeping training, and the safety of plant-based foods are three more topics covered in this Special Issue. Finally, new sustainable functional foods, bioactive skin grapes, renewable products, and essential oil compositions are the subjects included in the topic of new foods within this Special Issue.

**Dimitris Skalkos**
*Editor*

 *sustainability*



*Editorial*

# Prospects, Challenges and Sustainability of the Agri-Food Supply Chain in the New Global Economy II

Dimitris Skalkos

Laboratory of Food Chemistry, Department of Chemistry, University of Ioannina, 45110 Ioannina, Greece; dskalkos@uoi.gr; Tel.: +30-2651008345

In the new global era, the process "from farm to fork" as a holistic approach to the production and consumption of food will become a key factor for the sustainability and progress of the food industry. The subject of the agri-food supply chain (AFSC) is becoming more and more important not only from a scientific but also from a business point of view since it provides the means for a regular food supply worldwide. We initiated the series of special editions of AFSC a year ago with the first edition entitled *"Innovative Agrifood Supply Chain in the Post-COVID 19 Era"* [1]. This Special Issue [2] is focused on 11 selected topics from different parts of the A.S.C. in view of the post-COVID-19 era, expanding from innovative scientific insights and technological advances of natural resources, the identification of organic pollutants, new food product development, traceability, and packaging, chain management, to consumer's attitudes and eating motivations, aiming to tackle the foreseen changes of global economy and society.

The topic of AFSC is extremely interesting; therefore, major reviews publications were published both during and after the pandemic within the years 2001–2023 presented here:

*Artificial intelligence (AI)* will play a key role in the future of AFCS; therefore, scientific developments on the subject are critical challenges. Monteiro and Barata evaluated 18 papers highlighting mature areas for AI adoption, identifying opportunities for future research in the extended AFSC [3] The bibliometric analysis revealed that the AI in traditional stages of production need to be expanded using intelligent planning for demand uncertainty and personalized needs of end-customers, storage optimization, waste reduction in the post-production phase, and boundary-spanning analytics. For practice, the findings of the AI inspired startups dealing with AFSC ecosystems and incumbents in their projects for the intelligent and sustainable digital transformation of agri-food, with AI techniques contributing to closing the loop of sustainable agri-food supply chains.

*Circular economy (CE)* is a topic with potential solutions for social, economic, and environmental challenges, but with limited engagements yet to explore its initiatives in the AFSC. Mehmood et al. addressed the gap by critically reviewing the existing literature and identifying the drivers and barriers for implanting the CE in the AFSC [4]. They found that environmental (67%), policy and economic (47%), and financial benefits (43%) are the top three drivers. However, institutional (64%), financial (48%), and technological risks (40%) are the top three barriers in implementing CE practices in the AFSC. Indeed, there is the utmost need for international communities to introduce internationally accepted standards and framework for CE practices to be used globally to eliminate waste, particularly in the agriculture sector, and for government intervention to stimulate CE initiatives playing a critical role in the transition process.

*Harvest and production planning for food crops* is another key factor for AFSC sustainability and prospect. Tugce and Bilgen reviewed the optimization models used extensively in providing insights to decision makers on related issues [5]. Based on the reviews evaluated, a new classification scheme has been developed and analyzed via three sections: the problem scope, model characteristics, and modeling approach. These clearly show the gaps in the literature and determine research opportunities and future directions. The main conclusion of this review is the need for more studies on integrated decisions in AFSC and

**Citation:** Skalkos, D. Prospects, Challenges and Sustainability of the Agri-Food Supply Chain in the New Global Economy II. *Sustainability* **2023**, *15*, 12558. https://doi.org/10.3390/su151612558

Received: 14 August 2023
Accepted: 16 August 2023
Published: 18 August 2023

the need for a closer relationship between academia and stakeholders in order to generate more applied research.

*Blockchain technology (BKCT)* is a major parameter generating prospects in the future for AFSC. Srivastava and Dashora explored and analyzed the recent applications published on the subject matter, with a systematic in-depth literature analysis of papers from 2016 to 2021 [6]. The findings highlight the issue of food safety, traceability, transparency, eliminating intermediaries, and integrating the Internet of things with BKCT as prominent applications in the agrifood sector. The challenges of BKCT as identified in the review study are scalability, privacy, security, lack of regulations, and a lack of skills and training.

*Industry 4.0 (I 4.0)* is a paradigm adopted increasingly often by companies belonging to different industries, including the AFSC industry, thus providing challenges and prospects for the future. Bigliardi et al. explored the various applications of 4.0 technologies in the agri-food sector, reviewing the recent publications between 2018 and 2022, with the aim of understanding what are the new trends and changes in the sector [7]. The analysis led to the identification of three marco-areas, namely: (a) agribusiness technology transition, (b) supply chain management 4.0, and (c) sustainability and other trends. The incorporation of I 4.0 elements can help tackle many challenges facing the AFCS industry. These can help increase productivity and offer consumers more customized products. Concerning the challenge of sustainability, a deep focus on digital skills can favor the achievement of sustainable development goals, among which is the urgency to solve the problem of world hunger.

*Ecodesign concepts for sustainable food product development* across the supply chain reducing the environmental impact of AFCS products are reviewed by Silva et al. [8]. Based on their evaluation of the existing literature, they suggest that the relevant ecodesign principles fall into three main categories depending on the supply chain stage: "design for sustainable sourcing (DFSS)", "design for optimized resource use (DFORU)", and "design for end-of-life optimization (DFEO)". Applying this framework across the supply chain could significantly reduce the environmental impact of food production and indirectly contribute to dietary change.

A number of papers on various subjects of AFSC have already been published in 2023. Imran et al. investigated the deployment of specific knowledge management practices in the AFSC and found that firms' knowledge management practices work sequentially (knowledge acquisition, assimilation, and application) and develop a risk management culture in order to achieve supply chain resilience and minimize supply chain risks [9]. Zhao et al. showed that power and national culture are critical knowledge mobilization factors with the greatest ability to elicit other factors for focal companies of AFSCs and government [10]. Sharma et al. revealed that performance expectancy, effort expectancy, social influence facilitating conditions, interfirm trust and transparency are the drivers of blockchain adoption and have a significant impact on the behavioral intention of stakeholders of the AFSC companies [11]. Yontar investigated critical success factor analysis of blockchain technology in AFSC management and found that the "ability to prevent food waste"; "increased food security"; and "product life-cycle tracking" are factors that take priority in their ranking among the 12 factors studied [12]. Stevens and Teal developed revenue-based measures of firms' vertical (across the supply chain segment) and horizontal (within the supply chain segment) diversification, and found that diversification increases firms' resilience within the AFSC [13]. Fornes et al. studied the management of quality, supplier selection, and cold-storage contracts in AFSC and found that based on different scenarios, the value of the stochastic solutions shows that modeling and solving the proposed stochastic model minimizes costs by an average of around 6.4%, and the expected value of perfect information demonstrates that using a proactive strategy could cost up to 9% [14]. Pardaev et al. assessed the impact of risk on economic integration between entities in AFSC and found that applying "written contract" and "insurance" to collaborative relationships to reduce risk levels has been shown to reduce risks to coefficients of 0.6 [15]. Esteso et al. proposed a tool based on a system dynamics model to determine the robustness of an already designed five-stage fresh AFSC and its planting planning to disruption in demand, supply, transport, and the operability of its nodes [16].

In this second special edition, selected subjects on the prospects, challenges, and sustainability of the AFCS in the new global economy which are emerging are presented. The driving force of the chain is no doubt the end users of the food namely the consumers and their preferences, characteristics, etc. Five papers cover topics relevant to this subject:

Skalkos and Kalyva reviewed recent findings on food choice motives by consumers based on 10 main key food motives, namely, health, convenience, sensory appeal, nutritional quality, moral concerns, weight control, mood and anxiety, familiarity, price, and shopping frequency behaviour. These motives continue to be significant in the post-pandemic era, and their findings indicate that it is too premature to give definite answers as to what food choice motives in the post-COVID-19 era will be like.

Guine et al. investigated the level of knowledge about edible insects (EIs) in a sample of people in thirteen countries. The questionnaire survey concluded that the level of knowledge about EIs is highly variable according to the individual characteristics, namely that the social and cultural influences of the different countries lead to distinct levels of knowledge and interpretation of information, thus producing divergent approaches to the consumption of insects.

Crawshaw and Piazza explored the views of the livestock farmers' attitudes, compared with no farmers population, regarding emerging protein alternatives in UK using four products (plant-based burgers; plant-based milk alternatives; cultured beef; animal-free dairy milk). Overall farmers rated the four products less appealing and less beneficial to the industry compared to non-farmers. Both groups tended to agree that the alternatives offered advantages, particularly for the environment, resource use, food security, and animal treatment, though agreement rates were lower for farmers. Farmers tended to perceive more barriers to acceptance than non-farmers, with 'threat to farmers' and 'lack of support to local farmers' being of paramount concern to both groups.

Ranga et al. explored the acceptance amongst consumers and farmers in Ireland of insect-based feed (IBF). The research proved showed that information on the benefits of using IBF increased its acceptance, which means that IBF acceptance might depend on dedicated educational interventions which include addressing the safety aspect of the feed even among those with higher level of education.

Skalkos et al. explored consumers' perception of semi-hard and hard cheeses in Greece in the new global era. Using a self-response questionnaire survey through Google, they found that there is no significant change in consumers' motives today for these types of cheese except for a significant decline in consumption, reaching up to 8.4%, and concluded that in order to maintain sustainability and growth, one should stick to the good practices of production, promotion, and sales developed before the pandemic, exploring. However, new avenues and practices to increase consumption have been explored, which are currently declining.

New, innovative product development is also a key challenging factor for both future prospects and sustainability of the AFSC. Two papers are presented on this subject:

Slabu et al. synthesized renewable products with potentially interesting properties and application by functionalizing linseed oil via epoxidation and epoxy ring opening with carboxylic acids and anhydrides. LDHs (Layered Double Hydroxides), a well-known class of materials, were used for a wide range of reactions; these are the catalysts used in this study, with the overall advantages of facile separation and reusability.

Matran et al. produced a sustainable food product for the special purpose of nutrivigilance, as an adjuvant in the repair of the gastric mucosa. Through the development of forestry for the cultivation of white or black mulberry (Morus alba and Morus nigra), the raising of silkworms (Bombyx mori), the processing of fibroin to obtain natural silk, and the processing of sericin as a residue in the textile industry, the new food product was developed in order to actively contribute to the global economy.

Two papers from this Special Issue explore selected issues related to the chain process and relationship between the key players:

Hsieh and Luh explored contract farming for the agriculture sector dominated by smallholder farms partnered with modern distributors for higher returns in Taiwan. The findings suggest that the marginal treatment effects are generally in an increasing trend as the quantile increases, implying that the economic effects of contract farming or partnership with modern distributors are more pronounced for higher returns among rice farmers.

Guine et al. investigated the gaps in the updated knowledge of beekeepers and how these can be filled through lifelong learning via a survey conducted in seven European countries. This work revealed valuable information that should be used to design professional training actions to help the professionals in the beekeeping sector enhance their competencies and be better prepared to manage their activities successfully.

Three more papers have been presented, each considering various subjects:

Piglowski and Niewczas–Dobrowolska examined rapid alert systems for food and feed (RASFF) notifications for products of plant origin with respect to hazard, year, product, notifying country, origin country, notification type, notification basis, distribution status, and actions taken in 1998–2020 in selected countries. The study proved that to ensure the safety of food of plant origin, it is necessary to adhere to good agricultural and manufacturing practices, involve producers in the control of farmers, ensure proper transport conditions (especially from Asian countries), ensure that legislative bodies set and update hazard limits, and ensure their subsequent control by the authorities of EU countries.

Michalaki et al. examined the bioactivity of grape skin from extracts of small-berried muscat and Augustiatis from the island of Samos, Greece. The total phenolic content, antiradical activity, the inhibition of plasma oxidation and platelet aggregation, and the phenolic profile were examined. The specialized bioactivities found in both wine grape skin extracts from Samos were significant, giving them added value for further use as bioactive food ingredients in other food products.

Tsoumani et al. investigated the potential interconnection between the place of cultivation of Greek oregano samples and the composition and properties of their essential oils (EOs), identifying characteristic chemical features that could differentiate between geographical origins with the use of chemometric tools. The application of the cross-validation method resulted in high correct classification rates in both geographical groups studied (93.3% and 82.7%, respectively), attesting to a strong correlation between location and oregano EO composition.

**List of Contributions**

1. Skalkos, D.; Kalyva, Z.C. Exploring the Impact of COVID-19 Pandemic on Food Choice Motives: A Systematic Review. *Sustainability* **2023**, *15*, 1606. https://doi.org/10.3390/su15021606.
2. Guiné, R.P.F.; Florença, S.G.; Costa, C.A.; Correia, P.M.R.; Ferreira, M.; Cardoso, A.P.; Campos, S.; Anjos, O.; Chuck-Hernández, C.; Sarić, M.M.; et al. Investigation of the Level of Knowledge in Different Countries about Edible Insects: Cluster Segmentation. *Sustainability* **2022**, *15*, 450. https://doi.org/10.3390/su15010450.
3. Crawshaw, C.; Piazza, J. Livestock Farmers' Attitudes towards Alternative Proteins. *Sustainability* **2023**, *15*, 9253. https://doi.org/10.3390/su15129253.
4. Ranga, L.; Noci, F.; Vale, A.P.; Dermiki, M. Insect-Based Feed Acceptance amongst Consumers and Farmers in Ireland: A Pilot Study. *Sustainability* **2023**, *15*, 11006. https://doi.org/10.3390/su151411006.
5. Skalkos, D.; Bamicha, K.; Kosma, I.S.; Samara, E. Greek Semi-Hard and Hard Cheese Consumers' Perception in the New Global Era. *Sustainability* **2023**, *15*, 5825, https://doi:10.3390/su15075825.
6. Slabu, A.I.; Banu, I.; Pavel, O.D.; Teodorescu, F.; Stan, R. Sustainable Ring-Opening Reactions of Epoxidized Linseed Oil in Heterogeneous Catalysis. *Sustainability* **2023**, *15*, 4197. https://doi.org/10.3390/su15054197.

7.  Matran, I.M.; Tarcea, M.; Rus, D.C.; Voda, R.; Muntean, D.-L.; Cirnatu, D. Research and Development of a New Sustainable Functional Food under the Scope of Nutrivigilance. *Sustainability* **2023**, *15*, 7634. https://doi.org/10.3390/su15097634.
8.  Hsieh, M.-F.; Luh, Y.-H. Is Contract Farming with Modern Distributors Partnership for Higher Returns? Analysis of Rice Farm Households in Taiwan. *Sustainability* **2022**, *14*, 15188. https://doi.org/10.3390/su142215188.
9.  Guiné, R.P.F.; Oliveira, J.; Coelho, C.; Costa, D.T.; Correia, P.; Correia, H.E.; Dahle, B.; Oddie, M.; Raimets, R.; Karise, R.; et al. Professional Training in Beekeeping: A Cross-Country Survey to Identify Learning Opportunities. *Sustainability* **2023**, *15*, 8953. https://doi.org/10.3390/su15118953.
10. Pigłowski, M.; Niewczas-Dobrowolska, M. Hazards in Products of Plant Origin Reported in the Rapid Alert System for Food and Feed (RASFF) from 1998 to 2020. *Sustainability* **2023**, *15*, 8091. https://doi.org/10.3390/su15108091.
11. Michalaki, A.; Iliopoulou, E.N.; Douvika, A.; Nasopoulou, C.; Skalkos, D.; Karantonis, H.C. Bioactivity of Grape Skin from Small-Berry Muscat and Augustiatis of Samos: A Circular Economy Perspective for Sustainability. *Sustainability* **2022**, *14*, 14576. https://doi.org/10.3390/su142114576.
12. Tsoumani, E.S.; Kosma, I.S.; Badeka, A.V. Chemometric Screening of Oregano Essential Oil Composition and Properties for the Identification of Specific Markers for Geographical Differentiation of Cultivated Greek Oregano. *Sustainability* **2022**, *14*, 14762. https://doi.org/10.3390/su142214762.

**Funding:** This research received no external funding.

**Institutional Review Board Statement:** Not applicable.

**Informed Consent Statement:** Not applicable.

**Data Availability Statement:** Not applicable.

**Acknowledgments:** As Guest Editor of the Special Issue **"Prospects, Challenges and Sustainability of the Agri-Food Supply Chain in the New Global Economy II"**, I would like to express my deep appreciation to all authors whose valuable work was published under this issue and thus contributed to the success of the edition.

**Conflicts of Interest:** The authors declare no conflict of interest.

# References

1.  Skalkos, D. Innovative Agrifood Supply Chain in the Post-COVID 19 Era. *Sustainability* **2022**, *14*, 5359. [CrossRef]
2.  Skalkos, D. (Ed.) *Innovative Agrifood Supply Chain in the Post-COVID 19 Era*; MDPI: Basel, Switzerland, 2022; ISBN 978-3-0365-4187-7.
3.  Monteiro, J.; Barata, J. Artificial Intelligence in Extended Agri-Food Supply Chain: A Short Review Based on Bibliometric Analysis. *Procedia Comput. Sci.* **2021**, *192*, 3020–3029. [CrossRef]
4.  Mehmood, A.; Ahmed, S.; Viza, E.; Bogush, A.; Ayyub, R.M. Drivers and Barriers towards Circular Economy in agri-food Supply Chain: A Review. *Bus. Strategy Dev.* **2021**, *4*, 465–481. [CrossRef]
5.  Taşkıner, T.; Bilgen, B. Optimization Models for Harvest and Production Planning in Agri-Food Supply Chain: A Systematic Review. *Logistics* **2021**, *5*, 52. [CrossRef]
6.  Srivastava, A.; Dashora, K. Application of Blockchain Technology for Agrifood Supply Chain Management: A Systematic Literature Review on Benefits and Challenges. *Benchmarking Int. J.* **2022**, *29*, 3426–3442. [CrossRef]
7.  Bigliardi, B.; Bottani, E.; Casella, G.; Filippelli, S.; Petroni, A.; Pini, B.; Gianatti, E. Industry 4.0 in the Agrifood Supply Chain: A Review. *Procedia Comput. Sci.* **2023**, *217*, 1755–1764. [CrossRef]
8.  Silva, B.Q.; Vasconcelos, M.W.; Smetana, S. Conceptualisation of an Ecodesign Framework for Sustainable Food Product Development across the Supply Chain. *Environments* **2023**, *10*, 59. [CrossRef]
9.  Ali, I.; Golgeci, I.; Arslan, A. Achieving Resilience through Knowledge Management Practices and Risk Management Culture in Agri-Food Supply Chains. *Supply Chain Manag. Int. J.* **2023**, *28*, 284–299. [CrossRef]
10. Zhao, G.; Chen, H.; Liu, S.; Dennehy, D.; Jones, P.; Lopez, C. Analysis of Factors Affecting Cross-Boundary Knowledge Mobilization in Agri-Food Supply Chains: An Integrated Approach. *J. Bus. Res.* **2023**, *164*, 114006. [CrossRef]
11. Sharma, A.; Sharma, A.; Singh, R.K.; Bhatia, T. Blockchain Adoption in Agri-Food Supply Chain Management: An Empirical Study of the Main Drivers Using Extended UTAUT. *Bus. Process Manag. J.* **2023**, *29*, 737–756. [CrossRef]

12. Yontar, E. Critical Success Factor Analysis of Blockchain Technology in Agri-Food Supply Chain Management: A Circular Economy Perspective. *J. Environ. Manag.* **2023**, *330*, 117173. [CrossRef] [PubMed]
13. Stevens, A.W.; Teal, J. Diversification and Resilience of Firms in the Agrifood Supply Chain. *Am. J. Agric. Econ.* **2023**. [CrossRef]
14. Mateo-Fornés, J.; Soto-Silva, W.; González-Araya, M.C.; Plà-Aragonès, L.M.; Solsona-Tehas, F. Managing Quality, Supplier Selection, and Cold-storage Contracts in Agrifood Supply Chain through Stochastic Optimization. *Int. Trans. Oper. Res.* **2023**, *30*, 1901–1930. [CrossRef]
15. Pardaev, K.; Hasanov, S.; Muratov, S.; Saydullaeva, F. Mitigating Impact of Risks on Economic Integration between Entities in Agrifood Supply Chain. *E3S Web Conf.* **2023**, *365*, 04002. [CrossRef]
16. Esteso, A.; Alemany, M.M.E.; Ottati, F.; Ortiz, Á. System Dynamics Model for Improving the Robustness of a Fresh Agri-Food Supply Chain to Disruptions. *Oper. Res.* **2023**, *23*, 28. [CrossRef]

*sustainability*

*Article*

# Insect-Based Feed Acceptance amongst Consumers and Farmers in Ireland: A Pilot Study

Leocardia Ranga [1], Francesco Noci [2], Ana P. Vale [3] and Maria Dermiki [1,*]

[1] Department of Health and Nutritional Sciences, Atlantic Technological University, F91 YW50 Sligo, Ireland; leocardia.ranga@research.atu.ie

[2] Department of Sports Exercise and Nutrition, Atlantic Technological University, H91 T8NW Galway, Ireland; francesco.noci@atu.ie

[3] UCD Veterinary Sciences Centre, University College Dublin, Belfield, Dublin 4, Ireland; ana.vale@ucd.ie

* Correspondence: maria.dermiki@atu.ie

**Abstract:** The potential of insect-based feed (IBF) as a sustainable alternative to conventional animal feed is widely reported, yet there is extremely limited information on its acceptance in Ireland, a country with a strong farming background. Therefore, this study aims to provide baseline data on factors affecting acceptance of IBF amongst a segment of consumers and farmers in Ireland. Quantitative and qualitative data were collected amongst 233 consumers, 73 of which were farmers. Non-parametric statistical tests revealed that the willingness to consume foods from animals fed with IBF depends on the type of food and is affected by a combination of consumer- and product-related factors. Consumers' age, gender, diet, and education level, the foods' packaging information, safety, and price, and whether insects are part of an animal's natural diet or environmentally friendly had a significant effect. Safety concern regarding use of IBF was the main factor affecting farmers' willingness to use it. Qualitative findings revealed concerns emanating from the bovine spongiform encephalopathy outbreak and a general need for more information. Accordingly, information on the benefits of using IBF increased its acceptance. Thus, IBF acceptance might depend on dedicated educational interventions which include addressing the safety aspect of the feed even among those with higher level of education.

**Keywords:** insects; animal feed; insect meal; sustainability; consumer acceptance

**Citation:** Ranga, L.; Noci, F.; Vale, A.P.; Dermiki, M. Insect-Based Feed Acceptance amongst Consumers and Farmers in Ireland: A Pilot Study. *Sustainability* **2023**, *15*, 11006. https://doi.org/10.3390/su151411006

Academic Editor: Dimitris Skalkos

Received: 17 June 2023
Revised: 5 July 2023
Accepted: 12 July 2023
Published: 13 July 2023

## 1. Introduction

As the production of livestock and aquaculture (with the exception of algae) continue to increase worldwide to meet growing consumer demands, so is the use of animal feed ingredients [1,2]. The production of animal feed, however, is currently exploiting approximately a third of global arable land, adding pressure to land and water resources, which, according to the Food and Agriculture Organization of the United Nations (FAO), are now at a "breaking point" [3]. Moreover, life cycle assessments (LCAs) conducted on several fish [4–6] and livestock [7–9] farming systems identified feed as being one of the major contributors to negative environmental impact of these systems. Animal feed (for both livestock and fed aquaculture species) is also reportedly responsible for the largest share of farming costs [1,2,10] as its demand continues to increase along with increased production [1,2]. Thus, methods to improve the sustainability of animal production continue being explored. To support this process, the European Commission (EC), as part of the European Union's (EU) green deal [11] Farm-to-Fork strategy [12], pledged to facilitate the approval of new sustainable feed alternatives for use on animals in the region [12]. Accordingly, the EC approved the use of processed insect protein in the feed given to aquaculture species [13], pigs, and poultry [14], after being risk assessed by the European Food Safety Authority (EFSA) panel [15].

Several LCA studies have reported the potential of insect protein as a sustainable feed ingredient relative to conventional protein feed sources [16] when employing insect-rearing technologies that use low energy [17–19] and substrates of low economic value [17,18,20–23]. Despite the variations in the nutrient profiles of insects depending on the species and the rearing conditions [24], the general inference has been that insect-based feed (IBF) is a suitable alternative to the conventional feeds such as soy and fishmeal [24–26]. In addition, the potential contribution of insects to a circular economy [27], low feed conversion ratios, and subsequent reductions in production costs has also been documented [18].

For the successful adoption of IBF, consumers/farmers' willingness to accept its use on animals is crucial [28]. As such, some studies have over the years been undertaken to understand the factors affecting willingness in this regard in order to provide recommendations for possible intervention pathways [29–37]. Based on findings from such studies, willingness to accept the use of IBF for animals seems to be intricately affected by a combination of factors associated with the characteristics of the participants (consumer-/farmer-related factors) [31–37] and those associated with the characteristics of the IBF itself or the end product (product-related factors) [32,37–39].

Participant characteristics such as previous knowledge of insects being used in feed [31,33], residential location [32], gender [33–37], age [32,35], and level of education [34] are some of the named factors found to be influential in the acceptance of IBF among consumers and farmers. Characteristics specific to the farmers (farmer-related factors) such as the type [37] and quantity of animals being reared, type of animal feed ingredients being used, and previous experience with using IBF [31] have been found to affect their willingness to use IBF. Moreover, the supposed benefits/risks of using IBF are among the product-related factors previous works have found to also be influential in consumers' and farmers' acceptance of IBF [32,37,38]. Accordingly, being informed of the sustainability benefits of using IBF tends to improve its acceptance [30,33,34,36]. How each factor affects willingness to accept IBF, however, tends to vary depending on the country of the participants under investigation [40].

In Ireland, where agricultural production is an integral part of the country's economy [41], achieving the Sustainable Development Goal 12 of responsible consumption and production by 2030 remains a significant challenge [42]. Yet, there is a scarcity in published studies that have been conducted to understand the factors affecting acceptance of IBF amongst consumers/farmers in this state. As such, this study aims to collect data on the factors affecting IBF acceptance amongst a segment of consumers and farmers in Ireland and how this differs from studies conducted in other countries. The specific main research questions of this study are:

1. Which factors affect the willingness of a segment of consumers in Ireland to consume food products derived from animals that have been fed with IBF?
2. Which factors affect the willingness of a segment of farmers in Ireland to use IBF for their animals?

## 2. Materials and Methods

### 2.1. Study Design and Sampling

A pragmatic paradigm was adopted to answer the research questions of this study. Therefore, a "convergent parallel mixed-methods design" [43] was employed in the form of an online survey created using Qualtrics$^{TM}$ (first release 2005, copyright year 2022, available at https://www.qualtrics.com). Closed-ended questions were developed using themes from a literature review on the field. However, since there was a scarcity in reports on consumers' or farmers' acceptability of IBF in Ireland, open-ended questions were also included. This was done to explore other factors specific to the consumers and farmers in Ireland, which may not be otherwise available from studies undertaken in other countries. This survey was approved by the Institute Research Ethics Committee of the Atlantic Technological University (ATU) in Sligo, Ireland (Ref No. 2022001).

The survey was disseminated to a convenience sample by gatekeepers from Atlantic Technological University (ATU) and University College Dublin (UCD) via email to their respective staff and students based on Sligo, Galway, and Dublin campuses. To attract participants outside of ATU and UCD, a link to the survey was also shared on the researchers' social media accounts (Twitter, LinkedIn, and Facebook) and on farmers' social pages. Furthermore, posters with the survey's QR code were physically distributed at the Sligo Farmers' Market to the farmers selling their produce and those visiting the market. In all the above instances, a request was made for people to share the link or poster with anyone who might also be interested in participating, thus generating a snowball effect [44]. The survey was kept live for a period of three months (April, May and June) in 2022. The survey items used in this study are provided in Table S1.

### 2.2. Survey Construction

To ensure content validity of the survey, the main research questions of this study were broken down into embedded research questions to facilitate the construction of survey items (see Table S2). The survey was divided into four sections. The first section assessed consumer-related and farmer-related characteristics common for all participants such as their sociodemographic information (age, gender, level of education, location of residence, and workplace if applicable) and previous knowledge of insects being used in animal feed. In addition, participants were asked if they followed a particular diet (Yes/No) and if they did, they were requested to specify the type of diet. In the second section, using 5-point Likert scales, participants were asked to indicate how willing they would be to consume different food products derived from animals fed with IBF. The extent to which participants agreed with provided statements used to complete the sentences: "I am willing to eat food derived from animals that have been fed with insect-based feed if … " and "I am NOT willing to eat food products derived from animals that have been fed with insect-based feed if … ", was used to capture product-related reasons behind their willingness to consume these products. These questions were also asked on a 5-point Likert scale and two of the statements provided were adapted from past studies [34,37]. Participants were given an opportunity through an open-ended question to state "other" reasons (if any) behind their willingness to consume food products from animals fed with IBF. This section of the survey ended with the question "Do you participate in farming activities related to poultry, fish and/or livestock production?" (Yes/No). Those participants who selected "Yes" to this question moved to the third section of the survey, whilst those who selected "No" were automatically directed to the fourth and final section of the survey.

In the third section, farmer-related characteristics such as the type of farming activities, farm size, number of animals being reared, type of feed ingredients currently used, and prior experience with IBF were ascertained. In addition, farmers were asked to indicate their level of agreement with the use of IBF for different animals and how likely they would be to use it (5-point Likert scales). The extent to which farmers agreed with the provided product-related reasons behind their willingness to use IBF was also assessed. Five of these "reasons" were adapted from past studies [32,37]. An open-ended question explored "other" reasons (if any) behind farmers' willingness to use IBF. As a final question to this section, farmers were firstly provided with information that the use of insect protein in feed for pigs and poultry had been recently authorised in the EU. Thereafter, they were asked if there would be any other factors they would consider prior to using IBF for their animals. In the fourth and final section of the survey, participants' acceptance of IBF was assessed again (5-point Likert scales) after they were provided with information on its environmental and nutritional benefits.

A total of 284 participants completed the survey. However, 51 of them did not complete the first section of the survey; therefore, these were excluded from the analysis. The remaining participants (N = 233) who either answered all or at least 75% of the questions in the survey were included in the analysis.

*2.3. Data Analysis*

The statistical software SPSS (IBM® version 28.0) was used to analyse the quantitative data obtained. Descriptive statistics were used to outline the profile of the participants. The 5-point scales were collapsed to three groups each for analysis to have at least 5 counts in each cell of the cross-tabulation tables in order to run the chi-square test. In line with previous work [33], willingness 5-point scales were collapsed to 1: "unwilling", 2: "uncertain", and 3: "willing". Likewise, degree of likelihood was collapsed to 1: "unlikely", 2: "uncertain", and 3: "likely", while level of agreement was collapsed to 1: "disagree", 2: "neutral", and 3: "agree".

Non-parametric statistical tests (Mann–Whitney U, Kruskal–Wallis, and chi-square) were used to analyse the effect of (1) the consumer-related factors on the willingness of all participants to consume food products from animals fed with IBF and (2) the farmer-related factors on the willingness to use IBF amongst farmers, as shown in Table S3. Spearman's correlation was used to determine the correlation between participants' level of agreement to statements on factors relating to the characteristics of the IBF or end product (product-related) and their willingness to accept it. Lastly, the sign test was used to determine the differences between participants' willingness to accept the use of IBF for animals before and after being provided with information on its environmental and nutritional benefits. Significance for all statistical tests was established at $p < 0.05$.

The question on gender had four options for the participants to choose from ("male", "female", "other", and "prefer not to say"). Since less than one percent of the total participants (N = 233) selected "other" and "prefer not to say", respectively, these two categories were not included when analysing the effect of participants' gender. However, all four categories were included for the rest of the analysis. Moreover, since less than a fifth of the participants were adhering to a specific diet (vegan, vegetarian, calorie-restricted, or other), all participants were divided into those that followed a particular diet and those that did not when analysing the effect of diet.

To analyse the qualitative data collected through open-ended questions, each participant response was coded using an inductive approach [45]. Codes that linked together were then sorted into sub-themes and themes [46]. This process conducted by one researcher was appraised by a second researcher to ensure accurate reporting of results. Qualitative results are presented according to their themes, together with the quantitative results related to that theme. Participants' quotes are provided along with a participant's (consumer [C]/farmer [F]) number.

## 3. Results

*3.1. Participants' Profile*

In reporting the results of this study, the term "consumers" is used to refer to the total participants (N = 233) whether they were involved in farming activities or not, whilst the term "farmers" is used for those participants who answered Yes to the question "Do you participate in farming activities related to poultry, fish and/or livestock production?" (n = 73).

An overview of the participants' profile (N = 233) is presented in Table 1. Although more consumers worked in Connaught province (37.3%) than those who worked in other provinces, most of them resided in Leinster (42.1%). Connaught was, however, the province where more farmers resided (41.1%) and worked (38.4%) than other provinces. Just above half of the participants were female (58.8% consumers and 54.8% farmers) while those in the 18 to 29 age group (31.8% consumers and 45.2% farmers) numbered more than those in other age groups. All consumers had education at junior certificate level, with 47.6% reporting to have either a masters or a PhD degree. Nevertheless, the percentage of farmers with either a masters or a PhD degree (34.2%) was comparable to those farmers whose highest level of education was an honours degree (37.0%). When this study was conducted, most of the consumers (82%) and farmers (89%) were not adhering to any particular diet and most of them (67.0% consumers and 76.7% farmers) had prior knowledge of insects being used in animal feed.

**Table 1.** Participants' profile (N = 233 all participants consumers and farmers, n = 73 only farmers).

| Profile | Total Consumers (Including Farmers) (N = 233) | Farmers Only (n = 73) |
|---|---|---|
| | % | % |
| **Location of residence:** | | |
| Connaught | 39.5 | 41.1 |
| Leinster | 42.1 | 30.1 |
| Munster | 7.7 | 5.5 |
| Ulster | 10.7 | 23.3 |
| **Location of workplace:** | | |
| Connaught | 37.3 | 38.4 |
| Leinster | 36.5 | 26.0 |
| Munster | 8.2 | 8.2 |
| Ulster | 9.4 | 24.7 |
| Not Applicable | 8.6 | 2.7 |
| **Gender:** | | |
| Male | 39.5 | 45.2 |
| Female | 58.8 | 54.8 |
| Other | 0.9 | 0.0 |
| Prefer not to say | 0.9 | 0.0 |
| **Age:** | | |
| 18–29 | 31.8 | 45.2 |
| 30–39 | 18.5 | 15.1 |
| 40–49 | 22.3 | 12.3 |
| 50–59 | 17.6 | 20.5 |
| 60 and above | 9.9 | 6.8 |
| **Level of education:** | | |
| No formal education | 0.0 | 0.0 |
| Junior Certificate | 0.0 | 0.0 |
| Leaving Certificate | 6.9 | 8.2 |
| Advanced certificate | 9.0 | 12.3 |
| Bachelor's degree | 9.4 | 8.2 |
| Honours degree | 27.0 | 37.0 |
| Master's or PhD | 47.6 | 34.2 |
| **Follow a specific diet:** | | |
| Yes | 18.0 | 11.0 |
| No | 82.0 | 89.0 |
| **Previous knowledge of insects being used in feed:** | | |
| Yes | 67.0 | 76.7 |
| No | 33.0 | 23.3 |

Almost half (49.3%) of the farmers were involved in beef farming, followed by those involved in "sheep, goats and other grazing livestock" (41.1%), poultry (30.1%), dairy (19.2%), "mixed crops and livestock" (8.2%), pigs (6.8%), and fish (1.4%) production. When asked to state the sizes of their farms in hectares (Ha), 4.1% of farmers indicated having more than 100 Ha, 20.5% between 51 and 100 Ha, 17.8% between zero and ten Ha, and 17.8% between 31 to 50 Ha. Most farmers (40.3%) reared between 101 and 500 farm animals, followed by 26.4% with zero to 50 animals and 11.1% with 51 to 100 animals. Although none of the farmers reared between 500 and 1000 animals, 7% of them had more than a thousand animals on their farms. To feed these animals, most farmers used grass (80.8%), silage (71.2%), and "cereals including maize, wheat and rice" (79.5%). Approximately a third (34.2%) of the farmers fed their animals soymeal. Other farmers also used molasses

(19.2%), brewer's/distiller's grain (16.4%), rapeseed meal (12.3%), sunflower seeds (6.8%), layers pellets (5.5%), and palm kernel (5.4%) in their animal feed. However, only 2.8% of the farmers reported to have used IBF before.

*3.2. Willingness to Consume Foods Derived from Animals Fed with Insect-Based Feed (IBF)*

Participants (N = 233) were mostly willing to consume eggs (75.1%), chicken (73%), and dairy products (70%) derived from animals fed with IBF. Around three-fifths of the participants were willing to consume fish (64.4%), beef (62.7%), pork (62.7%), and lamb/mutton (56.7%). Figure 1 shows that although some participants expressed unwillingness to consume these food products, some were uncertain of their willingness.

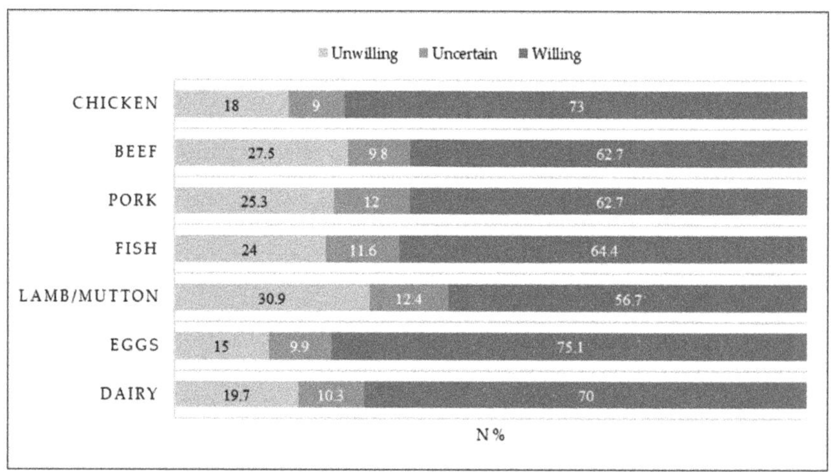


**Figure 1.** The willingness of participants (N = 233) to consume different food products derived from animals fed with insect-based feed (IBF).

*3.3. Factors Affecting Willingness to Consume Foods Derived from Animals Fed with IBF*

The effect of consumer-related factors on the willingness to consume foods derived from animals fed with IBF is outlined in Table S4. Province of residence, being involved in farming, or previous knowledge of insects being used in animal feed had no significant effect ($p > 0.05$). On the other hand, level of education had a significant effect only on the willingness to consume fish (H(4) = 10.761, $p = 0.029$). Those who had attained at least an honours degree were significantly more willing to consume fish fed with IBF compared to those who had not.

The province where consumers worked had no significant effect ($p > 0.05$) on their willingness to consume chicken, beef, pork, fish, and lamb/mutton fed with IBF. However, it had a significant effect on consumers' willingness to consume eggs (H(4) = 10.996, $p = 0.027$) and dairy products (H(4) = 10.974, $p = 0.027$). Those working in Connaught province were significantly more willing to consume eggs and dairy products than those who selected the "not applicable" option when asked about the province in which they worked. In contrast, whether consumers were on a particular diet or not had no significant effect ($p > 0.05$) on their willingness to consume eggs and dairy products, but it significantly affected ($p < 0.05$) their willingness to consume the other products. Consumers who were not on any specific diet were significantly more willing to consume chicken (U = 5483, $p < 0.001$), beef (U = 5893.5, $p < 0.001$), pork (U = 5838, $p < 0.001$), fish (U = 5570.5, $p < 0.001$), and lamb/mutton (U = 5698.5, $p < 0.01$) than those who were adhering to specific diets. Qualitative findings revealed that those who followed diets that either restricted or excluded meat were unwilling to eat animal-based products regardless of what the animals were fed:

*"I don't eat beef, pork, lamb or fish—hence my reply to those. It is not the objection to the insect feed"* (C18)

*"As a vegan I don't eat animals no matter what they are fed"* (C2)

Gender had no significant effect ($p > 0.05$) on consumers' willingness to consume chicken, eggs, and dairy products. Nevertheless, it significantly affected ($p < 0.05$) their willingness to consume beef (U = 5680, $p = 0.048$), pork (U = 5576, $p = 0.027$), fish (U = 5550, $p = 0.022$), and lamb/mutton (U = 5342, $p = 0.008$), such that female consumers were significantly less willing to consume these products compared to the male consumers. In contrast, the age of the consumers had no significant effect ($p > 0.05$) on their willingness to consume beef, pork, fish, and lamb/mutton but had a significant effect ($p < 0.05$) on their willingness to consume chicken (H(4) = 10.555, $p = 0.032$), eggs (H(4) = 14.958, $p = 0.005$), and dairy products (H(4) = 15.739, $p = 0.003$). Those aged between 18 and 29 were significantly the most willing, while those in the 40 to 49 age range were significantly the most unwilling, to consume chicken, eggs, and dairy products.

More than 75% of the participants agreed with the statements that they would be willing to consume food derived from animals fed IBF "if insects are naturally part of the animals' diet" (79%) and "if feeding animals with insect-based feed has a positive impact on the environment" (79.4%). This was followed by those willing to consume such food products "if the price of the food products is comparable to the existing food products in the market" (61.8%) and "if the information is specified on the food product packaging" (60.9%). Almost half of the consumers (46.8%) agreed that they would be willing to consume foods from animals fed insect-based feed "if the food products are cheaper" (see Figure 2).

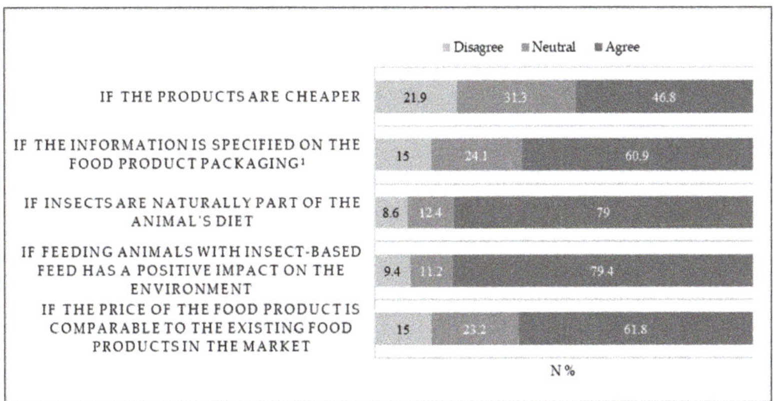


**Figure 2.** The level of participants' (N = 233) agreement to the statements "I am willing to eat food products derived from animals that have been fed with insect-based feed … " [1] Statements adapted from [34].

Consumers' level of agreement to all these statements (insects being a natural part of the animal's diet, insect-based feed having a positive impact on environment, price of the food products, and information on food packaging) was positively correlated with their willingness to consume the different type of foods from animals feeding on insect-based feed (see Table 2). However, consumers' willingness to consume lamb/mutton was not significantly affected ($p > 0.05$) by whether insects were naturally part of a sheep's diet or not.

**Table 2.** Correlation coefficients between product-related factors and participants' (N = 233) willingness to consume foods from animals fed with IBF.

| Reasons for Willingness to Consume Food Derived from Animals Fed IBF: | Willingness to Eat the following If the Animals Were Fed IBF: | | | | | | |
|---|---|---|---|---|---|---|---|
| | Chicken | Beef | Pork | Fish | Lamb/Mutton | Eggs | Dairy |
| | Correlation Coefficients [1] | | | | | | |
| If the products are cheaper | 0.471 ** | 0.510 ** | 0.443 ** | 0.383 ** | 0.445 ** | 0.437 ** | 0.465 ** |
| If the information is specified in the food packaging | 0.396 ** | 0.373 ** | 0.290 ** | 0.310 ** | 0.284 ** | 0.389 ** | 0.378 ** |
| If insects are naturally part of the animal's diet | 0.267 ** | 0.157 * | 0.154 * | 0.213 ** | 0.072 | 0.270 ** | 0.190 ** |
| If feeding animals with insect-based feed has a positive impact on the environment | 0.433 ** | 0.396 ** | 0.361 ** | 0.444 ** | 0.384 ** | 0.482 ** | 0.495 ** |
| If the price of the food products is comparable to the existing food products in the market | 0.379 ** | 0.429 ** | 0.357 ** | 0.344 ** | 0.386 ** | 0.399 ** | 0.435 ** |

[1] * Correlation significant at the 0.05 level; ** Correlation significant at the 0.01 Level.

About half of the consumers disagreed with the provided statements that they would be unwilling to eat food products derived from animals fed with IBF "because I am concerned whether I might have allergic reactions after eating these food products" (53.9%), "because I am concerned about the sensory appeal (i.e., taste, aroma & texture) of the food product" (46.1%) and "because I am concerned about the safety of the food products" (45.6%). Almost half of them, nevertheless, agreed that they would be unwilling to consume "if the food products are more expensive" (49.3%). Nearly a third of the consumers could neither agree nor disagree with all these statements (see Figure 3).

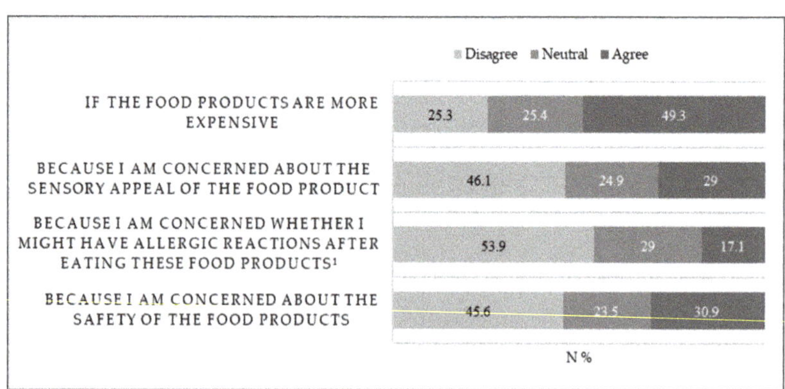


**Figure 3.** The level of participants' (N = 233) agreement (disagree, neutral and agree) to the statements "I am NOT willing to eat food products derived from animals that have been fed with insect-based feed … " [1] Statements adapted from a past study [37].

Consumers' level of agreement to most of the provided statements was not significantly correlated ($p > 0.05$) with their willingness to consume foods derived from animals fed with IBF. However, consumers' concern about the sensory appeal of eggs obtained from animals on an IBF diet was negatively correlated with their willingness to consume such eggs (r (231) = −0.158, $p < 0.05$). Similarly, consumers' concern regarding the safety of beef (r (231) = −0.158, $p < 0.05$), pork (r (231) = −0.210, $p < 0.01$), fish (r (231) = −0.179, $p < 0.01$), lamb/mutton (r (231) = −0.189), $p < 0.01$), eggs (r (231) = −0.169, $p < 0.05$), and dairy (r (231) = −0.217), $p < 0.01$) was negatively correlated with their willingness to consume those food products.

Analysis of qualitative data related to the research question "Which factors affect the willingness of consumers in Ireland to consume food products derived from animals that have been fed with IBF?" generated two main themes: "consumer-related factors" and "product-related factors". In relation to the product-related factors, the answers of 59 participants revealed that most of the participants' concern regarding the safety was further linked to the type of animal and unnatural animal diet. Consumers did not think it was natural for herbivores to feed on IBF and questioned the safety of such a practice for both humans and animals. This concern was linked to consumers' recollection of the bovine spongiform encephalopathy (BSE) outbreak that was first detected in cattle in the United Kingdom in 1986 and later spread to humans through consumption of meat that was infected with prions [47]:

> *"I worry about forcing animals to eat an unnatural diet that may cause problems for that animal and may once again cause problems to humans as it did with the Bovine spongiform encephalopathy (BSE) in cows … " (C19)*

> *"I would only eat products if it forms part of the natural diet of the animal, like chickens and fish. I think feeding insects to herbivores isn't healthy or natural" (C46)*

Consumers' willingness to consume foods from animals fed with IBF was also dependent on the sustainability of the IBF, nutritional value, and the sensory attributes of the foods:

> *"If the taste and appearance of the food was not significantly altered, if the insects were produced sustainably … " (C26)*

> *"If the eating quality and nutritional value of the products remain consistent" (C32)*

Most of the consumers required further information regarding the benefits/risks of feeding animals with IBF:

> *"I would need more information about whether insects are reasonably part of the animals' natural diet … I understand this would not be natural for cows, sheep, cattle and so more information about this would help my decision making" (C58)*

> *"I would like to have more information about the pros and cons of the differences in food products that have and have not had insects" (C51)*

Some consumers had "no reason not to eat" while others had "no reason to eat" foods from animals fed with insect-based feed.

### 3.4. Willingness to Use IBF Amongst Farmers

Most farmers agreed with the use of IBF for poultry (81.7%), fish (80.3%), and pigs (71.8%), as seen in Figure 4. Nearly 60% of the farmers agreed with the use of IBF for pets whilst slightly over half agreed with its use for cattle (53.5%) and sheep (52.1%). Regarding the use of IBF for cattle and sheep, the other half of the farmers was almost evenly divided between those who disagreed (23.9% and 22.5%, respectively) and those who were unsure (22.6% and 25.4%, respectively) (see Figure 4).

When asked how likely they would be to use IBF for their own animals prior to being provided with information on its benefits, 56.3% of the farmers indicated that they would likely use it while just 18.3% declared that they were unlikely to use it. About 25%, however, were uncertain ("neither likely nor unlikely").

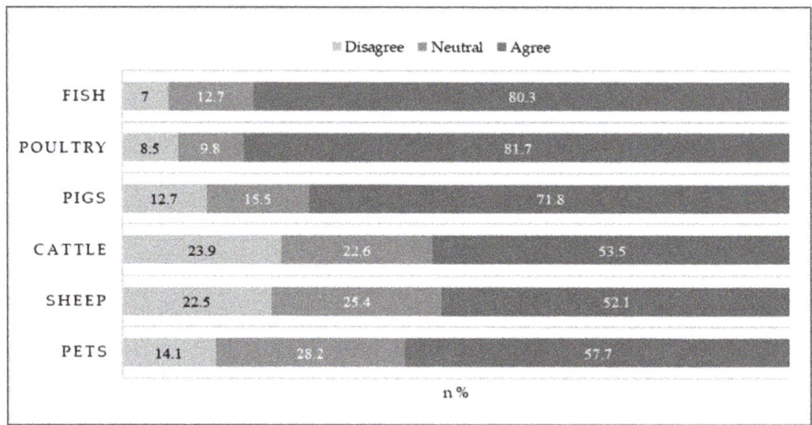



**Figure 4.** The extent to which farmers (n = 71; two farmers did not complete this question) agreed with the use of IBF for different animals.

### 3.5. Factors Affecting Willingness of Farmers to Use IBF

The farmers' willingness to use IBF for their animals was not significantly affected ($p > 0.05$) by any of the farmer-related factors investigated (location of residence/workplace, gender, age, level of education, previous knowledge of insects being used in animal feed, farm size, number of animals being reared, type of feed ingredients being used, type of farming, and previous experience with using IBF) in this study (see Table S5).

Figure 5 shows that most farmers agreed with all the provided statements that they would be willing to use IBF for their animals "if the feed is of high nutritional value" (91.5%), "if it is safe for animal consumption" (90.1%), "if consumers will purchase products of animals fed with insect-based feed" (88.7%), "if it reduces the price of feed and animal production" (88.7%), and "if the animals will grow faster" (76.1%). Even though most of the farmers agreed to these statements, it was their level of agreement to the statement "if it is safe for animal consumption" that was significantly correlated ($r (69) = 0.307, p < 0.01$) with their willingness to use IBF.

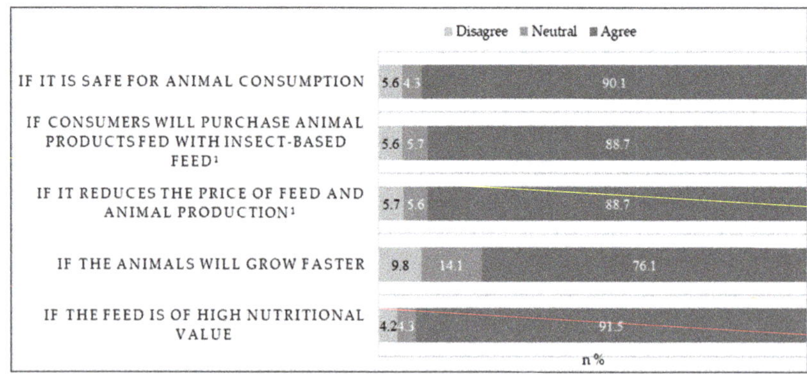



**Figure 5.** The level of farmers' (n = 71; two farmers did not complete this question) agreement (disagree, neutral, agree) with the statements "I am willing to use insect-based feed for my animals … "; [1] Statements provided adapted from a past study [37].

Almost all the farmers agreed that they would not be willing to use IBF for their animals "if it introduces microbial contamination or chemical residues to the food chain"

(92.6%) and "if it causes allergic reactions in animals or/and humans" (92.6%). This was followed by those who agreed that they would not be willing to use IBF "if it reduces consumer acceptance of food resulting from animal production" (73.5%) or "because I do not have enough information regarding the benefits/risks" (67.6%). About a fifth (20.6%) of the farmers could neither agree nor disagree to the statement "because I do not have enough information regarding benefits/risks" as their reason for unwillingness to use IBF (see Figure 6). No significant correlations ($p > 0.05$) were observed between the farmers' willingness to use insect-based feed and their level of agreement with any of these statements.

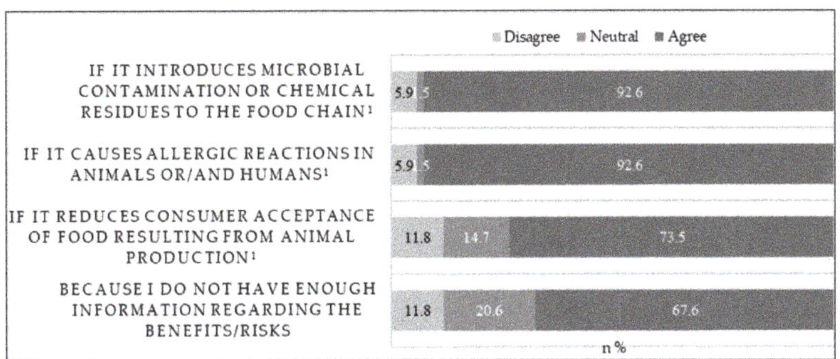


**Figure 6.** The level of farmers' (n = 68; five farmers did not complete this question) agreement (disagree, neutral, agree) to the statements "I am NOT willing to use insect-based feed for my animals … "; [1] Statements adapted from past studies [32,37].

Less than a quarter of the farmers provided "other" reasons for their willingness (21.1%) or unwillingness (11.8%) to use IBF through the open-ended questions, which were mainly "product-related factors" as revealed from the analysis of the qualitative data. Among these factors, sustainability and safety of the IBF were most frequently mentioned, while fewer participants named factors related to its nutritional value or availability. Though some only mentioned the term "sustainable" without further elaboration, the sustainability sub-theme (from the theme "product-related factors") was mostly linked to the economic and environmental pillars of sustainability [48]:

> *"There is a lot of wheat used in the chicken meal on our farm … If we could feed insects to the chickens, we may be able to use the wheat to make other products which will earn more than the insects cost to produce. Thereby increasing Ireland's net agricultural outputs"* (F6)

> *"If there is less impact on the environment from using insects as a source of feed"* (F5)

Farmers' concern regarding the safety of IBF was linked to the type of animal. Concerns were raised over herbivores being fed IBF, subsequently causing some to expect its safety to be substantiated through "extensive" research before they are willing to use it:

> *"If enough research has been done on the environmental impact of the insect production and alterations to the food chain, if it is proven that it is safe for herbivores to eat insects"* (F9)

The availability of the IBF was another factor pointed out by some farmers as having an impact on their willingness to use it:

> *"If it is not available to buy at local stores, I wouldn't be special ordering in insect meal"* (F1)

Others, however, associated IBF with a high protein content and favourable sensory properties in eggs; hence their willingness to use it for their animals:

*"Higher protein content than grain, deeper yolk colour and more flavour in eggs"* (F13)

Upon being provided information that the use of insect protein in feed given to pigs and poultry had been recently authorised in the EU, some farmers indicated that they would still consider other factors prior to using IBF for their animals. These factors were related to the environmental impact (11.8%), availability of supply (11.1%), amount of research conducted to back any benefits or risks of using IBF (9.5%), cost (7.4%), nutritional value (4.4%), and palatability for animals (4.4%) and the people (4.4%) who would consume the food from animals fed with IBF. Most of these farmers however, still wanted more information, as seen by how they mostly asked questions in their responses:

*"Is it likely to cause allergic reaction to individuals with hayfever?"* (F24)

*"Would there be a way to grow the insects using the waste products from the chicken house? Currently the chicken manure goes to the tillage farmers who plough it in and grow grain to be sold back to us as more feed. If we replace grain with insect protein the tillage men might not take our manure … "* (F9)

*3.6. Effect of Providing Information on Participants' Willingness to Accept IBF*

After being provided with the benefits of using IBF, 74.6% of farmers indicated that they would likely use it for their animals compared to 56.3% who had done so before. In addition, there was a decrease in the percentage of farmers who were unlikely (5.6%) or neither likely nor unlikely (12.7%) to use IBF after knowing its benefits. The exact sign test confirmed that these differences were significant ($p = 0.011$), as seen in Table 3. Similarly, providing information on the benefits of using IBF for animals induced a significant increase in consumers' willingness to consume beef ($p < 0.001$), pork ($p < 0.001$), fish ($p < 0.001$), lamb/mutton ($p < 0.001$), and dairy products ($p = 0.005$) from animals that were fed with IBF. However, it did not significantly affect their willingness to consume chicken ($p = 0.127$), or eggs ($p = 0.185$) (see Table 3).

**Table 3.** Effect of knowing the benefits of IBF on participants' (N = 207) willingness to consume food from animals fed with such feed and on farmers' (n = 63) willingness to use it.

| Willingness to Consume the following Products from Animals Fed IBF (N = 207 [1]): | Positive Differences | Negative Differences | Tiers | Sign Test |
|---|---|---|---|---|
| | % | % | % | *p*-Value [2] |
| Chicken | 15.5 | 9.7 | 74.9 | 0.127 |
| Beef | 27.1 | 4.8 | 68.1 | <0.001 * |
| Pork | 25.1 | 6.3 | 68.6 | <0.001 * |
| Fish | 23.2 | 8.7 | 68.1 | <0.001 * |
| Lamb/mutton | 31.4 | 4.8 | 63.8 | <0.001 * |
| Eggs | 13.5 | 8.7 | 77.8 | 0.185 |
| Dairy | 18.4 | 7.2 | 74.4 | 0.005 * |
| **Willingness of farmers to use insect-based feed for their animals (n = 63 [3])** | 28.6 | 7.9 | 63.5 | 0.011 * |

[1] Twenty-six consumers did not complete this question; [2] *p*-value significant when * $p < 0.05$; [3] Ten farmers did not complete this question.

## 4. Discussion

This study explored the factors affecting the acceptance of IBF amongst a segment of consumers and farmers in Ireland. Figure 7 shows a summary of the consumer- and product-related factors found from the analysis of the quantitative data as having an influence in that regard, while Table S6 shows a summary of the factors generated through the analysis of qualitative data.

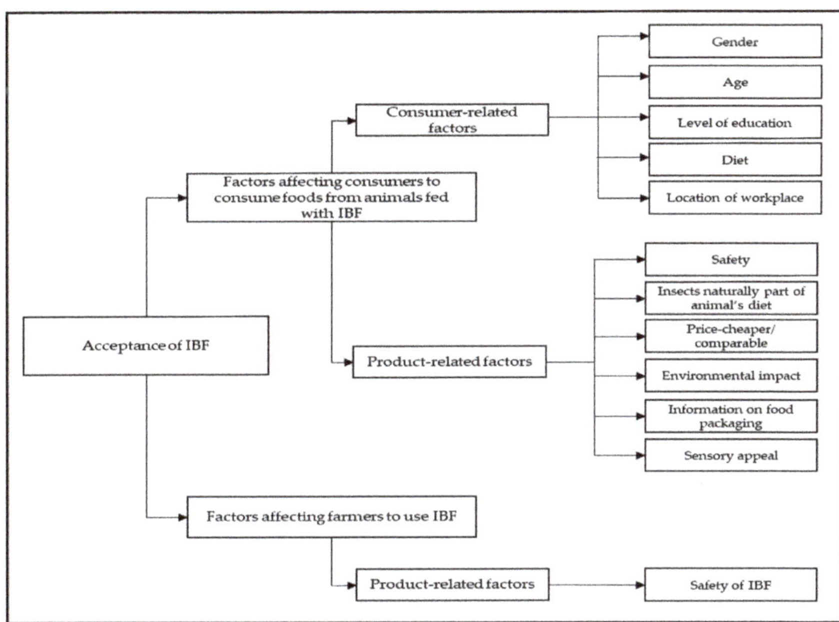

**Figure 7.** Summary of the consumer- and product-related factors found to influence acceptance of IBF in this study (based on the analysis of the quantitative data).

Several consumer-related factors affected the willingness of participants in this study to consume foods from animals fed with IBF. While gender had no influence on the willingness to consume chicken, dairy, and eggs, men were more willing to consume beef, pork, fish, and lamb/mutton. Several past studies have also reported men to be more willing to consume foods from animals fed with IBF [29,30,33,36]. Moreover, men have been found to generally consume beef, pork, fish, and lamb/mutton significantly more than females [49,50]. Comparably, in Ireland, men's overall animal protein intake is reportedly higher than that of females, who instead tend to consume more plant-based protein than men [51]. All this could have played a role in the gender effect on willingness to consume foods from animals fed with IBF in the present study. Furthermore, females have been reported to likely be more concerned than men about the safety aspect of these foods [30]. In the present study, the safety concern surrounding use of IBF was linked to consumers' recollection of the BSE outbreak in cows and subsequent transmission to humans [47]. The risk of getting infected from consuming dairy products from BSE infected animals was, however, found to be very rare [52,53], which could have led female consumers in this study to view dairy as being relatively safe to consume, whilst the greater acceptance of chicken and eggs from animals fed IBF may be explained by insects being part of the natural diet of poultry [24]. Consumers in the youngest age group (18–29) were significantly more willing than those in other age groups to consume these three products. Other researchers had also found young consumers to be more accepting of the foods from animals fed with IBF [29,32–34]. Most of the farmers in the present study were in the 18–29 age group and there were more female farmers than were male farmers (see Table 1), which could explain why this age group was more willing to consume chicken, dairy, and eggs for the above reasons.

Consumers' diet influenced their willingness to consume meat from animals fed with IBF, with those adhering to a particular diet being less willing in that regard. However, it had no effect on their willingness to consume eggs and dairy products. This could be attributed to the fact that this study did not exclude vegetarians, vegans, or those on selective meat diets, and most (88%) of the consumers who were adhering to a particular diet (18% of the total consumers: see Table 1), were on diets that did not exclude eggs and dairy (as revealed from their comments). Level of education influenced consumers' willingness to consume fish fed with IBF. Those who had completed an honours degree were more willing to consume IBF-fed fish. This agrees with previous studies that showed having a university degree to positively influence one's willingness to consume fish [30], duck [34], or animal products in general [36] from animals fed with IBF. In a study where less than a third of the participants had a university degree [32], level of education had no influence on willingness to consume foods from poultry, cattle, pigs, and fish fed with IBF. The use of insect protein in EU aquafeed was authorised in 2017 [13], which might have provided ample time for the most educated group of consumers in the present study to get acquainted with this information.

Consumers' willingness to consume foods from animals fed with IBF was also affected by several product-related factors. Concern regarding the safety of beef, pork, fish, lamb/mutton, eggs, and dairy from these animals significantly decreased consumers' willingness to consume them. Accordingly, and in line with results from the qualitative analysis, insects being naturally part of an animal's diet significantly increased consumers' willingness to consume these products. The safety of these foods and insects being a natural part of an animal's diet were also noted as contributing factors in past studies [30,34,36]. However, in the present study, these factors had no effect on the willingness to consume lamb/mutton, possibly because in Ireland, lamb/mutton is generally consumed far less yearly at 3.0 kg/capita compared to beef (19.8 kg/capita), poultry (24.7 kg/capita), or pork (31.1 kg/capita) [54].

The price of the foods derived from animals fed IBF influenced consumers' willingness to consume them. Consumers' willingness increased if the food products were cheaper or comparable to the existing alternatives, but it was not affected if the products were expensive. In a study conducted in Spain, participants were willing to buy fish fed with IBF even if it was more expensive than the alternatives [55]. These participants also believed IBF to be environmentally friendly compared to conventional aquafeeds, which could have contributed to their willingness to pay more for this type of fish [55]. In the present study, the willingness of consumers to consume all the foods (chicken, beef, pork, fish, lamb/mutton, eggs, and dairy) significantly increased when feeding animals with IBF had a positive impact on the environment, possibly explaining why their willingness was not affected by how expensive the products were.

The willingness to use IBF amongst farmers in the present study was not affected by any of the farmer-related factors explored as seen in Figure 7. Similarly, the intention to use IBF amongst farmers in France and the Netherlands was not significantly affected by age, gender, type of feed, and country location [56]. In a study conducted outside the EU, older poultry farmers who might have had more experience with using IBF were found to be more willing to use it than the younger farmers [35]. In the present study, however, most farmers were in the youngest age group (18–29), as they were more willing to complete an online survey. Moreover, almost all (97.2%) the farmers reported having no previous experience with using IBF. This was not surprising considering that the EU regulation allowing the use of insect protein in feed for poultry and pigs [14] came into force relatively recently in Ireland and the EU in general (36.9% were poultry and pig farmers, compared to just 1.4% of fish farmers).

IBF product-related factors, such as its safety, availability, sustainability, consumer acceptability, potential to reduce production costs, nutritional value, and improved growth performance of animals, were all generally important to the farmers in this study. However, safety significantly affected their willingness to use it on their animals. The more the farmers agreed to the statement that they would be willing to use IBF "if it is safe for animal consumption", the more willing they were to use it. Perceived risks associated with using IBF were also found to significantly reduce the willingness of farmers to accept its use in Belgium [37]. However, in that study [37], as well as in another conducted in France and the Netherlands [56], perceptions regarding the benefits associated with use of IBF had a higher impact on the farmers' willingness to use IBF compared to the perceived risks, which was different from what the present study found. Although important to the farmers, quantitative analysis in the present study revealed no significant correlations between three of the safety aspects of IBF investigated in the present study, i.e., microbial, chemical, and allergenic risks, and willingness to use IBF. However, qualitative analysis of the farmers' comments revealed that the safety aspect most crucial to them was the one dependent on the type of animal. Most responders questioned the safety of IBF for animals that are naturally herbivores, possibly explaining why more farmers agreed with its use for fish, poultry, and pigs than those who agreed with its use for cattle and sheep, as highlighted in the closed-ended questions. Moreover, 2.8% of farmers who had previous experience with IBF were involved in egg production and yet, while they were willing to use IBF, they could not agree with its use for cattle and sheep. It can be assumed that the safety of the IBF was a much higher priority to the farmers than their consideration of the number of animals being reared, farm size, or any farmer-related factors; hence the lack of significant results from the latter.

Providing information about the environmental and nutritional benefits of using IBF increased its acceptance amongst the consumers and farmers in the present study, which agrees with the findings of studies conducted in France and Italy [30,34]. This information, however, did not influence consumers' willingness to consume chicken and eggs, which were already the two most preferred products before the information was provided. Still, some participants (consumers and farmers) in the present study were uncertain of their willingness to accept IBF after being provided with information on its benefits. This could be attributed to the type of information provided, which did not include specific information on the safety for herbivores and/or humans or if insects can feed on manure; these were all details that participants were interested in, according to their comments. In addition, along with responses to the open-ended questions, most participants asked some questions that would suggest a general need for more information around the use of IBF. This need for information could explain the increased willingness to consume foods from animals fed with IBF "if the information is specified on the food packaging", as was also reported in another study [34]. Lack of information regarding the use of IBF has been found to cause uncertainties regarding its acceptance among consumers and stakeholders [57]. It can be assumed that the lack of significant results found on some participant-related factors in the present study, such as previous knowledge of insects being used in feed, for example, might have been influenced by this need for more information.

There were several strengths and limitations to this study. This is the first study to assess IBF acceptance amongst a segment of mostly younger and educated consumers and farmers in Ireland, a country with a substantial livestock production sector. Hence, the results could provide baseline data for IBF-related future studies in Ireland. This study also sheds light on the acceptance of IBF among an educated group of consumers and the future generation of farmers, as it mostly involved younger participants with a university education. Considering results from this study and other previous studies that found those younger [32–34] with a high level of education [36,58,59] to be more accepting of IBF, this group of consumers/farmers could potentially be among the early adopters of IBF. Identifying and understanding the factors affecting IBF acceptance among early adopters could aid targeted intervention strategies among this group of consumers/farmers. This is especially important since, according to several studies, attempts to introduce a novel practice might be best targeted at specific segments of the population who may be early adopters, in order to firstly establish some level of adoption before shifting the focus to the general population [60,61]. A limitation to the present study was that it did, however, prove challenging to recruit older farmers and those without a university education through online surveys; the latter was also observed in other studies [62].

## 5. Conclusions

This study aimed to provide baseline data on factors affecting the acceptance of IBF amongst a segment of younger and educated consumers and farmers in Ireland. The study found that a combination of consumer- and product-related factors affect consumers' willingness to consume foods from animals fed with IBF. This effect, nevertheless, depended on the type of food. Consumers' gender, age, level of education, and diet were the consumer-related factors found to significantly affect their willingness. On the other hand, safety, insects being a natural part of an animal's diet, environmental impact, price, and information reported on food packaging were the product-related factors found to influence consumers' willingness to consume foods from animals fed with IBF. Yet, the safety of the IBF for animal consumption, particularly herbivores, was a strong factor found to significantly affect the willingness of farmers to use it as feed on their own animals, unlike what was found in other studies conducted in the EU, where the benefits of using IBF were the stronger factors. The farmers in the present study were generally open to using IBF once its safety is substantiated through extensive research. This therefore calls for more research to be conducted to investigate the safety of IBF, particularly for ruminants, just as was recommended by the EFSA. In addition, providing information on the environmental and nutritional benefits of IBF increased its acceptance by both consumers and farmers. Future success on the adoption of IBF might depend on assuring farmers and consumers about its safety through enacting evidence-based educational strategies. Furthermore, EU public policy changes could be implemented to include a statement on the food packaging information that the food is from an animal fed with IBF, since this kind of transparency was shown to increase consumer acceptance in the present study. However, due to the relatively small sample size and participants' profile restrictions in this study, these conclusions may not be generalised for the entire Irish population. Therefore, for the future, this study could be extended using pen and paper questionnaires or face-to-face interviews with older farmers and those without university education, who were underrepresented in the current study. Recruitment for farmers could be focused on marts or the physical farmers markets across the different provinces of Ireland, although it is important to note that not all farmers in Ireland sell their produce at these markets. In addition to the farmers markets, recruitment for farmers could be undertaken via the different farmers' associations.

**Supplementary Materials:** The following supporting information can be downloaded at: https://www.mdpi.com/article/10.3390/su151411006/s1, Table S1: Survey items used in the present study; Table S2: How each main research question of this study was broken down (into embedded research questions) to facilitate survey construction and ensure content validity of the survey; Table S3: Statistical tests used to analyse the effect of various factors on the willingness of farmers and consumers to accept the use of insect-based feed on animals; Table S4: An outline of the effect of consumer-related factors (left) on the willingness of participants (N = 233) to consume foods (top) derived from animals fed with insect-based feed (IBF); Table S5: The effect of farmer-related factors on their willingness to use insect-based feed; Table S6: Themes, sub-themes, and corresponding quote examples generated from the qualitative analysis of the data from the open-ended questions.

**Author Contributions:** Conceptualization, M.D., F.N. and A.P.V.; methodology, L.R., M.D., F.N. and A.P.V.; validation, L.R., M.D. and F.N.; formal analysis, L.R. and M.D.; investigation, L.R., M.D., F.N. and A.P.V.; data curation, L.R. and M.D.; writing—original draft preparation, L.R.; writing—review and editing, M.D., F.N. and A.P.V.; visualization, L.R.; supervision, M.D., F.N. and A.P.V.; project administration, M.D.; funding acquisition, M.D. and F.N. All authors have read and agreed to the published version of the manuscript.

**Funding:** This study is part of a PhD project funded by the Connaught Ulster Alliance Bursary, grant number PCUAB024.

**Institutional Review Board Statement:** The study was conducted in accordance with the Declaration of Helsinki and approved by the Institute Research Ethics Committee of the Atlantic Technological University, Sligo, Ireland (reference number 2022001 approved on 9 February 2022).

**Informed Consent Statement:** Informed consent was obtained from all subjects involved in the study described in the research ethics application with reference number 2022001.

**Data Availability Statement:** The data presented in this study are available on request from the corresponding author. The data are not publicly available due to ethical reasons.

**Acknowledgments:** The authors would like to thank the respondents who participated in the piloting of the survey and those anonymous participants who took part in this study. We also thank the gatekeepers from ATU and UCD for disseminating the survey and Guy Marsden for allowing us to distribute survey posters at the Sligo Farmers' Market. Our appreciation also goes to the anonymous reviewers who contributed to the quality of this manuscript.

**Conflicts of Interest:** The authors declare no conflict of interest. The funders had no role in the design of the study; in the collection, analyses, or interpretation of data; in the writing of the manuscript; or in the decision to publish the results.

## References

1. Food and Agriculture Organisation of the United Nations (FAO). *The State of World Fisheries and Aquaculture 2022: Towards Blue Transformation*; FAO: Rome, Italy, 2022. [CrossRef]
2. The Organisation for Economic Co-operation and Development /Food and Agriculture Organisation of the United Nations. *OECD-FAO Agricultural Outlook 2022–2031*; OECD Publishing: Paris, France, 2022. [CrossRef]
3. Food and Agriculture Organisation of the United Nations (FAO). *The State of the World's Land and Water Resources for Food and Agriculture—Systems at Breaking Point*; FAO: Rome, Italy, 2022. [CrossRef]
4. Chen, X.; Samson, E.; Tocqueville, A.; Aubin, J. Environmental assessment of trout farming in France by life cycle assessment: Using bootstrapped principal component analysis to better define system classification. *J. Clean. Prod.* **2015**, *87*, 87–95. [CrossRef]
5. Maiolo, S.; Forchino, A.A.; Faccenda, F.; Pastres, R. From feed to fork–Life Cycle Assessment on an Italian rainbow trout (*Oncorhynchus mykiss*) supply chain. *J. Clean. Prod.* **2021**, *289*, 125155. [CrossRef]
6. Sanchez-Matos, J.; Regueiro, L.; González-García, S.; Vázquez-Rowe, I. Environmental performance of rainbow trout (*Oncorhynchus mykiss*) production in Galicia-Spain: A Life Cycle Assessment approach. *Sci. Total Environ.* **2023**, *856*, 159049. [CrossRef]
7. Asem-Hiablie, S.; Battagliese, T.; Stackhouse-Lawson, K.R.; Alan Rotz, C. A life cycle assessment of the environmental impacts of a beef system in the USA. *Int. J. Life Cycle Assess.* **2019**, *24*, 441–455. [CrossRef]
8. Mogensen, L.; Hermansen, J.E.; Halberg, N.; Dalgaard, R.; Vis, J.C.; Smith, B.G. Life cycle assessment across the food supply chain. In *Sustainability in the Food Industry*; Baldwin, C., Ed.; Wiley-Blackwell: Ames, IA, USA, 2009; pp. 115–144. [CrossRef]
9. Reckmann, K.; Traulsen, I.; Krieter, J. Life Cycle Assessment of pork production: A data inventory for the case of Germany. *Livest. Sci.* **2013**, *157*, 586–596. [CrossRef]
10. Pandey, A.K.; Kumar, P.; Saxena, M.J. Feed additives in animal health. In *Nutraceuticals in Veterinary Medicine*; Gupta, R., Srivasta, A., Lall, R., Eds.; Springer: Cham, Switzerland, 2019; pp. 345–362. [CrossRef]
11. European Commission. COM (2019) 640 Final. Communication from the Commission to the European Parliament, the Council, the European Economic and Social Committee, and the Committee of the Regions: The European Green Deal. Available online: https://eur-lex.europa.eu/legal-content/EN/TXT/?uri=CELEX%3A52019DC0640 (accessed on 7 June 2023).
12. European Commission. COM (2020). Communication from the Commission to the European Parliament, the Council, the European Economic and Social Committee, and the Committee of the Regions: A Farm to Fork Strategy for a Fair, Healthy and Environmentally Friendly Food System. Available online: https://eur-lex.europa.eu/legal-content/EN/TXT/?uri=CELEX%3A5 2020DC0381 (accessed on 7 June 2023).
13. EU Regulation (EU) 2017/893 of 24 May 2017 amending Annexes I and IV to Regulation (EC) No 999/2001 of the European Parliament and of the Council and Annexes X, XIV and XV to Commission Regulation (EU) No 142/2011 as regards the Provisions on Processed Animal Protein. *Off. J. Eur. Union.* **2017**, *L138*, 92–116. Available online: https://eur-lex.europa.eu/legal-content/EN/TXT/PDF/?uri=OJ:L:2017:138:FULL&from=EN (accessed on 7 June 2023).
14. EU Regulation (EU) 2021/1372 of 17 August 2021 amending Annex IV to Regulation (EC) No 999/2001 of the European Parliament and of the Council as regards the prohibition to feed non-ruminant farmed animals, other than fur animals, with protein derived from animals. *Off. J. Eur. Union.* **2021**, *L295*, 1–17. Available online: https://eur-lex.europa.eu/legal-content/EN/TXT/?uri=CELEX%3A32021R1372 (accessed on 7 June 2023).
15. European Food Safety Authority Scientific Committee. Risk profile related to production and consumption of insects as food and feed. *EFSA J.* **2015**, *13*, 4257. [CrossRef]
16. Halloran, A.; Hanboonsong, Y.; Roos, N.; Bruun, S. Life cycle assessment of cricket farming in north-eastern Thailand. *J. Clean. Prod.* **2017**, *156*, 83–94. [CrossRef]
17. Beyers, M.; Coudron, C.; Ravi, R.; Meers, E.; Bruun, S. Black soldier fly larvae as an alternative feed source and agro-waste disposal route–A life cycle perspective. *Resour. Conserv. Recycl.* **2023**, *192*, 106917. [CrossRef]

18. Thévenot, A.; Rivera, J.L.; Wilfart, A.; Maillard, F.; Hassouna, M.; Senga-Kiesse, T.; Le Féon, S.; Aubin, J. Mealworm meal for animal feed: Environmental assessment and sensitivity analysis to guide future prospects. *J. Clean. Prod.* **2018**, *170*, 1260–1267. [CrossRef]

19. Van Zanten, H.H.; Mollenhorst, H.; Oonincx, D.G.; Bikker, P.; Meerburg, B.G.; de Boer, I.J. From environmental nuisance to environmental opportunity: Housefly larvae convert waste to livestock feed. *J. Clean. Prod.* **2015**, *102*, 362–369. [CrossRef]

20. Féon, S.; Thévenot, A.; Maillard, F.; Macombe, C.; Forteau, L.; Aubin, J. Life Cycle Assessment of fish fed with insect meal: Case study of mealworm inclusion in trout feed, in France. *Aquaculture* **2019**, *500*, 82–91. [CrossRef]

21. Modahl, I.S.; Brekke, A. Environmental performance of insect protein: A case of LCA results for fish feed produced in Norway. *SN Appl. Sci.* **2022**, *4*, 183. [CrossRef]

22. Smetana, S.; Palanisamy, M.; Mathys, A.; Heinz, V. Sustainability of insect use for feed and food: Life Cycle Assessment perspective. *J. Clean. Prod.* **2016**, *137*, 741–751. [CrossRef]

23. Vauterin, A.; Steiner, B.; Sillman, J.; Kahiluoto, H. The potential of insect protein to reduce food-based carbon footprints in Europe: The case of broiler meat production. *J. Clean. Prod* **2021**, *320*, 128799. [CrossRef]

24. Van Huis, A.; Van Itterbeeck, J.; Klunder, H.; Mertens, E.; Halloran, A.; Muir, G.; Vantomme, P. *Edible Insects: Future Prospects for Food and Feed Security: FAO Forestry Paper no.171*; FAO: Rome, Italy, 2013.

25. Makkar, H.P.; Tran, G.; Heuzé, V.; Ankers, P. State-of-the-art on use of insects as animal feed. *Anim. Feed Sci. Technol.* **2014**, *197*, 1–33. [CrossRef]

26. Hong, J.; Kim, Y.Y. Insect as feed ingredients for pigs. *Anim. Biosci.* **2022**, *35*, 347. [CrossRef]

27. Cappellozza, S.; Leonardi, M.G.; Savoldelli, S.; Carminati, D.; Rizzolo, A.; Cortellino, G.; Terova, G.; Moretto, E.; Badaile, A.; Concheri, G.; et al. A first attempt to produce proteins from insects by means of a circular economy. *Animals* **2019**, *9*, 278. [CrossRef] [PubMed]

28. Final Report Summary—PROTEINSECT (Enabling the Exploitation of Insects as a Sustainable Source of Protein for Animal Feed and Human Nutrition). Available online: https://cordis.europa.eu/project/id/312084/reporting (accessed on 29 June 2023).

29. Baldi, L.; Mancuso, T.; Peri, M.; Gasco, L.; Trentinaglia, M.T. Consumer attitude and acceptance toward fish fed with insects: A focus on the new generations. *J. Insects Food Feed.* **2022**, *8*, 1249–1263. [CrossRef]

30. Bazoche, P.; Poret, S. Acceptability of insects in animal feed: A survey of French consumers. *J. Consum. Behav.* **2021**, *20*, 251–270. [CrossRef]

31. Chia, S.Y.; Macharia, J.; Diiro, G.M.; Kassie, M.; Ekesi, S.; Van Loon, J.J.; Dicke, M.; Tanga, C.M. Smallholder farmers' knowledge and willingness to pay for insect-based feeds in Kenya. *PLoS ONE* **2020**, *15*, e0230552. [CrossRef] [PubMed]

32. Domingues, C.H.D.F.; Borges, J.A.R.; Ruviaro, C.F.; Gomes Freire Guidolin, D.; Rosa Mauad Carrijo, J. Understanding the factors influencing consumer willingness to accept the use of insects to feed poultry, cattle, pigs and fish in Brazil. *PLoS ONE* **2020**, *15*, e0224059. [CrossRef]

33. Laureati, M.; Proserpio, C.; Jucker, C.; Savoldelli, S. New sustainable protein sources: Consumers' willingness to adopt insects as feed and food. *Ital. J. Food Sci.* **2016**, *28*, 652–668. [CrossRef]

34. Menozzi, D.; Sogari, G.; Mora, C.; Gariglio, M.; Gasco, L.; Schiavone, A. Insects as Feed for Farmed Poultry: Are Italian Consumers Ready to Embrace This Innovation? *Insects* **2021**, *12*, 435. [CrossRef]

35. Sebatta, C.; Ssepuuya, G.; Sikahwa, E.; Mugisha, J.; Diiro, G.; Sengendo, M.; Fuuna, P.; Fiaboe, K.K.M.; Nakimbugwe, D. Farmers' acceptance of insects as an alternative protein source in poultry feeds. *Int. J. Agric. Res. Innov. Techol.* **2018**, *8*, 32–41. [CrossRef]

36. Szendrő, K.; Nagy, M.Z.; Tóth, K. Consumer acceptance of meat from animals reared on insect meal as feed. *Animals* **2020**, *10*, 1312. [CrossRef]

37. Verbeke, W.; Spranghers, T.; De Clercq, P.; De Smet, S.; Sas, B.; Eeckhout, M. Insects in animal feed: Acceptance and its determinants among farmers, agriculture sector stakeholders and citizens. *Anim. Feed Sci. Technol.* **2015**, *204*, 72–87. [CrossRef]

38. Giotis, T.; Drichoutis, A.C. Consumer acceptance and willingness to pay for direct and indirect entomophagy. *Q Open* **2021**, *1*, qoab015. [CrossRef]

39. Popoff, M.; MacLeod, M.; Leschen, W. Attitudes towards the use of insect-derived materials in Scottish salmon feeds. *J. Insects Food Feed* **2017**, *3*, 131–138. [CrossRef]

40. Ribeiro, J.C.; Gonçalves, A.T.S.; Moura, A.P.; Varela, P.; Cunha, L.M. Insects as food and feed in Portugal and Norway—cross-cultural comparison of determinants of acceptance. *Food Qual. Prefer.* **2022**, *102*, 104650. [CrossRef]

41. Department of Agriculture, Food and the Marine. Policy: Agriculture and Food. Government of Ireland. 2020. Available online: https://www.gov.ie/en/policy/268a7-agriculture-and-food/ (accessed on 9 June 2023).

42. Sachs, J.; Kroll, C.; Lafortune, G.; Fuller, G.; Woelm, F. *Sustainable Development Report 2022*; Cambridge University Press: Cambridge, UK, 2022. [CrossRef]

43. Creswell, J.W. *Research Design: Qualitative, Quantitative, and Mixed Methods Approaches*, 4th ed.; SAGE: Thousand Oaks, CA, USA, 2014.

44. Goodman, L.A. Snowball sampling. *Ann. Math. Stat.* **1961**, *32*, 148–170. Available online: https://www.jstor.org/stable/2237615 (accessed on 9 June 2023).

45. Thomas, D.R. A general inductive approach for analyzing qualitative evaluation data. *Am. J. Eval.* **2006**, *27*, 237–246. [CrossRef]

46. Vaismoradi, M.; Jones, J.; Turunen, H.; Snelgrove, S. Theme development in qualitative content analysis and thematic analysis. *J. Nurs. Educ. Pract.* **2016**, *6*, 100–110. [CrossRef]

47. Nathanson, N.; Wilesmith, J.; Griot, C. Bovine spongiform encephalopathy (BSE): Causes and consequences of a common source epidemic. *Am. J. Epidemiol.* **1997**, *145*, 959–969. [CrossRef] [PubMed]
48. United Nations Conference for Sustainable Development (UNDP). Available online: https://sdgs.un.org/statements/united-nations-conference-sustainable-development-undp-14186 (accessed on 9 June 2023).
49. Rosenfeld, D.L.; Tomiyama, A.J. Gender differences in meat consumption and openness to vegetarianism. *Appetite* **2021**, *166*, 105475. [CrossRef]
50. Uzmay, A.; Cinar, G. The likelihood of sheep meat consumption in Turkey. *Ital. J. Food Sci.* **2017**, *29*, 209. [CrossRef]
51. Hone, M.; Nugent, A.P.; Walton, J.; McNulty, B.A.; Egan, B. Habitual protein intake, protein distribution patterns and dietary sources in Irish adults with stratification by sex and age. *J. Hum. Nutr. Diet.* **2020**, *33*, 465–476. [CrossRef] [PubMed]
52. Everest, S.J.; Thorne, L.T.; Hawthorn, J.A.; Jenkins, R.; Hammersley, C.; Ramsay, A.M.; Hawkins, S.A.; Venables, L.; Flynn, L.; Sayers, R.; et al. No abnormal prion protein detected in the milk of cattle infected with the bovine spongiform encephalopathy agent. *J. Gen. Virol.* **2006**, *87*, 2433–2441. [CrossRef] [PubMed]
53. Ryser, E.T. Safety of Dairy Products. In *Microbial Food Safety: Food Science Text Series*; Oyarzabal, O., Backert, S., Eds.; Springer: New York, NY, USA, 2012. [CrossRef]
54. FAO. FAOSTAT: Food Balances (2010–). Available online: https://www.fao.org/documents/card/fr/c/cb9574en/ (accessed on 14 June 2023).
55. Llagostera, P.F.; Kallas, Z.; Reig, L.; De Gea, D.A. The use of insect meal as a sustainable feeding alternative in aquaculture: Current situation, Spanish consumers' perceptions and willingness to pay. *J. Clean. Prod.* **2019**, *229*, 10–21. [CrossRef]
56. von Jeinsen, T.; Weinrich, R. Acceptance of insects as protein feed–evidence from pig and poultry farmers in France and in the Netherlands. *J. Insects Food Feed* **2023**, *9*, 707–719. [CrossRef]
57. Rumbos, C.I.; Mente, E.; Karapanagiotidis, I.T.; Vlontzos, G.; Athanassiou, C.G. Insect-based feed ingredients for aquaculture: A case study for their acceptance in Greece. *Insects* **2021**, *12*, 586. [CrossRef]
58. Sogari, G.; Menozzi, D.; Mora, C.; Gariglio, M.; Gasco, L.; Schiavone, A. How information affects consumers' purchase intention and willingness to pay for poultry farmed with insect-based meal and live insects. *J. Insects Food Feed* **2022**, *8*, 197–206. [CrossRef]
59. Naranjo-Guevara, N.; Fanter, M.; Conconi, A.M.; Floto-Stammen, S. Consumer acceptance among Dutch and German students of insects in feed and food. *Food Sci. Nutr.* **2021**, *9*, 414–428. [CrossRef] [PubMed]
60. House, J. Consumer acceptance of insect-based foods in the Netherlands: Academic and commercial implications. *Appetite* **2016**, *107*, 47–58. [CrossRef] [PubMed]
61. Ho, I.; Gere, A.; Chy, C.; Lammert, A. Use of Preference Analysis to Identify Early Adopter Mind-Sets of Insect-Based Food Products. *Sustainability* **2022**, *14*, 1435. [CrossRef]
62. Goel, S.; Obeng, A.; Rothschild, D. Non-Representative Surveys: Fast, Cheap, and Mostly Accurate. Work Pap 2015. Available online: https://researchdmr.com/FastCheapAccurate.pdf (accessed on 10 June 2023).

*Article*

# Livestock Farmers' Attitudes towards Alternative Proteins

Chloe Crawshaw and Jared Piazza *

Department of Psychology, Lancaster University, Lancaster LA1 4YF, UK
* Correspondence: j.piazza@lancaster.ac.uk

**Abstract:** New food technologies such as cultured meat, precision fermentation, and plant-based alternatives may one day supplant traditional modes of animal farming. Nonetheless, very little is known about how traditional animal farmers perceive these new products, despite being directly impacted by their advance. The present study explored the views of livestock farmers regarding emerging protein alternatives. We used a comparison group of omnivorous non-farmers as a frame of reference. Forty-five UK-based livestock farmers and fifty-three non-farmers read an informative vignette about emerging food technologies that reviewed their advantages vis-à-vis intensive animal agriculture. Afterwards, participants rated four products (plant-based burgers; plant-based milk alternatives; cultured beef; animal-free dairy milk) in terms of their personal appeal and how much they represented a positive change to the market. Participants furthermore voiced their agreement or disagreement towards 26 statements representing potential facilitators or barriers to product acceptance. Overall, farmers rated the four products less appealing and less beneficial to the industry compared to non-farmers. Positive change ratings tended to be higher than personal appeal ratings for all products. Both groups tended to agree that the alternatives offered advantages, particularly for the environment, resource use, food security, and animal treatment, though agreement rates were lower for farmers. Farmers tended to perceive more barriers to acceptance than non-farmers, with 'threat to farmers' and 'lack of support to local farmers' of paramount concern to both groups. These findings highlight how farmers' attitudes towards alternative proteins are mixed and, ultimately, shaped by the perceived vulnerability of farming communities.

**Keywords:** farmers; plant-based alternatives; cultured meat; animal-free dairy; consumer attitudes

**Citation:** Crawshaw, C.; Piazza, J. Livestock Farmers' Attitudes towards Alternative Proteins. *Sustainability* **2023**, *15*, 9253. https://doi.org/10.3390/su15129253

Academic Editors: Dario Donno and Dimitris Skalkos

Received: 25 April 2023
Revised: 24 May 2023
Accepted: 5 June 2023
Published: 8 June 2023

## 1. Introduction

The domestication of animals and the raising of livestock for food has been a normal feature of society for roughly 13,000 years. Throughout history, individuals and organizations have at times questioned the ethicality of this practice. The rise of intensive animal-agricultural systems in the twentieth century, as a solution to feeding larger populations at lower costs, have cast the issue in a new light, requiring ever-greater ethical scrutiny [1,2]. Large-scale "factory" farms or concentrated animal feeding operations, where vast numbers of animals are kept on feedlots or housed in cages and/or indoors, have a number of associated risks to human and planetary health, including causing animals undue stress and illness, increased risk of microbial resistance and zoonotic diseases, biodiversity and habitation loss, and elevated production of greenhouse gases [3–7].

Despite these issues, meat production continues to rise globally [8], and farming practices continue to intensify. In the UK, for example, the number of intensive pig and poultry farms increased by about 26% between 2011 and 2017 [9] and this trend does not appear to be reversing [10]. It was estimated in 2017 that over 70% of farmed animals in the UK are kept on industrialized "mega-farms" [11]. The growth of industrialized farming has also put pressure on small-scale farmers, who today struggle to make a profit and stay in business due to the rising costs of animal feed, fuel, and fertilizer, as well as labor shortages [12]. Furthermore, environmental demands to reduce herd sizes relative to land use have pushed farmers towards utilizing more concentrated forms of farming to maintain profits [13]. Even though consumer behavior and the demand for inexpensive foods is largely driving this trend, the vast majority of consumers say they are opposed to factory farming. For example, a recent YouGov survey by Open Cages [14] found that 78% of British respondents were opposed to it. Clearly, there is an appetite among consumers for more ethical forms of animal agriculture if practical concerns regarding retail costs and the availability of viable alternatives can be effectively addressed.

### 1.1. Alternative Proteins

Two major sources of alternative proteins include plant-based alternatives (PBAs) and cultured (i.e., 'animal free' or lab-based) animal products [15,16]. PBAs are made from plant proteins such as wheat, legumes, and fungi, and they are designed to mimic the taste and texture of traditional animal products [17]. For example, plant-based milk alternatives such as soy, almond, and oat-based beverages are made from pulse protein concentrate and water. Other components, such as vegetable oils, sugars and flavorings, can be added during production to more closely match the protein and fat content of cow's milk or alter the taste profile (for details, see, e.g., [18]). The market for PBAs has grown rapidly—for example, UK sales of meat-free foods increased by 40% between 2014 and 2019—and is estimated to reach over GBP 1.1 billion by 2024 [19]. Plant-based milk alternative sales are predicted to double in the UK by 2025 and be worth over GBP 705 million [20].

Cultured ('in vitro' or 'clean') meat and 'animal free' dairy products are comprised of animal proteins using cell-culturing technology. The principal benefit of such technology is that they do not require animal husbandry or slaughter. Cultured meat involves extracting cells from an animal and growing them in a laboratory [17,21]. The tissue produced can be turned into a wide variety of animal products. It is projected that, by 2030, cultured meat will reach mainstream production and account for up to 12% of the UK's consumer demand for meat [22] and be cost competitive with traditional meat [23]. Different from cultured meat, animal-free dairy is produced using precision fermentation, which uses microflora (e.g., yeast) to synthesize proteins, similar to methods used to produce insulin and rennet. The synthesized proteins can then be added to plant fats and water to create milk [24].

Both plant-based products and cultured products are envisioned as alternatives to intensive animal husbandry due to their substantially smaller impact on the environment and animals. In a review of over 57,000 UK products, plant-based meat alternatives were found to have roughly 1/5th to 1/15th the environmental impact compared to their corresponding animal products, and plant-based dairy alternatives had about half to 1/10th the impact [3]. It has been projected that cultured meat will produce up to 98% less greenhouse gas emissions, use 96% less water, and 99% less land than conventional beef [25] (cf. [22]). Additionally, the embrace of alternative proteins will likely substantially reduce the risk of zoonotic-disease transmission and environmental contamination, as animal–human interactions will be kept to a minimum, and lab production will eliminate the need for antibiotics, pesticides and other problematic substances [24].

Though alternative proteins arguably offer "a kinder, greener protein for a sustainable future" (Perfect Day slogan), they have not been embraced without concern. Consumer reactions to alternative proteins have been mixed, and there are practical, ontological, regulatory and legal obstacles facing animal-free meat and dairy [17,24,26]. Furthermore, there is growing concern about the health profile of some *processed* PBAs [27]. At the present, there is greater support among consumers for PBAs, though this may change as lab-grown animal products become more widely available and affordable. Research into initial attitudes towards cultured meat by Verbeke et al. [28] revealed a healthy skepticism about the new technology, though some consumers warmed up to the idea when learning about its potential benefits (e.g., to reduce the carbon footprint of red meat production). Since this initial work, research into alternative proteins has exploded. There have been several literature reviews targeting research into cultured meat acceptance (e.g., [29,30]). These reviews have observed a consistent trend, with the majority of consumers reporting a willingness to at least try cultured meat (e.g., [31]). Most studies comparing cultured meat to PBAs have found higher rates of interest for PBAs (e.g., [32,33]). For example, one recent study found that UK-based meat eaters rated (images of) plant-based burgers equally pleasant as conventional burgers, though they thought conventional burgers would be more filling [34]. By contrast, the same UK meat eaters thought cultured beef burgers would be less pleasant to eat compared to conventional burgers. Though consumer interest in cultured meat and animal-free dairy is mixed, studies show that consumers consistently recognize the potential for these alternatives to improve animal welfare and food safety (e.g., reducing the risk of contamination and disease transmission) and to reduce the impact of animal agriculture on the environment [29,35,36]. At the same time, impediments to acceptance extend beyond concerns about taste to include the perceived "unnaturalness" of cultured meat, its nutritional profile, affordability, safety concerns (e.g., distrust in food companies and regulatory bodies), and ethical concerns (e.g., possible harm to animals [e.g., calves] from whom cells and sera are taken) [17,26,29,30,37,38].

*1.2. Moderators of Alternative-Protein Acceptance and the Case of Farmers*

To date, most studies of alternative proteins have focused on the attitudes of everyday consumers. These studies have yielded crucial insights into the moderators of consumer acceptance. For instance, studies have observed greater interest in PBAs and cultured meat among younger rather than older adults [39,40]. Women are more likely to purchase PBAs than men [39,41,42], whereas men are more receptive to cultured meat [43]. Locality also seems to be an important moderator, with urban dwellers being more receptive to lab-grown products than rural dwellers [44,45] and more likely to purchase PBAs [41,46].

Arguably, farmers are a segment of the population with the greatest potential to be impacted by the rise of protein alternatives. Yet, currently, little is known about their attitudes towards these products. There is reason to believe that they may be more pessimistic and resistant than the average consumer, since their livelihoods could be adversely affected if the demand for alternative proteins is sufficiently disruptive to projected markets. Additionally, farmers, on average, tend to endorse more conservative values related to conformity and tradition and score lower on openness to change than the general population [47]. Furthermore, animal agriculture is an integral part of farmers' heritage, with farmers tending to live and work in the same location for multiple generations [48]. Though the food industry has largely controlled the narrative regarding animal-free meat and dairy, and as a result media coverage has largely been positive [49], this optimism about cultured animal products may not be shared by farmers. At least one study by Bryant and van der Weele [50] held focus groups with Dutch farmers about cultured meat and meat production. Among some of the concerns raised, farmers mentioned the precarity of their economic position due to increased governmental regulations and the lack of appreciation they felt from consumers and the government.

Concern about the impact of lab-grown animal products on farmer livelihoods might also be shared among the wider population of consumers. For instance, rural Irish consumers expressed concern about the economic impact of cultured meat on Irish farmers, noting the dependence of the Irish economy on beef production [44]. At the same time, farmers (at least in the US) do not appear to be very concerned about the rising demand for PBAs. When asked in 2021 to estimate the expected market share of PBAs, 31% of the US farmers surveyed expected a market share less than 1%, and another 55% expected it to be less than 10% [51]. Though some farmers anticipate some reduction in income due to the growth of alternative proteins, interviews with expert informants with ties to animal agriculture seem to reveal more pressing concerns expressed over the declining number of farmers due to increased pressure to intensify production, waning interest in the profession, and the expense of new technologies further alienating small-scale farmers [52]. Moreover, some livestock farmers may welcome the development of alternative proteins as an opportunity to diversify and for rural communities to host new infrastructures, machinery (e.g., bioreactors for cultured meat production) and jobs [13,52].

*1.3. Present Study and Research Questions*

The aim of the present study was to explore the attitudes of UK livestock farmers towards emerging protein alternatives, particularly cultured meat and animal-free dairy products and plant-based alternatives. We preregistered two sets of research questions (https://aspredicted.org/kz88s.pdf). The first set involved farmers' support for emerging meat and dairy alternatives. We presented UK farmers, recruited primarily from southwest England, with PBAs (specifically, plant-based burgers and milk alternatives) and cultured meat and dairy alternatives (specifically, cultured beef burgers and fermentation-derived dairy milk). Support for each product was assessed in terms of its personal appeal and whether it was perceived as a positive change within the food industry. We asked:

**RQ1**: *Of these four protein alternatives, which are farmers most supportive of?*

We contrasted farmers' attitudes with those of a comparable sample of UK-based non-farmers. The groups were matched on diet, age and gender. We asked:

**RQ2**: *How do farmers and non-farmers differ with respect to the alternative proteins they support?*

Our second set of research questions related to the perceived benefits offered by meat/dairy alternatives, and the perceived barriers to acceptance. The four products were presented as alternatives to factory-farmed beef and dairy products. Though most of the farmers surveyed for this study worked (or had previously worked) on small-scale farms, we used factory-farmed products as the contrasting referent because the majority of animal agricultural products (e.g., over 70% in the UK) are derived from mega-farms or concentrated animal feeding operations [11]. Participants were presented background information about alternative proteins. Subsequently, participants expressed their agreement or disagreement towards 26 statements reflecting potential advantages or disadvantages of the four target products. We asked:

**RQ3**: *What are the biggest perceived facilitators of product acceptance, and how do farmers and non-farmers differ in this respect?*

**RQ4**: *What are the biggest perceived barriers to product acceptance, and how do farmers and non-farmers differ in this respect?*

We made no specific hypotheses regarding which facilitators or barriers would be most commonly endorsed. However, we were particularly interested in discovering the extent to which "threat to farmers" (i.e., their livelihood and farming traditions) would be endorsed as a disadvantage. Additionally, we were interested to see to what extent farmers endorsed the animal welfare benefits of alternative proteins, given that farmers tend to believe that the welfare standards of animal agriculture are much higher than asserted by critics outside of the farming industry [53–55].

## 2. Method

*2.1. Participant Recruitment, Exclusions, and Demographics*

We pre-registered a plan to recruit a minimum of 40 participants per group (https://aspredicted.org/kz88s.pdf). This was a target we believed we could reliably recruit based on prior work with this population and the limited incentives we had at our disposal. Between July and August 2022, two groups of participants were recruited: adults with experience in livestock farming (farmers), and a comparison group of adults who were animal product consumers without livestock-farming experience (non-farmers). Farmers make up a very small percentage (0.2%) of the UK population [56], so we included in our recruitment criteria current farmers, retired farmers, farm workers, and members of farming families. Thirty-two livestock farmers predominately living in Gloucestershire were recruited using snowball sampling. Farmers that were known to the first author were contacted via telephone or social media, or visited in-person. Interested participants were provided a brief description of the study, a URL link to the online questionnaire, and a request to forward the information to other individuals in the farming community.

Individuals without internet access received a paper copy of the questionnaire and completed it at their convenience, and the first author later collected it from them. Because we struggled to collect the target sample of farmers via snowball sampling, we turned to Prolific to augment our strategy. A further 23 farmers were recruited using Prolific by pre-screening for UK-based workers in the 'Agriculture, Food, and Natural Resources' employment sector. Prolific participants were paid GBP 3.75. The description of the study encouraged participation among those with "experience of working with farmed animals." Non-farmers were recruited through convenience sampling. Contacts of the first author were emailed or contacted via social media and sent a URL link to the online survey to complete; ten non-farmers were recruited incidentally via Prolific (i.e., participants who reported not being involved in farming). As farmers are typically older males [57], we attempted, as much as possible, to match the ages and gender of the non-farmers to the farmers. To qualify for the study, farmers and non-farmers had to be omnivore, as the study focuses on the perspective of those who regularly consume meat and animal products, not those who are practicing plant-forward diets.

A total of 130 participants consented and completed the study: 55 farmers and 61 non-farmers. Following our preregistered exclusion criteria, 18 participants who reported dietary restrictions related to meat or animal products—including "semi-vegetarian or reducetarian", "pescatarian", "lacto- or ovo-vegetarian", "strict vegetarian", "dietary vegan" and "lifestyle vegan"—were excluded (10 farmers, 8 non-farmers). The final sample consisted of 45 farmers (14 "meat lover", 31 "omnivore"; 98% British) and 53 non-farmers (9 "meat lover", 44 "omnivore"; 96% British). Farmers consisted of 12 working farmers, 4 farm workers, 4 retired farmers, 24 members of farming families, and 1 unclassified.

Table 1 presents the age and gender profiles of the two groups. Farmers were somewhat older on average; however, a Kruskal-Wallis test showed that age did not significantly differ between farmers ($M_{\mathrm{rank}}$ = 52.82, $SD$ = 17.75, 95% CI [42.87, 53.53]), and non-farmers ($M_{\mathrm{rank}}$ = 46.68, $SD$ = 18.80, 95% CI [39.14, 49.50]), $H(1)$ = 1.14, $p$ = 0.28, $\varepsilon^2$ = 0.001. Thus, age was not treated as a covariate. The distribution of males to females was somewhat balanced within and between groups, and a Pearson's Chi-square test showed that gender distributions did not significantly differ between groups, $\chi^2$ (2, $N$ = 98) = 1.77, $p$ = 0.413, Cramer's $V$ = 0.13.

**Table 1.** Demographics by group.

| | $N$ | Age (In Years) | | Gender | | |
|---|---|---|---|---|---|---|
| | | $M$ | $SD$ | Male | Female | Non-Binary |
| Farmers | 45 | 48.36 | 16.18 | 25 | 19 | 1 |
| Non-Farmers | 53 | 41.86 | 20.31 | 26 | 27 | 0 |
| Total | 98 | | | | | |

## 2.2. Design

A 2 (Group) × 4 (Product Type) mixed design was used. Group was a between-subjects factor (Farmers, Non-Farmers), and Product type was within-subjects (Plant-based burgers, Cultured-beef burgers, Plant-based milk alternatives, Cultured [i.e., fermentation-derived] cows' milk).

## 2.3. Procedure

The study was approved by Lancaster University's Department of Psychology Ethics Committee on 30 May 2022. It was advertised to participants as a study about "existing and emerging food innovations". Participation was anonymous, and farmers were not asked to disclose the name or location of their farm. Participants were assigned a unique numerical ID that they could later use if they wished to withdraw their data. Before consenting to the study (by agreeing to several statements), they were first provided information about the study and researcher contact details, and they had an opportunity to ask questions. Next, participants read a vignette that outlined the problems with intensive farming, introduced plant-based and lab-cultured alternatives, and described the advantages of these alternatives relative to industrial-farmed animal products. Attention checks were embedded in this opening segment. Afterwards, participants answered measures that assessed (a) consumer appeal of the products, (b) perceived facilitators and barriers to product acceptance, (c) previous experiences with plant-based alternatives, purchase intentions, and shopping behavior, (d) farming experiences and demographics. Most participants completed the survey within 25 min ($M$ = 14.64, $SD$ = 7.45, excluding those who spent more than an hour). On completion, participants were debriefed, reminded of their right to withdraw their data, thanked and compensated.

### 2.4. Materials and Measures

### 2.4.1. Vignette about Alternatives to Animal Products

There were four main sections to the vignette, each section on a separate page, with attention checks at the bottom of three of the pages to encourage careful engagement. The full vignette and images can be found in Supplements A, along with the sources used in generating the vignette. Section 1 presented information about the widespread use of intensive farming practices and "concerns" with this way of producing animal products (e.g., "Animals are closely housed together and remain inside for the whole of their productive lives"). Section 2 introduced plant-based products as an alternative to industrial-farmed animal products and highlighted their "advantageous qualities" (e.g., "They mimic the taste and appearance of animal products"; "They are nutritionally similar to animal products, however, no animal slaughter is required"). Section 3 introduced cultured animal products and briefly explained cell culturing and precision fermentation methods of producing cultured beef and animal-free dairy. It included a helpful figure and highlighted "several expected advantageous qualities" (e.g., "Animals are only required for the initial cell or DNA samples, then no further animals are needed for production").

In the Section 4, a comparison table adapted from Van Loo et al. [58] was displayed, which presented the environmental impact of a plant-based soya burger and a cultured-beef burger relative to a factory-farmed beef burger. As a visual memory aide, participants were also presented a summary table comparing the key beneficial features of plant-based and cultured alternatives, vis-à-vis factory farming, discussed in the vignette. Participants were not excluded on the basis of failing the attention checks. Rather, if they selected the incorrect response, the correct answer was highlighted, and an explanation was provided as to why their response was incorrect.

### 2.4.2. Product Acceptance

Product acceptance was measured in two ways: how *appealing* the product is to them as a consumer, and the extent to which they view the product as a *positive change* within the wider food system. Participants were asked to "Imagine yourself in a future time when plant-based alternatives and cultured animal products are widely available and fairly equivalent in price to organic animal products". We had participants consider the products at an equivalent price since price is a major determinant of consumer behavior [46,59]. Many consumers assume that plant-based and cultured products are more expensive than conventional products, which can negatively impact their acceptance [43,60]. For each product, they considered "In such a future reality, how appealing would each of these products be to you?" rated on a 1–7 scale (1 = *Not at all appealing*; 7 = *Very appealing*). Likewise, for each product, they considered "To what extent do you consider the introduction of these products a positive change in the global food system?" (1 = *Not at all positive*, 7 = *Very positive*). Spearman's correlations revealed that the appeal and positive change measures related significantly (medium to large relationships for all measures), underscoring their reliability as dual aspects of product acceptance—see Table S1 for correlation table.

### 2.4.3. Facilitators and Barriers to Acceptance

For each product, participants viewed a list of statements (facilitators first, barriers second) and were asked to tick those they agreed with. The list of potential facilitators (10 items) and barriers (16 items) were generated by the authors' review of the literature (e.g., [29,59])—see Table 2 for the item labels and descriptions. An image of the product was displayed alongside the statements. Participants could also enter an additional comment underneath the list of statements to account for potentially overlooked factors. This task was repeated for all four products. The order of plant-based and cultured products was randomized to reduce potential order effects; however, to minimize participant confusion, the burger product was always presented first.

**Table 2.** Facilitator and barrier items: labels and descriptions.

| Label | Description |
| --- | --- |
| *Facilitators* | |
| Good taste | "I will enjoy the taste of [product]" |
| Good texture | "I will enjoy the texture of [product]" |
| Nutrition | "[product] is high in nutritional value (e.g., vitamins, minerals, fibre)" |
| Protein | "[product] is a good source of protein" |
| Health | "[product] is healthier than [traditional animal product] (e.g., lower in saturated fats, reduced risk of heart disease)" |
| Environment | "[product] is better for the environment (e.g., produces fewer greenhouse gases) than traditional [animal product]" |
| Resource use | "[product] is a better use of natural resources (e.g., requires less water or land to produce) than traditional [animal product]" |
| Animal treatment | "[product] involves better animal treatment than traditional [animal product]" |
| Food security | "[product] provides greater food security (e.g., reduced transmission of animal-derived diseases) than traditional [animal product]" |
| Curiosity | "I've never tried [product], but I'm curious to know how it would taste" |
| *Barriers* | |
| Bad taste | "I will not like the taste of [product]" |
| Bad texture | "I will not like the texture of [product]" |
| Satiety | "I don't think [product] would satisfy my hunger in the same way as traditional [animal product]" |
| No need | "I don't perceive a need for developing alternatives to traditional [animal product]" |
| Unwilling | "I'm not willing to give up traditional [animal product] as I enjoy it so much" |
| Does not support local farmers | "[product] does not support local farmers or the local economy" |
| Against culture or values | "[product] does not align with my culture or values" |
| Threatens tradition | "[product] threatens important food traditions we need to preserve" |
| Threatens farmers | "[product] threatens to put farmers out of business" |
| Unnatural to produce | "[product] is less natural to produce than the traditional [animal product]" |
| Unnatural content | "[product] contains more artificial ingredients than the traditional [animal product]" |
| Do not trust companies | "I don't trust food companies to produce [product] that is safe for consumers" |
| Food neophobia | "I'm not interested in [product] because I don't like changing my food habits" |
| Usability | "I would find it difficult to use [product]" |
| Cannot convince others | "It would be difficult to convince my partner or family to consume [product]" |
| Criticism | "My friends and/or family would criticise or tease me for consuming [product]" |

### 2.4.4. Experiences with Plant-Based Alternatives

Participants were asked about their experience of PBAs. Items included *plant-based . . . burgers, sausages, mince, chicken products e.g., nuggets, fish, milk, dairy products e.g., cheese, yoghurt* and *none of the above*. They selected which item(s) they had "previously tried or bought". They also indicated their level of intention towards buying plant-based products from the options: "No, I do not currently buy plant-based products and would never buy them", "No, I do not currently buy plant-based products, but I intend to start buying plant-based alternatives", "Yes, I do currently buy plant-based products occasionally", or "Yes, I do currently buy plant-based products regularly." Participants were further asked whether they were primarily responsible for household purchases (Yes/No) and where they typically shopped for animal products from the following options: *supermarkets, local shops such as local butchers, farmers market or local market,* and *I don't buy animal products.*

### 2.4.5. Farming Experiences and Demographics

Participants were asked whether they or members of their family had been involved in farming (Yes/No). Those that did classified themselves by selecting from the following options: *farmer, farm worker, retired farmer, part of a farming family,* and *other (please specify).* Participants provided information on their length of involvement in farming and classified their farm type and farm size using DEFRA [61] classifications (see Supplements B, Table S2, for categories and counts). Participants provided their age, gender, ethnicity, and nationality and self-classified their diet from a list of options adapted from Crawshaw and Piazza [53].

### 2.5. Analysis Plan

All statistical analyses were conducted using SPSS [62]. An anonymized version of the dataset is available at [https://osf.io/a7f9k]. The analysis plan involved, first, testing the distributions of each variable. Shapiro-Wilk tests showed that the assumption of normality was violated for all variables (see Table S3). The distributions of *product appeal* and *positive change* ratings show that farmers had bimodal distributions, and non-farmers had highly negatively skewed distributions (see Supplements C, Figures S1–S8). For non-parametric descriptive statistics of the key variables, see Table S4. Figure 1 presents means and standard deviations. Further (exploratory) analysis of the distributions for the different types of farmers showed that occupational farmers (farmers, farm workers and retired farmers) had more negative views of the products, compared to members of farming families (see Figures S1–S8). As stated in our preregistered analysis plan, because of the non-normal distributions, non-parametric tests were used, with one exception—we were unable to identify a suitable non-parametric alternative for the 2 × 4 MANOVA that was preregistered to test main effects and interactions on *product acceptance* (treating product appeal and positive change as two aspects of product acceptance). We report Pillai's Trace alongside the test statistics for the MANOVA as it is considered a robust test for violations of normality [63].

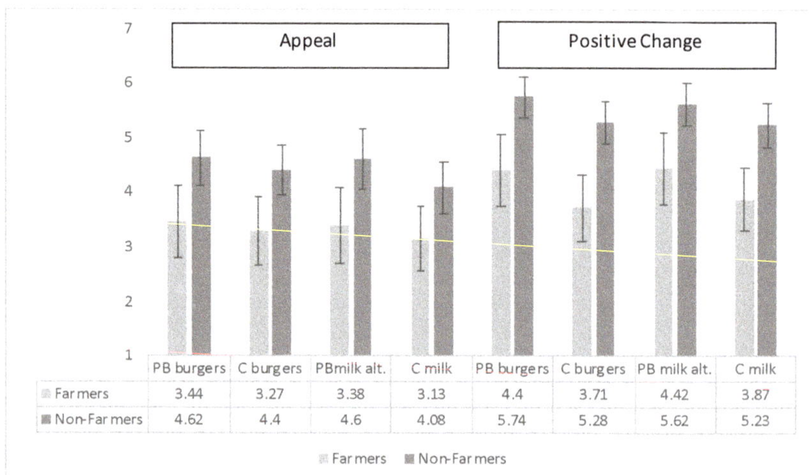



| | PB burgers | C burgers | PB milk alt. | C milk | PB burgers | C burgers | PB milk alt. | C milk |
|---|---|---|---|---|---|---|---|---|
| Farmers | 3.44 | 3.27 | 3.38 | 3.13 | 4.4 | 3.71 | 4.42 | 3.87 |
| Non-Farmers | 4.62 | 4.4 | 4.6 | 4.08 | 5.74 | 5.28 | 5.62 | 5.23 |

**Figure 1.** Product Acceptance: Means for Farmer and Non-Farmer. PB = Plant-based. C = Cultured. Error bars represent ±1 *SE.*

To explore the first set of research questions about product support (RQ1–2), pairwise Wilcoxon Signed Rank tests were used as follow-up tests, comparing the acceptance of each product (we did this first for farmers, then for non-farmers). A Bonferroni correction of alpha was applied based on the number of product comparisons, i.e., $p = 0.05/6 = 0.0083$. Further, Mann–Whitney U tests were conducted to contrast product acceptance ratings for farmers vs. non-farmers. A Bonferroni correction was applied based on the number of group comparisons, i.e., $p = 0.05/4 = 0.0125$. To explore the second set of research questions (RQ3–4), we conducted Chi-square tests comparing group differences in endorsement for each facilitator statement, and, separately, each barrier statement. A Bonferroni correction was applied based on the number of statements (facilitators, $p = 0.05/10 = 0.005$; barriers, $p = 0.05/16 = 0.0031$). Age was not treated as a covariate in any analysis—age correlated with only one measure (cultured burger appeal, with older participants finding cultured burgers less appealing)—see Table S1.

## 3. Results

### 3.1. Experience with Plant-Based Alternatives, Purchase, and Shopping Behavior: Farmers vs. Non-Farmers

Non-farmers were somewhat more likely to have tried plant-based alternatives, though there were comparably low rates for farmers—see Table S5. Non-farmers also appeared somewhat more willing to purchase or had already purchased PBAs (see Table S6), though not at statistically significant levels, $\chi^2(3) = 4.85$, $p = 0.183$, Cramer's V = 0.226. Most participants shopped at supermarkets (farmers = 80%; non-farmers = 98%), though at higher rates for non-farmers, $\chi^2(1) = 8.71$, $p = 0.003$, Cramer's V = 0.298. Farmers shopped at local shops more than non-farmers (71% vs. 28%, respectively), $\chi^2(1) = 17.87$, $p < 0.001$, Cramer's V = 0.427. Most participants reported being the primary food purchaser in their household (farmers = 67%; non-farmers = 70%).

### 3.2. RQ1–2: Which Animal-Product Alternatives Do Farmers and Non-Farmers Support?

The $2 \times 4$ MANOVA yielded a significant main effect for Group, $F(2, 95) = 8.86$, $p < 0.001$, partial-$\eta^2 = 0.16$, Pillai's Trace = 0.16, and Product, $F(6, 91) = 4.15$, $p = 0.001$, partial-$\eta^2 = 0.22$, Pillai's Trace = 0.22, on product acceptance, but no significant Group x Product interaction, $F(2, 95) = 0.73$, $p = 0.625$, partial-$\eta^2 = 0.05$, Pillai's Trace = 0.05.

#### 3.2.1. Farmers

Pairwise Wilcoxon Signed Rank tests revealed no significant differences in farmers' appeal ratings for the four products. A pattern emerged for perceptions of positive change, whereby farmers rated the plant-based products as a significantly more positive change for the industry than the cultured products—see Table S7. The only exception was between plant-based burgers vs. cultured milk, which did not reach adjusted significance levels. The positive change ratings for the two plant-based products were not significantly different, nor were the positive change ratings for the two cultured products. Thus, in answer to RQ1, farmers did not exhibit a personal preference for one product over another; however, they rated the plant-based products as a more positive change for the food industry than lab-grown products.

#### 3.2.2. Non-Farmers

There were no differences between non-farmer appeal ratings for the four products, see Table S8. Plant-based burgers were rated as a significantly more positive change in the industry than cultured products. However, there was no significant difference between ratings of plant-based milk and the cultured products, nor between the plant-based products themselves, nor between the cultured products themselves. Thus, non-farmers exhibited a similar response pattern as farmers, not preferring any one product over another, and tending to see plant-based products as a more positive change than cultured products.

### 3.2.3. Farmers vs. Non-Farmers

Mann–Whitney U tests revealed that non-farmers perceived the plant-based burgers, cultured burgers, and plant-based milk alternatives as significantly more appealing than farmers—see Table 3. There was no significant group difference in the appeal of animal-free milk. Non-farmers perceived all products as a significantly more positive change to the food industry compared to farmers. Thus, relative to farmers, non-farmers exhibited greater overall support for both the plant-based and lab-cultured alternative proteins (RQ2).

**Table 3.** Mann–Whitney U tests: product acceptance ratings for farmers vs. non-farmers.

| | Farmer $M_{rank}$ | Non-Farmer $M_{rank}$ | $U$ | $Z$ | $\eta^2$ | $p$ |
|---|---|---|---|---|---|---|
| *Appeal* | | | | | | |
| PB Burgers | 41.00 | 56.72 | 1575.00 | 2.78 | 0.08 | 0.005 * |
| C Burgers | 40.74 | 56.93 | 1586.50 | 2.86 | 0.08 | 0.004 * |
| PB Milk Alt. | 41.29 | 56.47 | 1562.00 | 2.67 | 0.07 | 0.008 * |
| C Milk | 42.21 | 55.69 | 1520.50 | 2.38 | 0.06 | 0.017 |
| *Pos. Change* | | | | | | |
| PB Burgers | 40.19 | 57.41 | 1611.50 | 3.06 | 0.10 | 0.002 * |
| C Burgers | 37.34 | 59.82 | 1739.50 | 3.99 | 0.16 | 0.000 * |
| PB Milk Alt. | 41.69 | 56.13 | 1544.00 | 2.56 | 0.07 | 0.010 * |
| C Milk | 38.57 | 58.78 | 1684.50 | 3.59 | 0.13 | 0.000 * |

Note. $N_{farmer}$ = 45, $N_{non\text{-}farmer}$ = 53. For *SD* and 95% CI see Table S4. * $p$ < 0.0125. PB = Plant-based. C = Cultured.

### 3.3. RQ3–4: Perceived Facilitators and Barriers of Product Acceptance

Tables 4 and 5 present the percentage of farmers and non-farmers that endorsed the ten facilitators and sixteen barriers to acceptance for the four products.

**Table 4.** Facilitator and barrier endorsement for plant-based burgers and cultured burgers: ordered by farmer endorsement (%) for plant-based burgers. Group-comparison statistics (Chi-square test) provided.

| | Plant-Based Burger | | | Cultured Burger | | |
|---|---|---|---|---|---|---|
| | Farmer | Non-Farmer | $\chi^2$ (V) | Farmer | Non-Farmer | $\chi^2$ (V) |
| *Facilitators* | | | | | | |
| Environment | 58% | 92% | 16.29 * (0.408) | 53% | 87% | 13.35 * (0.369) |
| Health | 56% | 85% | 10.27 * (0.324) | 40% | 55% | 2.11 (0.147) |
| Resource Use | 53% | 81% | 8.70 * (0.298) | 51% | 83% | 11.46 * (0.342) |
| Animal Treatment | 51% | 81% | 9.97 * (0.319) | 40% | 79% | 15.79 * (0.401) |
| Food Security | 51% | 79% | 8.63 * (0.297) | 44% | 83% | 15.98 * (0.404) |
| Nutrition | 47% | 75% | 8.59 * (0.296) | 38% | 57% | 3.46 (0.188) |
| Protein | 47% | 57% | 0.96 (0.099) | 44% | 72% | 7.48 (0.276) |
| Good Taste | 40% | 60% | 4.04 (0.203) | 40% | 64% | 5.70 (0.241) |
| Good Texture | 33% | 30% | 0.11 (0.034) | 27% | 51% | 5.99 (0.247) |
| Curiosity | 24% | 42% | 3.17 (0.180) | 42% | 81% | 15.85 * (0.402) |
| *Barriers* | | | | | | |
| Does Not Support Local Farmers | 67% | 34% | 10.42 * (0.326) | 67% | 45% | 4.50 (0.214) |
| Threatens Farmers | 58% | 45% | 1.52 (0.125) | 60% | 49% | 1.17 (0.109) |
| Unnatural Content | 38% | 25% | 2.01 (0.143) | 40% | 23% | 3.45 (0.188) |
| Cannot Convince Others | 33% | 36% | 0.07 (0.026) | 42% | 25% | 3.47 (0.188) |
| Unnatural to Produce | 33% | 17% | 3.52 (0.189) | 56% | 45% | 1.03 (0.102) |
| No Need | 31% | 8% | 9.01 * (0.303) | 31% | 8% | 9.01 * (0.303) |
| Against Culture or Values | 31% | 9% | 7.32 (0.273) | 33% | 9% | 14.83 * (0.389) |
| Threatens Tradition | 29% | 11% | 4.81 (0.221) | 33% | 19% | 8.56 * (0.296) |

**Table 4.** *Cont.*

| | Plant-Based Burger | | | Cultured Burger | | |
| --- | --- | --- | --- | --- | --- | --- |
| | **Farmer** | **Non-Farmer** | **$\chi^2$ (V)** | **Farmer** | **Non-Farmer** | **$\chi^2$ (V)** |
| Satiety | 27% | 28% | 0.03 (0.018) | 18% | 8% | 2.37 (0.156) |
| Do Not Trust Companies | 22% | 11% | 2.12 (0.147) | 33% | 19% | 2.68 (0.165) |
| Bad Taste | 20% | 11% | 1.41 (0.120) | 9% | 9% | 0.01 (0.009) |
| Unwilling | 20% | 11% | 1.41 (0.120) | 20% | 6% | 4.66 (0.218) |
| Bad Texture | 18% | 23% | 0.35 (0.060) | 13% | 9% | 0.37 (0.062) |
| Criticism | 18% | 8% | 2.37 (0.156) | 18% | 8% | 2.37 (0.156) |
| Usability | 13% | 11% | 0.09 (0.031) | 11% | 9% | 0.08 (0.028) |
| Food Neophobia | 9% | 11% | 0.16 (0.040) | 11% | 11% | 0.00 (0.003) |

Note. V = Cramer's V (effect size estimate). For facilitators, * $p < 0.005$. For barriers, * $p < 0.0031$.

**Table 5.** Facilitator and barrier endorsement for plant-based milk alternatives and animal-free milk: ordered by farmer endorsement (%) for plant-based milk. Group comparison statistics (Chi-square test) provided.

| | Plant-Based Milk Alt. | | | Animal-Free Milk | | |
| --- | --- | --- | --- | --- | --- | --- |
| | **Farmer** | **Non-Farmer** | **$\chi^2$ (V)** | **Farmer** | **Non-Farmer** | **$\chi^2$ (V)** |
| *Facilitators* | | | | | | |
| Nutrition | 60% | 72% | 1.49 (0.123) | 40% | 60% | 4.04 (0.203) |
| Environment | 60% | 70% | 1.03 (0.103) | 51% | 81% | 9.97 * (0.319) |
| Food Security | 53% | 70% | 2.81 (0.169) | 51% | 60% | 0.85 (0.093) |
| Animal Treatment | 49% | 79% | 9.90 * (0.318) | 40% | 68% | 7.67 (0.280) |
| Health | 47% | 55% | 0.63 (0.080) | 29% | 38% | 0.85 (0.093) |
| Resource Use | 40% | 66% | 6.64 (0.260) | 49% | 81% | 11.33 * (0.340) |
| Protein | 36% | 62% | 6.94 (0.266) | 36% | 66% | 9.06 * (0.304) |
| Good Taste | 33% | 51% | 3.08 (0.177) | 27% | 47% | 4.35 (0.211) |
| Good Texture | 27% | 42% | 2.37 (0.155) | 22% | 43% | 4.89 (0.223) |
| Curiosity | 22% | 25% | 0.07 (0.027) | 42% | 72% | 8.69 * (0.298) |
| *Barriers* | | | | | | |
| Does Not Support Local Farmers | 62% | 36% | 6.78 (0.263) | 73% | 49% | 5.99 (0.247) |
| Threatens Farmers | 53% | 51% | 0.06 (0.024) | 64% | 53% | 1.35 (0.117) |
| Unnatural Content | 47% | 23% | 6.29 (0.253) | 44% | 25% | 4.32 (0.210) |
| Cannot Convince Others | 40% | 36% | 0.18 (0.043) | 40% | 23% | 3.45 (0.188) |
| Unwilling | 33% | 23% | 1.39 (0.119) | 22% | 11% | 2.12 (0.147) |
| Against Culture or Values | 33% | 6% | 12.43 * (0.356) | 38% | 8% | 13.21 * (0.367) |
| Usability | 31% | 23% | 0.90 (0.096) | 20% | 17% | 0.15 (0.039) |
| Unnatural to Produce | 31% | 32% | 0.01 (0.010) | 58% | 42% | 2.58 (0.162) |
| Threatens Tradition | 29% | 11% | 4.81 (0.221) | 33% | 13% | 5.66 (0.240) |
| Bad Taste | 27% | 28% | 0.03 (0.018) | 16% | 13% | 0.11 (0.033) |
| Bad Texture | 27% | 19% | 0.85 (0.093) | 13% | 8% | 0.89 (0.095) |
| No Need | 27% | 8% | 6.51 (0.258) | 27% | 11% | 3.82 (0.197) |
| Criticism | 22% | 6% | 5.80 (0.243) | 22% | 4% | 7.71 (0.280) |
| Satiety | 20% | 8% | 3.28 (0.183) | 18% | 6% | 3.59 (0.191) |
| Food Neophobia | 18% | 9% | 1.47 (0.123) | 13% | 6% | 1.72 (0.132) |
| Do Not Trust Companies | 18% | 8% | 2.37 (0.156) | 40% | 19% | 5.33 (0.233) |

Note. V = Cramer's V (effect size estimate). For facilitators, * $p < 0.005$. For barriers, * $p < 0.0031$.

### 3.3.1. Facilitators

In general, non-farmers tended to agree with the facilitator statements at rates higher than farmers (RQ3). The rank order of facilitators was fairly similar between groups. For all products, apart from plant-based milk alternatives, 'Environment' and 'Resource Use' were the most popular facilitators for both groups. 'Food Security' was also an important facilitator for non-farmers for all products, but less so for farmers. For plant-based milk alternatives, 'Nutrition' and 'Environment' were the most important facilitators for farmers, whereas 'Animal Treatment' and 'Nutrition' were the most important for non-farmers. Regarding group-level differences in facilitator endorsement, the most consistent difference emerged with regards to 'Animal Treatment' (all products except animal-free milk) with non-farmers seeing animal welfare as a benefit at rates higher than farmers. 'Environment' and 'Resource Use' also tended to be viewed as facilitators by non-farmers more so than by farmers. For cultured products, 'Curiosity' was endorsed as a facilitator by non-farmers more so than farmers. Finally, non-farmers were more optimistic than farmers about the 'Protein' content of plant-based milk alternatives and animal-free milk.

### 3.3.2. Barriers

In general, farmers tended to endorse most barriers to acceptance at rates higher than non-farmers (RQ4), though the rank order of barriers was quite similar between groups—see Tables 4 and 5. For all products, 'Does Not Support Local Farmers' and 'Threatens Farmers' were the highest endorsed barriers for both groups. 'Unnatural to Produce' was an important barrier for both groups for the cultured products, and 'Cannot Convince Others' specifically for farmers. Regarding group-level differences in agreement, 'Against Culture or Values' was endorsed at a significantly higher rate by farmers than non-farmers as a barrier to cultured products and plant-based milk alternatives. 'Does not Support Local Farmers' was a significantly greater concern of farmers, particularly for animal-free milk and plant-based products. Farmers perceived there to be 'No Need' for plant-based and cultured burgers at rates higher than non-farmers.

## 4. General Discussion

The need for more sustainable alternatives to livestock production has driven a number of important technological advances in cellular and acellular agriculture and plant-based alternatives. These technologies have the potential to revolutionize the food industry, but they also have the power to transform lives and whole communities dependent on traditional forms of animal agriculture [52]. With this in mind, the present study explored the attitudes of UK-based livestock farmers towards emerging alternative proteins—individuals likely to be directly impacted by the advance of such technologies. We utilized a comparison group of non-farmers of similar nationality, age, gender and dietary profile as a frame of reference. The study returned a number of important insights that help expand what little we know regarding this segment of producer–consumers.

As might be expected, relative to non-farmer consumers, farmers generally considered the four alternatives—plant-based burgers and milk alternatives, cultured burgers and animal-free milk—significantly less appealing and less of a positive change to the food system. A subset of the farmers was strongly opposed to these meat and dairy alternatives; however, not all farmers were as pessimistic about them. In general, occupational farmers tended to be more pessimistic than farming family members. Consistent with some previous findings (e.g., [34]), plant-based alternatives were generally considered by both groups as more appealing than cultured products. PBAs were furthermore seen as a more positive change to the food industry than cultured products.

Investigation into the reasons for the less positive reception of cultured alternatives revealed that both groups were highly concerned about the impact that lab-cultured products would have on traditional farmers. This is a concern that has been raised in at least two other studies. Shaw and Iomaire [44] surveyed the views of urban and rural consumers in Ireland regarding cultured meat, and in their qualitative analysis they uncovered a theme related to the detrimental impact of cultured meat for Irish farmers and the Irish economy (among other themes). By contrast, Newton and Blaustein-Rejto [52] interviewed 37 experts with ties to the meat and plant-based meat sectors regarding the challenges of the alternative-protein market. Most respondents expressed little concern about alternative proteins supplanting traditional animal products. Instead, respondents expected that, at least in the near future, meat alternatives would form a complementary market that would help manage the growing demand for meat. Nonetheless, some concerns were voiced around the potential for new technologies to shift cultural narratives around food, such that traditional farming might appear outdated, inefficient, and, potentially, unethical.

In the present study, we observed high levels of concern among UK livestock farmers regarding the potential for protein alternatives to impact the lives of farmers. Unexpectedly, 'support for farmers' was also of paramount concern for non-farmers—in fact, it was the most commonly perceived barrier for all four products, though at rates consistently below farmers. Among the alternatives investigated here, farmers were most concerned about the economic impact of animal-free milk: 73% agreed that this technology does not support farmers, and 64% agreed that it will put dairy farmers out of business (compared to 62% and 53%, respectively, for plant-based milk alternatives). Farmers in our study were largely dairy farmers; thus, the somewhat elevated attitudes towards fermentation-derived milk may partly reflect a personal concern for their own livelihood. Plant-based milk alternatives are purchased by a third of British adults [64], which is far from market saturation. Dairy farmers may be somewhat more apprehensive about animal-free milk as it is expected to be identical in taste and texture to cow's milk [24,35,65]. However, it is worth pointing out that, at least among our sample of UK farmers, there was some doubt about whether the 'taste' and 'texture' profile of animal-free milk would be a strong facilitator of its acceptance (see Table 5). It may be that modern consumers are increasingly recognizing that the taste/texture profile of alternative proteins are not significant barriers to their appeal. Yet, they also recognize that what is *most appealing* about these alternatives has more to do with their ecological, health, safety, and welfare advantages than their taste (see also [59]).

Consistent with this theorizing, both farmers and non-farmers thought that alternative proteins represented a positive change in the market more than they found the products personally appealing. The relatively high positive-change ratings are likely explained by the perception that alternative proteins provide improvements upon industrial farming related to food security, resource use, environmental impact, animal treatment, and, for PBAs, nutrition. All of these positives were endorsed more strongly by non-farmers than farmers (though most consistently with regards to animal treatment). The lower ratings of personal appeal are likely explained by the perceived threat they pose to farmers and the perceived "unnaturalness" of the products and processes by which they are generated. These perceptions were higher for cultured animal products than PBAs (see Tables 4 and 5). Unnaturalness perceptions have been documented in previous work on alternative proteins, particularly cultured meat (e.g., [37,38,43,66]). In the present study, unnaturalness concerns were endorsed more widely by farmers than non-farmers. However, endorsement rates rarely surpassed 50%. Thus, even farmers were divided regarding the perceived "unnaturalness" of alternative proteins as a barrier to their acceptance.

In addition, a small percentage of consumers felt that there was 'No need' to replace traditional animal products, did not trust the companies involved, viewed the products as threatening to cultural traditions, and/or believed that people in their community would be against them. All of these concerns tended to be reported at higher rates by farmers than non-farmers (especially concern for the loss of cultural traditions). Lack of trust in companies and/or regulatory bodies is a concern that has been documented in previous studies on cultured meat acceptance [29], for example, among Irish [44] and Chinese consumers [67]. The concern regarding cultural traditions likely connects to the perceived threat to farmers. If farming communities are hurt by the growth of the alternative protein market, this will likely entail a corresponding loss of rural landscapes, communities, and traditions [28]. Though our participants seemed to focus on the *immediate* economic implications of alternative proteins for farmers (e.g., going out of business), a few participants (particularly farmers) also considered the more distal impact this could have in terms of eroding rural cultural traditions.

As a whole, the UK farmers we sampled had somewhat less experience purchasing and consuming PBAs than the non-farmers. They also appeared less interested in trying them (33% said they would 'never' buy them vs. 21% of non-farmers; see Tables S5 and S6).

However, group differences in PBA consumption and purchase intentions were not statistically significant. Overall, the results portray a diverse picture of farmers, who appear to fall along a spectrum of product "rejection", "contemplation", and even some "action", as described by the transtheoretical model of behavioral change [68]. Rejection involves not adopting a practice *despite* being aware of the arguments in favor of it, as opposed to simply being uninformed. Though some farmers rejected the positive claims about animal product alternatives, about 40% to 60% agreed that such products entailed improvements in animal treatment, resource use, and a reduced impact on the environment. A small percentage had even purchased plant-based alternatives. By contrast, UK-based non-farmers provided responses that might better typify a range of "contemplation", "preparation" and "action" stages of change [68]. Non-farmers tended to accept the positive arguments made about plant-based and cultured alternatives at rates above farmers, were more curious than farmers about lab-based alternatives, and they were less likely to perceive barriers to their adoption. Arguably, the differences we observed between farmers and non-farmers may partly be due to farmers' greater *personal* investment in the production of traditional animal products (see also [53]). However, the farmers in our sample also tended to shop locally and support small-scale farms more so than the non-farmers. Thus, the farmers' greater investment in farming *communities* and rural culture, which they perceived as under threat, likely also played a role in their heightened resistance to these emerging alternatives.

### 4.1. Limitations and Future Directions

There were several limitations to this research. Firstly, a fairly small number of occupational livestock farmers were recruited, with nationalities restricted to the UK. Livestock farmers make up a very small percentage of the UK population (e.g., 0.5% in England [56]), which poses challenges to sampling. Continued effort should be made by researchers to explore the views of farming communities on this topic, particularly in under-researched regions (e.g., Africa, Asia). The present findings suggest that addressing concerns about the future of farmers and rural communities may be paramount to promoting acceptance of these new food technologies. Secondly, to mitigate the length of the questionnaire, we had to restrict our focus to four animal-product alternatives. Future research would benefit from widening the scope to include, for example, plant-based and animal-free egg replacers. Future studies should also probe participants' views of animal product alternatives using more extensive, open-ended questions, and by having participants consider the use of different PBAs within specific meals or beverages (e.g., in coffee [69]). Though we allowed participants to volunteer comments at the end of the survey, this method yielded limited insights, as only a handful of participants made use of this opportunity. It would also be of value to systematically compare the attitudes held by occupational farmers and members of farming families, as our data suggested there may be graded differences that relate to a person's level of involvement in animal agriculture. Finally, future work with farmers would benefit from probing in more detail the specific apprehensions farmers have regarding the 'threat' protein alternatives pose to farmers and farming traditions, and how these perceived threats may be mitigated.

### 4.2. Applications

Our findings have direct implications for how alternative proteins might be better positioned to be more broadly appealing to consumers, particularly from rural backgrounds. Alternative proteins represent both opportunities and challenges for farming communities [52]. While it remains to be seen whether the popularity of alternative proteins will be significant enough to supplant the steady demand for conventional animal products [70], our findings suggest that farmers and non-farmers alike are concerned about the implications of protein alternatives for farming communities. Thus, it would be to the benefit of stakeholders and proponents of emerging alternatives to consider ways to center the investment in alternative proteins around rural life, for example, through the repurposing of existing land and infrastructures to support crop production for PBAs or cultured meat facilities [13]. Guarantees to invest in rural communities and support farming transitions would help mitigate concerns about farmers being further alienated by changes to the market and send a message to consumers that rural traditions have a valued role in the future food system.

*4.3. Conclusions*

The present study revealed that although farmers were less enthusiastic than non-farmers about the growth of emerging protein alternatives, they did recognize many of the advantages of these products when contrasted with traditional, large-scale animal agriculture. The farmers' biggest apprehension about these food technologies was not their perceived "unnaturalness", but their potential to threaten and damage the livelihoods of farmers and farming communities. This was a concern echoed by non-farmers to a high degree as well. Thus, a key take-away from this research is the need to address anxieties about what it means for the future of rural communities if protein alternatives and emerging food technologies continue to expand and reshape the animal-product market. Ongoing work in this area should consider ways that stakeholders can manage these anxieties via the commitments and assurances they make to producers and consumers about the progress of alternative proteins.

**Supplementary Materials:** The following supporting information can be downloaded at: https://www.mdpi.com/article/10.3390/su15129253/s1. [5,24,25,71–76].

**Author Contributions:** Conceptualization, C.C. and J.P.; Methodology, C.C. and J.P.; Formal analysis, C.C.; Investigation, C.C.; Writing—original draft, C.C. and J.P.; Supervision, J.P.; Project administration, C.C. All authors have read and agreed to the published version of the manuscript.

**Funding:** This research received no external funding.

**Institutional Review Board Statement:** The study was approved by Lancaster University's Department of Psychology Ethics Committee for MSc Student Projects on 30 May 2022.

**Informed Consent Statement:** Informed consent was obtained from all subjects involved in the study.

**Data Availability Statement:** The data presented in this study are openly available at https://osf.io/a7f9k/.

**Conflicts of Interest:** The authors declare no conflict of interest.

## References

1. Hartung, J. A short history of livestock production. In *Livestock Housing: Modern Management to Ensure Optimal Health and Welfare of Farm Animals*; Wageningen Academic Publishers: Wageningen, The Netherlands, 2013; pp. 81–146. Available online: https://www.academia.edu/download/36706331/Lecture_3_livestock_Population_and_their_history.pdf (accessed on 18 April 2023).
2. Reese, J. *The End of Animal Farming: How Scientists, Entrepreneurs, and Activists Are Building an Animal-Free Food System*; Beacon Press: Boston, MA, USA, 2018.
3. Clark, M.; Springmann, M.; Rayner, M.; Scarborough, P.; Hill, J.; Tilman, D.; Macdiarmid, J.I.; Fanzo, J.; Bandy, L.; Harrington, R.A. Estimating the environmental impacts of 57,000 food products. *Proc. Natl. Acad. Sci. USA* **2022**, *119*, e2120584119. [CrossRef] [PubMed]
4. Jones, B.A.; Grace, D.; Kock, R.; Alonso, S.; Rushton, J.; Said, M.Y.; McKeever, D.; Mutua, F.; Young, J.; McDermott, J.; et al. Zoonosis emergence linked to agricultural intensification and environmental change. *Proc. Natl. Acad. Sci. USA* **2013**, *110*, 8399–8404. [CrossRef] [PubMed]
5. Poore, J.; Nemecek, T. Reducing food's environmental impacts through producers and consumers. *Science* **2018**, *360*, 987–992. [CrossRef] [PubMed]
6. Ritchie, H.; Roser, M.; Rosado, P. CO$_2$ and Greenhouse Gas Emissions. Our World in Data. 2020. Available online: https://ourworldindata.org/co2-and-other-greenhouse-gas-emissions (accessed on 18 April 2023).
7. Willett, W.; Rockström, J.; Loken, B.; Springmann, M.; Lang, T.; Vermeulen, S.; Garnett, T.; Tilman, D.; DeClerck, F.; Wood, A.; et al. Food in the Anthropocene: The EAT–Lancet Commission on healthy diets from sustainable food systems. *Lancet* **2019**, *393*, 447–492. [CrossRef]

8.  Revell, B. Meat and Milk Consumption 2050: The Potential for Demand-side Solutions to Greenhouse Gas Emissions Reduction. *Eurochoices* **2015**, *14*, 4–11. [CrossRef]
9.  Davies, M.; Wasley, A. Intensive Farming in the UK, by Numbers. *The Bureau of Investigative Journalism*, 17 July 2017. Available online: https://www.thebureauinvestigates.com/stories/2017-07-17/intensive-numbers-of-intensive-farming (accessed on 18 April 2023).
10. Colley, C.; Wasley, A. Industrial-Sized Pig and Chicken Farming Continuing to Rise in UK. *The Guardian*, 7 April 2020. Available online: https://www.theguardian.com/environment/2020/apr/07/industrial-sized-pig-and-chicken-farming-continuing-to-rise-in-uk (accessed on 18 April 2023).
11. Compassion in World Farming. UK Factory Farming Map. 2017. Available online: https://assets.ciwf.org/factory-farm-map/ (accessed on 18 April 2023).
12. Wood, Z. 'It Is Unsustainable': Soaring Inflation Squeezes Budgets of UK Dairy Farmers. *The Guardian*, 25 March 2022. Available online: https://www.theguardian.com/environment/2022/mar/25/dairy-farmers-milk-production-inflation-prices-costs (accessed on 18 April 2023).
13. Proveg International. Amplifying Farmers Voices: Farming Perspectives on Alternative Proteins and a Just Transition. 2022. Available online: https://corporate.proveg.com/wp-content/uploads/2022/06/Amplifying_Farmers_Voices_Report.pdf (accessed on 18 April 2023).
14. Open Cages. Majority of Brits Strongly Oppose the Use of Cruel Farming Practises to Produce Cheap Food, Poll Shows. *Pressat*, 12 February 2022. Available online: https://pressat.co.uk/releases/majority-of-brits-strongly-oppose-the-use-of-cruel-farming-practises-to-produce-cheap-food-poll-shows-57867bb4edf8d6346685aa0b4f10bdcc/ (accessed on 18 April 2023).
15. Bryant, C.J. Plant-based animal product alternatives are healthier and more environmentally sustainable than animal products. *Future Foods* **2022**, *6*, 100174. [CrossRef]
16. Santo, R.E.; Kim, B.F.; Goldman, S.E.; Dutkiewicz, J.; Biehl, E.M.B.; Bloem, M.W.; Neff, R.A.; Nachman, K.E. Considering Plant-Based Meat Substitutes and Cell-Based Meats: A Public Health and Food Systems Perspective. *Front. Sustain. Food Syst.* **2020**, *4*, 134. [CrossRef]
17. Lee, H.J.; Yong, H.I.; Kim, M.; Choi, Y.-S.; Jo, C. Status of meat alternatives and their potential role in the future meat market—A review. *Asian-Australas. J. Anim. Sci.* **2020**, *33*, 1533–1543. [CrossRef]
18. Vogelsang-O'dwyer, M.; Zannini, E.; Arendt, E.K. Production of pulse protein ingredients and their application in plant-based milk alternatives. *Trends Food Sci. Technol.* **2021**, *110*, 364–374. [CrossRef]
19. Mintel. Plant-Based Push: UK Sales of Meat-Free Foods Shoot Up 40% between 2014–19. 17 January 2020. Available online: https://www.mintel.com/press-centre/food-and-drink/plant-based-push-uk-sales-of-meat-free-foods-shoot-up-40-between-2014-19 (accessed on 18 April 2023).
20. The Vegan Society. UK Plant Milk Market. 2019. Available online: https://www.vegansociety.com/news/market-insights/dairy-alternative-market/european-plant-milk-market/uk-plant-milk-market (accessed on 18 April 2023).
21. Humbird, D. Scale-up economics for cultured meat. *Biotechnol. Bioeng.* **2021**, *118*, 3239–3250. [CrossRef] [PubMed]
22. Oxford Economics. The Socio-Economic Impact of Cultivated Meat in the UK. *Oxford Economics*, 30 September 2021. Available online: https://www.oxfordeconomics.com/resource/the-socio-economic-impact-of-cultivated-meat-in-the-uk/ (accessed on 18 April 2023).
23. Watson, E. When Will Cell-Cultured Meat Reach Price Parity with Conventional Meat? Food Navigator USA. 15 March 2021. Available online: https://www.foodnavigator-usa.com/Article/2021/03/15/When-will-cell-cultured-meat-reach-price-parity-with-conventional-meat# (accessed on 18 April 2023).
24. Mendly-Zambo, Z.; Powell, L.J.; Newman, L.L. Dairy 3.0: Cellular agriculture and the future of milk. *Food Cult. Soc.* **2021**, *24*, 675–693. [CrossRef]
25. Tuomisto, H.L.; de Mattos, M.J.T. Environmental Impacts of Cultured Meat Production. *Environ. Sci. Technol.* **2011**, *45*, 6117–6123. [CrossRef] [PubMed]
26. Post, M.J.; Levenberg, S.; Kaplan, D.L.; Genovese, N.; Fu, J.; Bryant, C.J.; Negowetti, N.; Verzijden, K.; Moutsatsou, P. Scientific, sustainability and regulatory challenges of cultured meat. *Nat. Food* **2020**, *1*, 403–415. [CrossRef]
27. Macdiarmid, J.I. The food system and climate change: Are plant-based diets becoming unhealthy and less environmentally sustainable? *Proc. Nutr. Soc.* **2022**, *81*, 162–167. [CrossRef]
28. Verbeke, W.; Marcu, A.; Rutsaert, P.; Gaspar, R.; Seibt, B.; Fletcher, D.; Barnett, J. 'Would you eat cultured meat?': Consumers' reactions and attitude formation in Belgium, Portugal and the United Kingdom. *Meat Sci.* **2015**, *102*, 49–58. [CrossRef] [PubMed]
29. Bryant, C.; Barnett, J. Consumer Acceptance of Cultured Meat: An Updated Review (2018–2020). *Appl. Sci.* **2020**, *10*, 5201. [CrossRef]
30. Hartmann, C.; Siegrist, M. Consumer perception and behaviour regarding sustainable protein consumption: A systematic review. *Trends Food Sci. Technol.* **2017**, *61*, 11–25. [CrossRef]
31. Bryant, C.; Szejda, K.; Parekh, N.; Deshpande, V.; Tse, B. A Survey of Consumer Perceptions of Plant-Based and Clean Meat in the USA, India, and China. *Front. Sustain. Food Syst.* **2019**, *3*, 11. [CrossRef]
32. Circus, V.E.; Robison, R. Exploring perceptions of sustainable proteins and meat attachment. *Br. Food J.* **2019**, *121*, 533–545. [CrossRef]

33. Gómez-Luciano, C.A.; de Aguiar, L.K.; Vriesekoop, F.; Urbano, B. Consumers' willingness to purchase three alternatives to meat proteins in the United Kingdom, Spain, Brazil and the Dominican Republic. *Food Qual. Prefer.* **2019**, *78*, 103732. [CrossRef]

34. Vural, Y.; Ferriday, D.; Rogers, P.J. Consumers' attitudes towards alternatives to conventional meat products: Expectations about taste and satisfaction, and the role of disgust. *Appetite* **2023**, *181*, 106394. [CrossRef] [PubMed]

35. Thomas, O.Z.; Bryant, C. Don't Have a Cow, Man: Consumer Acceptance of Animal-Free Dairy Products in Five Countries. *Front. Sustain. Food Syst.* **2021**, *5*, 678491. [CrossRef]

36. Weinrich, R.; Strack, M.; Neugebauer, F. Consumer acceptance of cultured meat in Germany. *Meat Sci.* **2020**, *162*, 107924. [CrossRef] [PubMed]

37. Siegrist, M.; Sütterlin, B.; Hartmann, C. Perceived naturalness and evoked disgust influence acceptance of cultured meat. *Meat Sci.* **2018**, *139*, 213–219. [CrossRef]

38. Wilks, M.; Phillips, C.J.C.; Fielding, K.; Hornsey, M.J. Testing potential psychological predictors of attitudes towards cultured meat. *Appetite* **2019**, *136*, 137–145. [CrossRef] [PubMed]

39. Alae-Carew, C.; Green, R.; Stewart, C.; Cook, B.; Dangour, A.D.; Scheelbeek, P.F. The role of plant-based alternative foods in sustainable and healthy food systems: Consumption trends in the UK. *Sci. Total. Environ.* **2022**, *807*, 151041. [CrossRef]

40. Slade, P. If you build it, will they eat it? Consumer preferences for plant-based and cultured meat burgers. *Appetite* **2018**, *125*, 428–437. [CrossRef] [PubMed]

41. Beacom, E.; Bogue, J.; Repar, L. Market-oriented Development of Plant-based Food and Beverage Products: A Usage Segmentation Approach. *J. Food Prod. Mark.* **2021**, *27*, 204–222. [CrossRef]

42. Modlinska, K.; Adamczyk, D.; Maison, D.; Pisula, W. Gender Differences in Attitudes to Vegans/Vegetarians and Their Food Preferences, and Their Implications for Promoting Sustainable Dietary Patterns–A Systematic Review. *Sustainability* **2020**, *12*, 6292. [CrossRef]

43. Wilks, M.; Phillips, C.J.C. Attitudes to in vitro meat: A survey of potential consumers in the United States. *PLoS ONE* **2017**, *12*, e0171904. [CrossRef] [PubMed]

44. Shaw, E.; Iomaire, M.M.C. A comparative analysis of the attitudes of rural and urban consumers towards cultured meat. *Br. Food J.* **2019**, *121*, 1782–1800. [CrossRef]

45. Tucker, C.A. The significance of sensory appeal for reduced meat consumption. *Appetite* **2014**, *81*, 168–179. [CrossRef] [PubMed]

46. Cuffey, J.; Chenarides, L.; Li, W.; Zhao, S. Consumer spending patterns for plant-based meat alternatives. *Appl. Econ. Perspect. Policy* **2022**, *45*, 63–85. [CrossRef]

47. Dobricki, M. Basic Human Values in the Swiss Population and in a Sample of Farmers. *Swiss J. Psychol.* **2011**, *70*, 119–127. [CrossRef]

48. Farming UK. Family Farm Businesses Help Generate £15bn to UK Economy. 20 April 2018. Available online: https://www.farminguk.com/news/family-farm-businesses-help-generate-15bn-to-uk-economy_49106.html (accessed on 18 April 2023).

49. Painter, J.; Brennen, J.S.; Kristiansen, S. The coverage of cultured meat in the US and UK traditional media, 2013–2019: Drivers, sources, and competing narratives. *Clim. Chang.* **2020**, *162*, 2379–2396. [CrossRef] [PubMed]

50. Bryant, C.J.; van der Weele, C. The farmers' dilemma: Meat, means, and morality. *Appetite* **2021**, *167*, 105605. [CrossRef] [PubMed]

51. Mintert, J.; Langemeier, M. Producers Bullish about Farmland Values Amid Strong Current Conditions. Ag Economy Barometer, Purdue University. 2 May 2021. Available online: https://ag.purdue.edu/commercialag/ageconomybarometer/producers-bullish-about-farmland-values-amid-strong-current-conditions/ (accessed on 18 April 2023).

52. Newton, P.; Blaustein-Rejto, D. Social and Economic Opportunities and Challenges of Plant-Based and Cultured Meat for Rural Producers in the US. *Front. Sustain. Food Syst.* **2021**, *5*, 624270. [CrossRef]

53. Crawshaw, C.; Piazza, J. How conflicted are farmers about meat? Livestock farmers' attachment to their animals and attitudes about meat. *Psychol. Hum.-Anim. Intergroup Relat.* **2022**, *1*, e8513. [CrossRef]

54. Taylor, N.; Fraser, H. The cow project: Analytical and representational dilemmas of dairy farmers' conceptions of cruelty and kindness. *Anim. Stud. J.* **2019**, *8*, 133–153. [CrossRef]

55. Velde, H.T.; Aarts, N.; Van Woerkum, C. Dealing with Ambivalence: Farmers' and Consumers' Perceptions of Animal Welfare in Livestock Breeding. *J. Agric. Environ. Ethics* **2002**, *15*, 203–219. [CrossRef]

56. DEFRA. DEFRA Statistics: Agricultural Facts—England Regional Profiles. 2021. Available online: https://www.gov.uk/government/statistics/agricultural-facts-england-regional-profiles (accessed on 18 April 2023).

57. DEFRA. Farm Structure Survey 2016. [Data Set]. 2019. Available online: https://assets.publishing.service.gov.uk/government/uploads/system/uploads/attachment_data/file/771494/FSS2013-labour-statsnotice-17jan19.pdf (accessed on 18 April 2023).

58. Van Loo, E.J.; Caputo, V.; Lusk, J.L. Consumer preferences for farm-raised meat, lab-grown meat, and plant-based meat alternatives: Does information or brand matter? *Food Policy* **2020**, *95*, 101931. [CrossRef]

59. Bryant, C.; Sanctorum, H. Alternative proteins, evolving attitudes: Comparing consumer attitudes to plant-based and cultured meat in Belgium in two consecutive years. *Appetite* **2021**, *161*, 105161. [CrossRef] [PubMed]

60. Michel, F.; Hartmann, C.; Siegrist, M. Consumers' associations, perceptions and acceptance of meat and plant-based meat alternatives. *Food Qual. Prefer.* **2021**, *87*, 104063. [CrossRef]

61. DEFRA. Farm Classification in the United Kingdom. 2014. Available online: https://farmbusinesssurvey.co.uk/DataBuilder/UK_Farm_Classification_2014_Final.pdf (accessed on 18 April 2023).

62. IBM Corp. *IBM SPSS Statistics for Windows*, Version 25.0; IBM Corp.: Armonk, NY, USA, 2017. Available online: https://www.ibm.com/support/pages/downloading-ibm-spss-statistics-25 (accessed on 18 April 2023).

63. Olson, C.L. Comparative robustness of six tests in multivariate analysis of variance. *J. Am. Stat. Assoc.* **1974**, *69*, 894–908. [CrossRef]

64. Mintel. The Cream of the Vegan Milk Crop: Sales of Oat Milk Overtake Almond in the UK. 17 September 2021. Available online: https://www.mintel.com/press-centre/food-and-drink/the-cream-of-the-vegan-milk-crop-sales-of-oat-milk-overtake-almond-in-the-uk (accessed on 18 April 2023).

65. Floris, R. Industry Insights from NIZO: Precision Fermentation—Making the Case for 'Animal Protein without the Animal'. Food Navigator Europe, 16 March 2022. Available online: https://www.foodnavigator.com/Article/2022/03/16/Industry-insights-from-NIZO-Precision-fermentation-Making-the-case-for-animal-protein-without-the-animal (accessed on 18 April 2023).

66. Laestadius, L.I.; Caldwell, M.A. Is the future of meat palatable? Perceptions of in vitro meat as evidenced by online news comments. *Public Health Nutr.* **2015**, *18*, 2457–2467. [CrossRef] [PubMed]

67. Zhang, M.; Li, L.; Bai, J. Consumer acceptance of cultured meat in urban areas of three cities in China. *Food Control* **2020**, *118*, 107390. [CrossRef]

68. Bryant, C.J.; Prosser, A.M.; Barnett, J. Going veggie: Identifying and overcoming the social and psychological barriers to veganism. *Appetite* **2022**, *169*, 105812. [CrossRef] [PubMed]

69. Gorman, M.; Knowles, S.; Falkeisen, A.; Barker, S.; Moss, R.; McSweeney, M.B. Consumer Perception of Milk and Plant-Based Alternatives Added to Coffee. *Beverages* **2021**, *7*, 80. [CrossRef]

70. Siegrist, M.; Hartmann, C. Why alternative proteins will not disrupt the meat industry. *Meat Sci.* **2023**, *203*, 109223. [CrossRef] [PubMed]

71. FAIRR. Factory Farming: Assessing Investment Risks. 2016. Available online: https://www.fairr.org/article/factory-farming-assessing-investment-risks/ (accessed on 18 April 2023).

72. Kustar, A.; Patino-Echeverri, D. A review of environmental life cycle assessments of diets: Plant-based solutions are truly sustainable, even in the form of fast foods. *Sustainability* **2021**, *13*, 9926. [CrossRef]

73. New Scientist. Accelerating the Cultured Meat Revolution. New Scientist. 2020. Available online: https://www.newscientist.com/article/mg24032080-400-accelerating-the-cultured-meat-revolution/ (accessed on 18 April 2023).

74. Tesco. Tesco Groceries. 2022. Available online: https://www.tesco.com/groceries/?icid=dchp_groceriesshopgroceries (accessed on 18 April 2023).

75. Viva. Beef Cattle. Available online: https://viva.org.uk/animals/cows/beef/ (accessed on 18 April 2023).

76. Viva. Dairy Cows. Available online: https://viva.org.uk/animals/cows/dairy-cows/ (accessed on 18 April 2023).

*Article*

# Professional Training in Beekeeping: A Cross-Country Survey to Identify Learning Opportunities

Raquel P. F. Guiné [1], Jorge Oliveira [1], Catarina Coelho [1,2,*], Daniela Teixeira Costa [1], Paula Correia [1], Helena Esteves Correia [1], Bjørn Dahle [3], Melissa Oddie [3], Risto Raimets [4], Reet Karise [4], Luis Tourino [5], Salvatore Basile [6], Emilio Buonomo [6], Ivan Stefanic [7] and Cristina A. Costa [1]

[1]   CERNAS Research Centre, Polytechnic Institute of Viseu, 3504-510 Viseu, Portugal; raquelguine@esav.ipv.pt (R.P.F.G.); joliveira@esav.ipv.pt (J.O.); daniela@esav.ipv.pt (D.T.C.); paulacorreia@esav.ipv.pt (P.C.); hecorreia@esav.ipv.pt (H.E.C.); amarocosta@esav.ipv.pt (C.A.C.)
[2]   CECAV, Animal and Veterinary Research Center, University of Trás-os-Montes e Alto Douro, Quinta de Prados, Apartado 1013, 5000-801 Vila Real, Portugal
[3]   Norwegian Beekeepers Association, 2040 Kløfta, Norway; bjorn.dahle@norbi.no (B.D.); melissa.oddie@nordi.no (M.O.)
[4]   Institute of Agricultural and Environmental Sciences, Estonian University of Life Sciences, 51014 Tartu, Estonia; ristorai@gmail.com (R.R.); reet.karise@emu.ee (R.K.)
[5]   Eosa Estrategia y Organización SA, 36202 Vigo, Spain; ltourino@eosa.com
[6]   Bio-Distretto Cilento, 84052 Salerno, Italy; presidente@ecoregions.eu (S.B.); emiliobuonomo@gmail.com (E.B.)
[7]   Tera Tehnopolis, 31000 Osijek, Croatia; istefanic@fazos.hr
*   Correspondence: ccoelho@esav.ipv.pt

**Abstract:** Habitat loss, climate change, and other environmental degradations pose severe challenges to beekeepers. Therefore, this sector needs to rely on updated information so that the intervening actors can deal with the problems. In this context, and assuming that professional training can greatly help those acting in the beekeeping sector, this work intended to investigate the gaps in the updated knowledge of beekeepers and how these can be filled through lifelong learning. The research was conducted in seven European countries (Croatia, Estonia, Finland, Italy, Norway, Portugal, and Spain). The data were collected through a questionnaire survey translated into the native languages of all participating countries. The results revealed that the topics of highest interest are apiary health and pest control and the management of the colonies throughout the year. The beekeepers update their knowledge through family, complemented by professional training, with participants preferring in-person courses as well as, in the workplace or in internships. The learning methodologies they consider most useful are project-based learning and learning through gamification. The videos and paper books or manuals are particularly valued as learning materials, and practical exercises are considered the most helpful assessment format. Finally, considering the effect of sociodemographic variables on the learning experiences and preferences of beekeeping actors, it was observed that the country was the most influential of the variables under study. In conclusion, this work revealed valuable information that should be used to design professional training actions to help the professionals in the beekeeping sector enhance their competencies and be better prepared to manage their activities successfully.

**Keywords:** distance learning; mobile learning; professional learning; beekeeping; survey

**Citation:** Guiné, R.P.F.; Oliveira, J.; Coelho, C.; Costa, D.T.; Correia, P.; Correia, H.E.; Dahle, B.; Oddie, M.; Raimets, R.; Karise, R.; et al. Professional Training in Beekeeping: A Cross-Country Survey to Identify Learning Opportunities. *Sustainability* **2023**, *15*, 8953. https://doi.org/10.3390/su15118953

Academic Editor: Giacomo Falcone

Received: 21 April 2023
Revised: 24 May 2023
Accepted: 31 May 2023
Published: 1 June 2023

## 1. Introduction

Beekeeping is a key sector from multiple perspectives. Sustainability is one of the relevant aspects linked with the roles of bees and, consequently, beekeeping activities. Bees are important pollinators for many crops and plants. It is estimated that bees and other pollinators are responsible for one-third of the food we eat. By maintaining healthy bee populations through beekeeping, the pollination of crops is ensured, leading to greater food security and more sustainable agriculture. On the other hand, bees play a critical role in maintaining biodiversity by pollinating wildflowers and other plants. This helps to maintain healthy ecosystems and support other wildlife. Keeping bees helps support local biodiversity and contributes to the preservation of ecosystems. Bee products such as honey, beeswax, propolis and royal jelly have antimicrobial properties. By using these products and other natural beekeeping techniques, beekeepers can reduce the need for synthetic pesticides and other chemical treatments that can harm the environment and human health. Finally, in the social sustainability dimension, beekeeping can provide sustainable livelihoods for people in rural and urban areas. It can be a low-cost, low-impact form of agriculture practised on a small scale. By providing a source of income and livelihood, beekeeping can help support local communities and contribute to sustainable development [1–6].

Beekeepers face many challenges in maintaining their apiaries in good equilibrium, increasing productivity, enhancing performance, and being more competitive in the context of globalization. Since many beekeepers' businesses are of a small dimension and greatly contribute to the social development of rural populations, providing them with proper training is essential to help them cope with the sector's challenges. Investing in lifelong learning (LL) and professional training (PT) might make the difference between a successful business or a failure [7–9].

The human s capacity to learn and accumulate knowledge from a wide amount of information that is considered relevant is enormous. The human brain's capacity to learn and accumulate knowledge is closely related to synaptic plasticity, which refers to the ability of the connections between neurons, called synapses, to change in response to experience. When learning new things or acquiring new information, the human brain forms new connections between neurons or strengthens existing ones, which is known as synaptic potentiation. This allows for people to encode and store new information in long-term memory. Synaptic plasticity is influenced by various factors, including the frequency, intensity, and duration of neuronal activity, as well as the age, genetics, and environmental factors of the individual. Thus, the capacity of the human brain to learn and accumulate knowledge is intimately linked to the ability of its synapses to adapt and change in response to experience, which is a fundamental aspect of synaptic plasticity. With increased learning and experience, more connections are formed in the brain, strengthening and accumulating more knowledge over time. It is the synaptic plasticity of the brain that enables it to learn new representations as well as to eliminate previously learned information, constituting a foundation for shaping memory and learning that culminates in the LL process [10–14].

LL refers to the ongoing and voluntary pursuit of knowledge and skills throughout one's life, beyond traditional classroom education. It is an attitude and approach to learning that recognizes that learning is not just limited to formal education but can take place through a variety of experiences and activities such as work, hobbies, personal interests, and social interactions. LL is becoming increasingly important in today's rapidly changing world, where new technologies, information, and industries are constantly emerging. It allows individuals to adapt to changing circumstances, keep up with the latest trends and developments, and improve their personal and professional prospects. LL involves a commitment to personal development and self-improvement, and can bring numerous benefits, such as increased knowledge, improved job performance, better social and communication skills, increased confidence, and a sense of personal fulfilment. LL encompasses different analytic perspectives: the social organization of learning and individual learning. These indicate the way in which past definitional concerns related to formal, non-formal and informal learning. The recognition of learning outcomes must be modern and consider eventually contrasting viewpoints, in the European context as well as from the global perspective [15–18].

PT encompasses the process of building knowledge, skills and competencies, either on an individual person or in a group or team. PT can have a significant impact on an individual's personal and professional development, as well as on his organization/company.

Some of the key impacts of professional training are improved job performance, career advancement, increased job satisfaction and motivation and enhanced organizational performance. Overall, PT can have a positive impact on both individuals and organizations, leading to improved job performance, career advancement, job satisfaction, and organizational performance. Effective training improves not only knowledge and skills but also attitudes and resilience [19–22].

The beeB project—Foster for beekeeping bridges through innovative and participative training, which was approved by the European Union under Ref. 2019-1-PT01-KA202-060782, aims to contribute to the technical training of beekeepers and other agents involved in the beekeeping sector, as well as providing appropriate tools in mobile-learning (m-learning) contexts, to improve beekeepers' ability to manage their businesses successfully. The project team integrates six partners from different European countries and encompasses the identification of needs and the development of training opportunities, facilitating beekeepers' access to distance learning courses, platforms and content. In this context, the aim of this work was to identify the needs of those acting in the beekeeping sector and understand how these needs can be fulfilled through lifelong learning. Additionally, differences will be identified according to the country or other sociodemographic variables of the participants. These elements will offer valuable information for the design of courses and other learning tools that will be easily available for use by all stakeholders in the beekeeping sector to enhance their knowledge and skills.

## 2. Materials and Methods

### 2.1. Instrument Used for the Research

The questionnaire used for the present research was divided into six sections:

I. Experience in beekeeping (10 questions);
II. Training needs (3 questions);
III. Experience in beekeeping training activities (3 questions);
IV. Use of distance learning technologies and tools (3 questions);
V. Distance learning tools (4 questions);
VI. Sociodemographic characterization (6 questions).

The questionnaire was first produced in Portuguese and validated through a pre-test with 50 participants through direct interviews. The final instrument was then obtained after this pre-test. Before general application, the questionnaire was translated into the native languages in the seven countries where the data were collected, following a back-translation methodology.

This research paper is focused on professional training and its relationship with the sociodemographic variables, addressing questions from parts II–VI.

### 2.2. Data Collection

The survey was applied to beekeepers in different countries (Croatia, Estonia, Finland, Italy, Norway, Portugal, Spain) as a part of the project Beeb—Foster for beekeeping bridges through innovative and participative training (2019-1-PT01-KA202-060782), approved and developed by the Polytechnic Institute of Viseu, as leading partner.

The sample was selected from all the potentially interested people in the different countries included in the study. The target group comprised people linked to the beekeeping sector, either beekeepers, academia members, or those dealing with bee products' transformation and commercialization, as examples. This also included people who participated in activities other than beekeeping, those who recently engaged in this sector or those who have beekeeping as a complementing activity to their other principal activities.

The questionnaire was delivered in paper form, face-to-face, during training or dissemination events organized by beekeepers' associations or companies in each country. Additionally, online tools were used to complement the data collection and reach a wider audience among those connected with the beekeeping sector.

A total of 313 valid responses were obtained considering the whole set of countries. In the case of variable age, cases in which the participants did not indicate their age were excluded. For variable sex, cases where the participants identified themselves explicitly with men or women were considered. In the case of variable education, and due to the very low representativeness of the group basic school (only 3%), basic classes plus secondary school were merged into a single class.

### 2.3. Data Analysis

The non-parametric tests U-Mann–Whitney and Kruskal–Wallis were used to compare quantitative variables between two groups or three or more groups, respectively. Non-parametric tests were used in the present study due to the low number of cases in some groups, inequality of group dimensions and non-verification of normality distribution. Chi-square tests were used to test the differences between some categorical variables. For all data analysis, the software SPSS, from IBM Inc. (version 28), was used, complemented by Excel 2016. The level of significance considered was 5%.

## 3. Results

### 3.1. Sample Characterization

Data were collected from six European countries participating in the European Erasmus+ project beeB (Foster for beekeeping bridges through innovative and participative training/Ref. 2019-1-PT01-KA202-060782), namely: Croatia, Estonia, Finland, Italy, Norway, Portugal, and Spain. Figure 1 shows how the participants were distributed among the countries included in this study. The percentages varied from 5% for participants from Finland (n = 15 out of 313 participants) to 24% for participants from Norway (n = 74 out of 313).

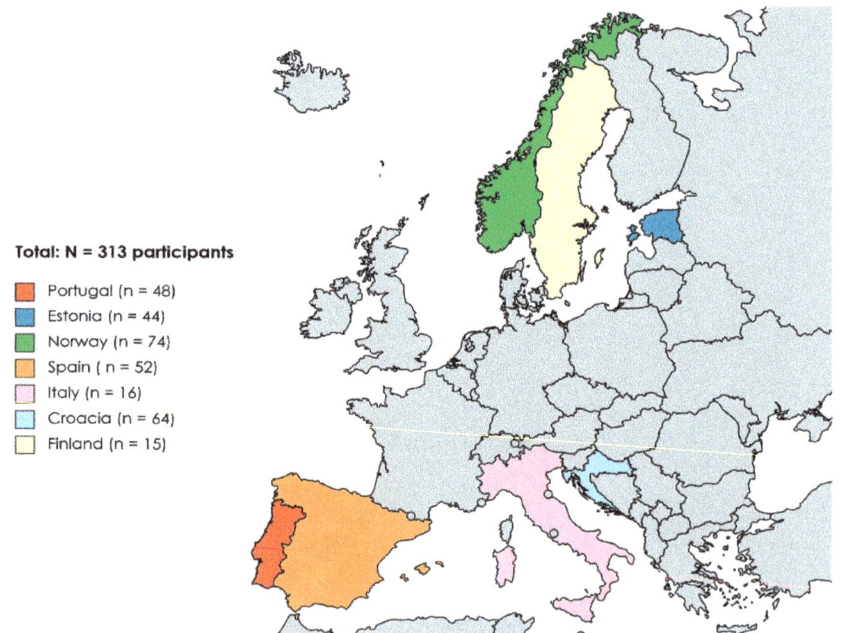


**Figure 1.** Geographical location of the countries included in the study and the corresponding number of participants.

Figure 2 shows that most of the participants in the survey (68%) had ages ranging from 31 to 59 years old, followed by those aged over 60 years (18%), and the class under 30 years had a lower expression (9%). The majority were male (74%), with only about one-fourth

(23%) females. Concerning the education level, 59% had a university degree, 35% had completed secondary school, and only 3% had a very low level of education (basic school). Concerning the income, the distribution by classes was more even, with 38% having an income between 15 and 50 thousand euros per year, 25% having an income lower than 15 thousand EUR/y and 23% over 50 thousand EUR/y.

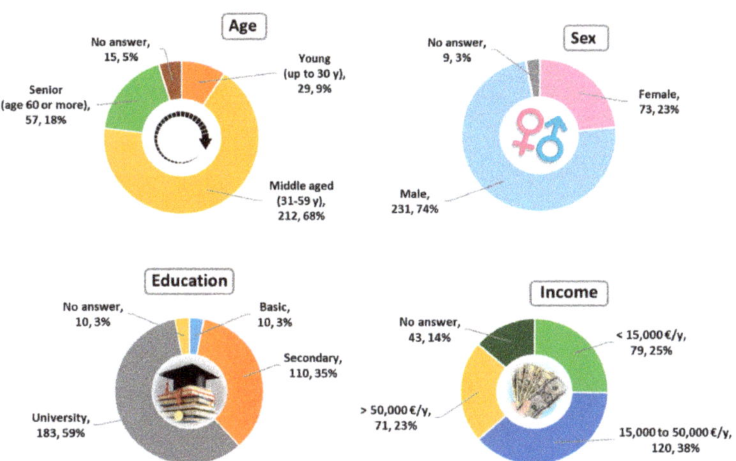


**Figure 2.** Sociodemographic characterization of the participants.

### 3.2. *Access to the Internet in the Apiaries*

The participants were questioned about whether they had access to internet in their apiaries, with the results presented in Figure 3. No significant differences were found between countries regarding access to the internet in the apiaries. Nevertheless, most participants in Italy (92.3%) and Finland (91.7%) have internet in all apiaries. Portugal and Spain are the countries with less internet access in the apiaries (69.8% and 68.6%, respectively). Considering all data (countries), mean access to the internet covers 78.9% of the apiaries.

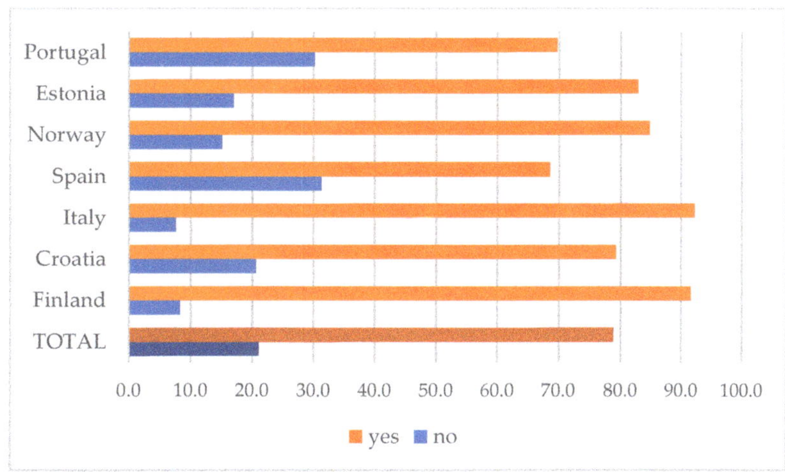


**Figure 3.** Country frequencies for access to internet in the apiaries (Chi-square test (level of significance $p < 0.05$); $p = 0.135$).

### 3.3. Use of Technologies and Purposes

The participants were questioned about how frequently they use mobile devices in their beekeeping activities, with the results presented in Table 1. Significant differences were found between countries for the frequency of utilization of mobile devices. Italy (62.5%), Croatia (61.7%) and Finland (58.3%) were the countries where the daily frequency of utilization of mobile devices was higher. Significant differences were also encountered between age groups for the frequency of utilization of mobile devices for beekeeping activities, with percentage of participants using them daily increasing as age decreased. Finally, the frequency of utilization of mobile devices was also found to vary significantly according to income, with increased daily usage for lower incomes.

**Table 1.** Group differences for frequency of mobile devices utilization.

| Variable | Group | Percentage and Significance | | | | |
| --- | --- | --- | --- | --- | --- | --- |
| | | Frequency of Utilization of Mobile Devices for Beekeeping Activities | | | | |
| | | Daily | 1-2x/Week | 1-2x/Month | Very Sporadically | Never |
| Country | Portugal | 37.2 | 27.9 | 18.6 | 7.0 | 9.3 |
| | Estonia | 10.8 | 13.5 | 5.4 | 56.8 | 13.5 |
| | Norway | 10.0 | 25.7 | 20.0 | 34.3 | 10.0 |
| | Spain | 50.0 | 13.9 | 8.3 | 27.8 | 0.0 |
| | Italy | 62.5 | 6.3 | 0.0 | 12.5 | 18.8 |
| | Croatia | 61.7 | 3.3 | 8.3 | 23.3 | 3.3 |
| | Finland | 58.3 | 25.0 | 8.3 | 8.3 | 0.0 |
| | Sig. [1] | <0.001 | | | | |
| Age | 18–30 y | 50.0 | 20.8 | 4.2 | 25.0 | 0.0 |
| | 31–59 y | 37.8 | 17.0 | 10.6 | 27.7 | 6.9 |
| | 60+ y | 20.4 | 14.3 | 24.5 | 28.6 | 12.2 |
| | Sig. [1] | 0.040 | | | | |
| Sex | Female | 25.0 | 15.6 | 12.5 | 31.3 | 15.6 |
| | Male | 38.9 | 17.2 | 12.3 | 26.1 | 5.4 |
| | Sig. [2] | 0.054 | | | | |
| Education | Secondary | 40.0 | 14.3 | 9.5 | 28.6 | 7.6 |
| | University | 32.9 | 18.0 | 13.0 | 28.0 | 8.1 |
| | Sig. [2] | 0.712 | | | | |
| Income | Low | 52.1 | 8.5 | 5.6 | 26.8 | 7.0 |
| | Medium | 35.6 | 17.3 | 12.5 | 27.9 | 6.7 |
| | High | 15.4 | 27.7 | 18.5 | 32.3 | 6.2 |
| | Sig. [1] | 0.001 | | | | |

[1] Chi-square test (level of significance $p < 0.05$). [2] Fisher's exact test (level of significance $p < 0.05$).

Table 2 shows the results of cross-tabulation between the sociodemographic variables and the reasons why the beekeepers use their mobile devices in beekeeping activities. Again, countries were shown to have significant differences for all the possible usages, while age and sex were variables that did not lead to significant differences. However, significant differences were found for the variable income, just like country. A higher income is associated with a higher percentage of utilization of mobile devices for all the tested reasons. Finally, significant differences were found between participants with a university degree from those with up to a secondary school education in the use of mobile devices to 'Take pictures', 'Make videos' and 'Use apps'.

### 3.4. Previous Knowledge and Experience in Training Activities

The results in Figure 4 show the mean value for the importance attributed to the different sources of information in previous knowledge. For the sources of information, the scale varied from 1 (most important) to 3 (least important), and the results indicated that the most important source was 'Family', with a mean score closest to 1, while the least important was 'Seminars', with the highest mean score of 1.55.

**Table 2.** Group differences for motivations to use mobile devices.

| Variable | Group | Take Pictures | | Make Videos | | Do Research | | Use Apps | | Browse Specialized Platforms | |
|---|---|---|---|---|---|---|---|---|---|---|---|
| | | No | Yes | No | Yes | No | Yes | No | Yes | No | Yes |
| Country | Portugal | 36.1 | 63.9 | 75.0 | 25.0 | 22.2 | 77.8 | 63.9 | 36.1 | 58.3 | 41.7 |
| | Estonia | 0.0 | 100.0 | 0.0 | 100.0 | 0.0 | 100.0 | 0.0 | 100.0 | 0.0 | 100.0 |
| | Norway | 0.0 | 100.0 | 6.3 | 93.8 | 2.8 | 97.2 | 8.3 | 91.7 | 10.0 | 90.0 |
| | Spain | 35.3 | 64.7 | 62.9 | 37.1 | 31.4 | 68.6 | 71.4 | 28.6 | 45.7 | 54.3 |
| | Italy | 0.0 | 100.0 | 23.1 | 76.9 | 23.1 | 76.9 | 23.1 | 76.9 | 46.2 | 53.8 |
| | Croatia | 22.4 | 77.6 | 55.2 | 44.8 | 51.7 | 48.3 | 77.6 | 22.4 | 63.8 | 36.2 |
| | Finland | 0.0 | 100.0 | 0.0 | 100.0 | 0.0 | 100.0 | 0.0 | 100.0 | 0.0 | 100.0 |
| | Sig. [1] | <0.001 | | <0.001 | | <0.001 | | <0.001 | | 0.014 | |
| Age | 18–30 y | 12.5 | 87.5 | 37.5 | 62.5 | 18.2 | 81.8 | 57.9 | 42.1 | 68.4 | 31.6 |
| | 31–59 y | 17.8 | 82.2 | 45.0 | 55.0 | 24.6 | 75.4 | 59.5 | 40.5 | 49.1 | 50.9 |
| | 60+ y | 15.6 | 84.4 | 60.0 | 40.0 | 40.0 | 60.0 | 65.0 | 35.0 | 50.0 | 50.0 |
| | Sig. [1] | 0.792 | | 0.313 | | 0.146 | | 0.880 | | 0.294 | |
| Sex | Female | 10.9 | 89.1 | 35.1 | 64.9 | 24.3 | 75.7 | 51.9 | 48.1 | 61.5 | 38.5 |
| | Male | 18.2 | 81.8 | 48.6 | 51.4 | 26.1 | 73.9 | 61.4 | 38.6 | 50.8 | 49.2 |
| | Sig. [2] | 0.237 | | 0.194 | | 1.000 | | 0.394 | | 0.390 | |
| Education | Secondary | 25.9 | 74.1 | 59.0 | 41.0 | 30.9 | 69.1 | 68.9 | 31.5 | 56.5 | 43.5 |
| | University | 9.2 | 90.8 | 35.0 | 65.0 | 24.1 | 75.9 | 51.8 | 48.2 | 46.3 | 53.7 |
| | Sig. [2] | 0.002 | | 0.002 | | 0.326 | | 0.036 | | 0.254 | |
| Income | Low | 16.1 | 83.9 | 52.5 | 47.5 | 40.0 | 60.0 | 72.7 | 27.3 | 67.3 | 32.7 |
| | Medium | 22.7 | 77.3 | 47.5 | 52.5 | 26.9 | 73.1 | 55.1 | 44.9 | 45.5 | 54.5 |
| | High | 4.1 | 95.9 | 20.8 | 79.2 | 8.3 | 91.7 | 38.9 | 61.6 | 37.5 | 62.5 |
| | Sig. [1] | 0.017 | | 0.027 | | 0.003 | | 0.021 | | 0.023 | |

[1] Chi-square test (level of significance $p < 0.05$). [2] Fisher's exact test (level of significance $p < 0.05$).

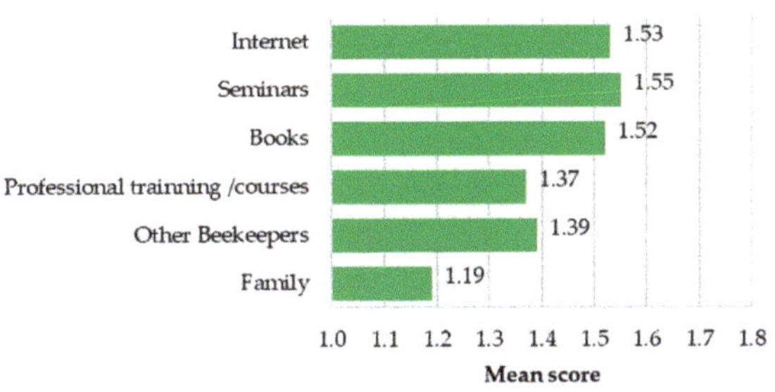


**Figure 4.** Level of importance of the sources of information.

Table 3 presents the results of the non-parametric statistical tests performed on the relations between the considered variables and the level of importance attributed to the sources of information in beekeeping. The results reveal that significant differences were found between countries for all sources of information'. For the other variables, differences were observed between groups for some of the sources of information.

Table 4 presents the results for cross-tabulation between the variables accounting for part experience in training in beekeeping and the sociodemographic variables under study. The results indicated significant differences between those who already participated in training activities and those who did not, according to country (higher participation in Norway—93.2%—and lower in Croatia—58.3%), age (higher percentage for older participants—83.6%—and lower for younger—51.7%), and income (higher participation for higher income—93.0%—and lower for lower income—71.1%). Regarding participation as a trainee, significant differences were found according to country, education, and income (a higher percentage of participants were trainees in Estonia, with a university education and high income). Concerning the participation as trainer/coordinator, significant differences were found for country and age (higher percentage of participants from Portugal and Estonia and in the age group of 60+ years (Table 4).

**Table 3.** Group differences for the level of importance of the sources of information.

| | | Percentage and Significance | | | | | |
| | | Sources of Information | | | | | |
| Variable | Group | Family | Other Beekeepers | Professional Training/Courses | Books | Seminars | Internet |
|---|---|---|---|---|---|---|---|
| Country | Portugal | 32.00 | 64.50 | 65.83 | 55.50 | 21.50 | 43.95 |
| | Estonia | 35.05 | 134.56 | 115.50 | 112.53 | 46.36 | 88.66 |
| | Norway | 46.30 | 124.53 | 97.77 | 123.94 | 52.17 | 97.68 |
| | Spain | 32.00 | 64.50 | 63.00 | 55.50 | 21.50 | 40.50 |
| | Italy | 50.10 | 101.58 | 97.50 | 90.00 | 45.79 | 73.25 |
| | Croatia | 32.00 | 64.50 | 63.00 | 55.50 | 21.50 | 40.50 |
| | Finland | 70.00 | 113.13 | 63.00 | 139.00 | 52.17 | 93.33 |
| | Sig. [1] | <0.001 | <0.001 | <0.001 | <0.001 | <0.001 | <0.001 |
| Age | 18–30 y | 29.50 | 68.77 | 99.10 | 74.00 | 28.00 | 52.56 |
| | 31–59 y | 34.58 | 89.24 | 78.78 | 81.61 | 27.11 | 53.70 |
| | 60+ y | 29.50 | 101.54 | 80.95 | 83.28 | 41.72 | 75.44 |
| | Sig. [1] | 0.301 | 0.022 | 0.219 | 0.719 | 0.017 | 0.008 |
| Sex | Female | 39.79 | 102.58 | 90.82 | 104.97 | 29.36 | 60.42 |
| | Male | 32.74 | 86.11 | 78.74 | 75.72 | 30.15 | 57.36 |
| | Sig. [2] | 0.030 | 0.019 | 0.058 | <0.001 | 0.898 | 0.628 |
| Education | Secondary | 32.86 | 79.09 | 79.50 | 69.06 | 27.30 | 51.77 |
| | University | 36.04 | 96.57 | 81.80 | 89.15 | 33.57 | 61.24 |
| | Sig. [2] | 0.260 | 0.006 | 0.698 | 0.002 | 0.101 | 0.072 |
| Income | Low | 29.11 | 66.93 | 68.97 | 54.61 | 27.83 | 42.39 |
| | Medium | 25.59 | 74.41 | 67.70 | 71.61 | 27.64 | 46.85 |
| | High | 35.17 | 101.19 | 83.35 | 91.50 | 35.83 | 79.06 |
| | Sig. [1] | 0.420 | <0.001 | 0.042 | <0.001 | 0.245 | <0.001 |

[1] Kruskal–Wallis test (level of significance $p < 0.05$). [2] U-Mann–Whitney test (level of significance $p < 0.05$).

**Table 4.** Group differences for participation in training activities.

| | | Past Experience (Percentage and Significance) | | | | | |
| | | Already Participated in Training Activities | | Role: Trainee | | Role: Trainer/Coordinator | |
| Variable | Group | No | Yes | No | Yes | No | Yes |
|---|---|---|---|---|---|---|---|
| Country | Portugal | 8.3 | 91.7 | 77.3 | 22.7 | 0.0 | 100.0 |
| | Estonia | 16.7 | 83.3 | 0.0 | 100.0 | 0.0 | 100.0 |
| | Norway | 6.8 | 93.2 | 11.8 | 88.2 | 63.6 | 36.4 |
| | Spain | 32.0 | 68.0 | 48.1 | 51.9 | 82.7 | 17.3 |
| | Italy | 31.3 | 68.8 | 31.3 | 68.8 | 93.8 | 6.3 |
| | Croatia | 41.7 | 58.3 | 8.6 | 91.4 | 77.1 | 22.9 |
| | Finland | 20.0 | 80.0 | 58.3 | 41.7 | 8.3 | 91.7 |
| | Sig. [1] | <0.001 | | <0.001 | | <0.001 | |
| Age | 18–30 y | 48.3 | 51.7 | 50.0 | 50.0 | 65.0 | 35.0 |
| | 31–59 y | 19.3 | 80.7 | 28.4 | 71.6 | 57.7 | 42.3 |
| | 60+ y | 16.4 | 83.6 | 30.6 | 69.4 | 38.1 | 61.9 |
| | Sig. [1] | 0.001 | | 0.118 | | 0.047 | |

**Table 4.** *Cont.*

| Variable | Group | Past Experience (Percentage and Significance) | | | | | |
| --- | --- | --- | --- | --- | --- | --- | --- |
| | | Already Participated in Training Activities | | Role: Trainee | | Role: Trainer/Coordinator | |
| | | No | Yes | No | Yes | No | Yes |
| Sex | Female | 22.5 | 77.5 | 20.7 | 79.3 | 65.3 | 34.7 |
| | Male | 20.9 | 79.1 | 34.4 | 65.6 | 51.9 | 48.1 |
| | Sig. [2] | 0.743 | | 0.054 | | 0.107 | |
| Education | Secondary | 23.5 | 76.5 | 40.6 | 59.4 | 53.7 | 46.3 |
| | University | 20.4 | 79.6 | 23.8 | 76.2 | 55.2 | 44.8 |
| | Sig. [2] | 0.564 | | 0.005 | | 0.893 | |
| Income | Low | 28.9 | 71.1 | 39.1 | 60.9 | 56.9 | 43.1 |
| | Medium | 23.1 | 76.9 | 37.4 | 62.6 | 51.7 | 48.3 |
| | High | 7.0 | 93.0 | 13.6 | 86.4 | 58.7 | 41.3 |
| | Sig. [1] | 0.003 | | 0.001 | | 0.667 | |

[1] Chi-square test (level of significance $p < 0.05$). [2] Fisher's exact test (level of significance $p < 0.05$).

### 3.5. Identification of Training Needs

The respondents were asked to classify several topics for possible training modules according to their level of interest on a scale from 1 (little interest) to 5 (much interest). Figure 5 presents the average scores for each option, calculated as the mean value of the classifications attributed by all participants. The topics considered of the highest global interest were 'Apiary health and pest control', followed by 'Colony management throughout the year'. Topics of the lowest interest are 'Organic production mode' and 'Beehive production'.

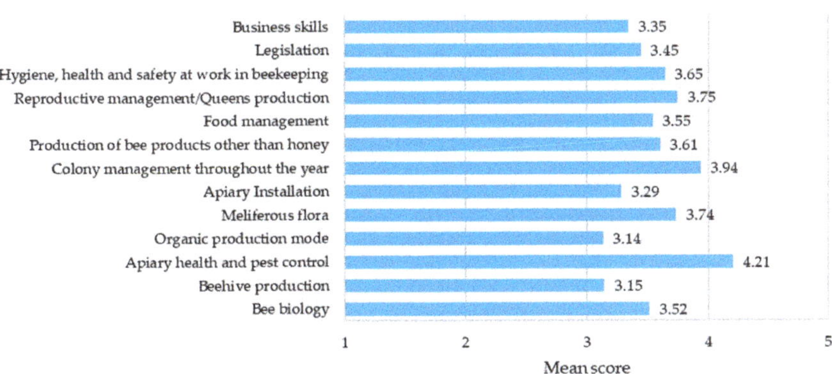


**Figure 5.** Level of interest in training subjects in beekeeping.

The results in Table 5 show that country is the variable for which significant differences were found for a higher number of training topics in beekeeping. Topics such as 'Beehive production', 'Organic production mode' and 'Food management' showed significant differences between countries, with $p < 0.001$, but topics such as 'Meliferous flora', 'Apiary Installation', 'Production of bee products other than honey', 'Hygiene, health and safety at work in beekeeping', 'Legislation', and 'Business skills' had a $p$-value below the significance level ($p < 0.05$). Additionally, the variable sex showed significant differences for many topics, five. Regarding age, significant differences were found for three of the topics, and income revealed differences for two topics. On the other hand, no significant differences were found between the participants with a university education and those with a secondary school education for any of the topics considered (Table 5).

Table 5. Group differences for the level of interest in training subjects in beekeeping.

| Variable | Group | | | | | | | | | | | | | |
|---|---|---|---|---|---|---|---|---|---|---|---|---|---|---|
| | | | | | | | | Mean Rank and Significance | | | | | | |
| | | Bee Biology | Beehive Production | Apiary Health and Pest Control | Organic Production Mode | Meliferous Flora | Apiary Installation | Colony Management throughout the Year | Production of Bee Products Other Than Honey | Food Management | Reproductive Management/ Queens Production | Hygiene, Health and Safety at Work in Beekeeping | Legislation | Business Skills |
| Country | Portugal | 142.62 | 201.49 | 176.81 | 164.65 | 173.38 | 166.15 | 167.33 | 158.71 | 181.07 | 157.84 | 174.40 | 174.43 | 165.68 |
| | Estonia | 119.75 | 122.25 | 158.72 | 136.78 | 114.98 | 404.43 | 127.96 | 126.34 | 112.93 | 138.23 | 136.30 | 129.76 | 155.47 |
| | Norway | 147.42 | 114.97 | 146.54 | 96.05 | 151.62 | 136.44 | 156.68 | 118.71 | 154.90 | 150.38 | 132.01 | 157.47 | 128.50 |
| | Spain | 122.00 | 142.68 | 133.63 | 151.81 | 130.69 | 146.75 | 128.02 | 130.20 | 138.58 | 142.83 | 122.12 | 112.68 | 118.60 |
| | Italy | 182.22 | 204.59 | 155.94 | 229.50 | 171.91 | 167.25 | 138.03 | 141.84 | 108.06 | 114.53 | 116.41 | 142.81 | 174.28 |
| | Croatia | 145.83 | 130.09 | 149.18 | 180.03 | 149.77 | 157.30 | 153.66 | 166.38 | 165.46 | 145.87 | 167.06 | 153.58 | 165.15 |
| | Finland | 116.04 | 177.08 | 131.23 | 126.35 | 111.73 | 116.73 | 143.69 | 145.38 | 102.27 | 154.12 | 147.65 | 117.08 | 169.77 |
| | Sig. [1] | 0.082 | <0.001 | 0.180 | <0.001 | 0.008 | 0.005 | 0.158 | 0.013 | <0.001 | 0.642 | 0.006 | 0.007 | 0.010 |
| Age | 18–30 y | 145.67 | 165.18 | 146.47 | 162.80 | 148.75 | 166.25 | 128.03 | 143.74 | 148.43 | 127.05 | 130.29 | 128.11 | 154.39 |
| | 31–59 y | 137.85 | 141.01 | 145.00 | 142.18 | 141.49 | 138.95 | 144.30 | 137.69 | 142.00 | 140.73 | 140.53 | 143.00 | 145.89 |
| | 60+ y | 109.77 | 117.73 | 141.62 | 123.13 | 129.81 | 111.45 | 135.85 | 113.59 | 133.47 | 138.87 | 137.95 | 137.38 | 123.24 |
| | Sig. [1] | 0.035 | 0.028 | 0.947 | 0.083 | 0.513 | 0.007 | 0.490 | 0.086 | 0.680 | 0.678 | 0.802 | 0.620 | 0.132 |
| Sex | Female | 156.50 | 133.16 | 153.51 | 160.45 | 160.28 | 153.42 | 160.77 | 133.82 | 144.13 | 153.51 | 154.93 | 162.47 | 146.06 |
| | Male | 129.40 | 143.45 | 145.55 | 137.06 | 137.01 | 134.58 | 138.59 | 135.40 | 143.96 | 137.54 | 136.64 | 136.21 | 143.34 |
| | Sig. [2] | 0.012 | 0.355 | 0.448 | 0.038 | 0.035 | 0.089 | 0.040 | 0.881 | 0.988 | 0.139 | 0.096 | 0.018 | 0.807 |
| Education | Secondary | 129.48 | 130.25 | 143.62 | 142.91 | 139.25 | 135.26 | 132.32 | 125.71 | 148.87 | 134.94 | 146.08 | 137.08 | 143.11 |
| | University | 138.32 | 146.06 | 148.32 | 141.43 | 142.90 | 139.74 | 149.81 | 139.46 | 139.42 | 144.62 | 136.16 | 145.03 | 143.74 |
| | Sig. [2] | 0.353 | 0.104 | 0.611 | 0.879 | 0.704 | 0.640 | 0.065 | 0.141 | 0.333 | 0.314 | 0.300 | 0.413 | 0.949 |
| Income | Low | 123.26 | 131.60 | 132.91 | 154.21 | 138.94 | 138.50 | 131.52 | 124.40 | 139.96 | 128.19 | 136.56 | 122.52 | 134.79 |
| | Medium | 115.33 | 132.30 | 130.28 | 133.39 | 118.64 | 119.22 | 122.75 | 121.69 | 119.63 | 124.00 | 125.44 | 122.12 | 131.66 |
| | High | 120.38 | 105.92 | 128.24 | 84.38 | 123.94 | 113.84 | 130.84 | 107.92 | 127.28 | 125.27 | 108.75 | 135.82 | 11.40 |
| | Sig. [1] | 0.735 | 0.035 | 0.917 | <0.001 | 0.153 | 0.087 | 0.633 | 0.283 | 0.169 | 0.925 | 0.058 | 0.397 | 0.100 |

[1] Kruskal–Wallis test (level of significance $p < 0.05$). [2] U-Mann–Whitney test (level of significance $p < 0.05$).

The results in Figure 6 present the mean value for the interest attributed to the different forms of training activities on a scale varying from 1 (little interest) to 5 (much interest). The results indicate that the activities carried out in person are preferred by the participants, with the highest means score (3.94), while the distance training received the lowest score (3.06).

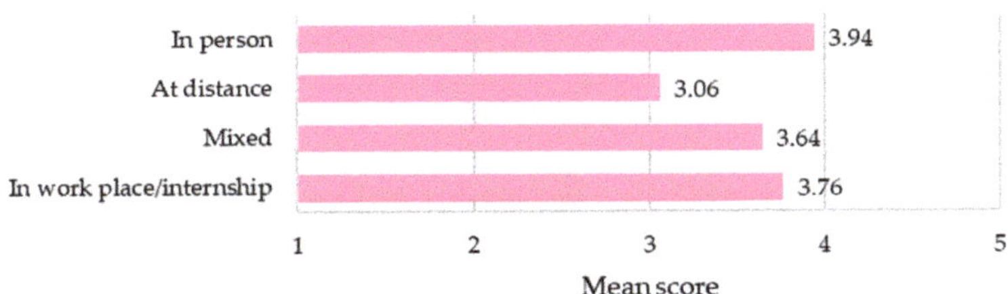


**Figure 6.** Level of interest according to the type of training activity.

Table 6 shows the results for the tests of group differences in training mode for the sociodemographic variables considered and reveals that country differences were statistically significant in all cases, i.e., for all types of training modes investigated. Higher means ranks were found for the 'In person' mode in Portugal (MR = 153.78), for the 'At distance' mode in Italy (MR = 167.44), for the 'Mixed' mode in Italy (MR = 152.75) and for more practical modes in Estonia (MR = 132.71 for the 'In work place/internship' mode). With respect to sex, significant differences were found for the 'At distance' and 'Mixed' modes. Finally, for education level, significant differences were encountered only for the 'At distance' mode, with this being preferred by people with a university degree.

**Table 6.** Group differences for preferences in training mode.

| Variable | Group | Mean Ranks and Significance | | | |
| --- | --- | --- | --- | --- | --- |
| | | In Person | At Distance | Mixed | In Workplace/ Internship |
| Country | Portugal | 153.78 | 103.42 | 85.88 | 136.54 |
| | Estonia | 122.89 | 106.65 | 127.38 | 132.71 |
| | Norway | 103.08 | 122.85 | 114.98 | 123.72 |
| | Spain | 117.25 | 86.65 | 122.55 | 84.10 |
| | Italy | 123.91 | 167.44 | 152.75 | 116.94 |
| | Croatia | 135.51 | 129.72 | 128.83 | 110.51 |
| | Finland | 102.82 | 130.41 | 118.00 | 93.00 |
| | Sig. [1] | 0.007 | 0.002 | 0.017 | 0.015 |
| Age | 18–30 y | 129.18 | 116.92 | 107.24 | 95.39 |
| | 31–59 y | 116.75 | 114.12 | 113.59 | 112.57 |
| | 60+ y | 115.50 | 103.73 | 113.31 | 122.63 |
| | Sig. [1] | 0.699 | 0.615 | 0.915 | 0.276 |
| Sex | Female | 130.66 | 138.78 | 131.79 | 123.75 |
| | Male | 117.26 | 105.89 | 110.01 | 111.42 |
| | Sig. [2] | 0.174 | <0.001 | 0.025 | 0.198 |
| Education | Secondary | 120.10 | 96.55 | 108.86 | 111.66 |
| | University | 119.13 | 123.91 | 118.18 | 115.45 |
| | Sig. [2] | 0.910 | 0.002 | 0.279 | 0.656 |
| Income | Low | 109.27 | 100.18 | 106.38 | 105.75 |
| | Medium | 116.18 | 99.11 | 106.26 | 104.03 |
| | High | 99.87 | 115.20 | 101.59 | 103.97 |
| | Sig. [1] | 0.243 | 0.217 | 0.871 | 0.981 |

[1] Kruskal–Wallis test ($p < 0.05$). [2] U-Mann–Whitney test ($p < 0.05$).

*3.6. Preferred Tools for Distance Learning*

When enquired whether the participants preferred digital or printed information about beekeeping for the purpose of lifelong learning and training activities, 177 said they preferred digital, and 136 preferred printed information. Countries where a higher number of participants prefer digital materials include Croatia (n = 44 against 17 who prefer printed), Finland (n = 9 against 5), Italy (n = 10 against 6), Portugal (n = 25 against 21) and Spain (n = 37 against 15). Contrarily, in Estonia and Norway, a higher number of participants prefer printed materials.

The participants' opinions about the usefulness of learning methodologies, materials, and assessment forms are presented in Figure 7. The mean scores were obtained as an average of all participants, and the measurement scale ranged from 1 (little useful) to 5 (very useful). The results indicate that project-based learning was the methodology considered most useful by the participants (mean score of 4.21), followed by the use of games and other challenges through gamification (means score of 3.82). The short courses were the least valued by the participants (with the lowest mean score of 2.44). With respect to the learning supports, the most valued were 'Videos' and 'Books/Paper manuals', with mean scores of 4.03 and 4.00, respectively. Strangely, the 'Educational games' came in last (with a mean score of 2.28), being considered a less useful learning support, somehow contradicting the results of the previous question, where gamification was a much-valued learning methodology. Finally, concerning the assessment formats, the 'Practical exercises' obtained the highest mean score (3.95), while 'Paper tests/questionnaires' obtained the lowest score (3.11) (Figure 7).

Tables 7 and 8 present the results obtained for the non-parametric tests performed to investigate possible significant differences between the groups regarding the sociodemographic variables studied in relation to the usefulness of learning methodologies, supports, and assessment formats. For the learning methodologies (Table 7), significant differences were observed between countries for practically all options, except for 'Monitoring of pilot farms'. For example, Italian participants attributed the lowest level of usefulness to 'Group work' (MR = 40.94) or 'Forum/Chat' (MR = 46.41), while attributing the highest usefulness to gamification (MR = 193.38) and to 'Short courses' (MR = 166.19). The differences according to age were only significant for 'Field visits' and for 'Short courses' (both options rated as less useful by older participants). Differences according to sex were also encountered for the same two learning methodologies, 'Field visits' and 'Short courses', which were preferred by female participants (mean ranks of 148.92 and 154.18, respectively). For education level, significant differences were found only for 'Gamification', with this methodology considered more useful by participants with a university degree (MR = 148.03). Finally, for the classes of income, significant differences were observed only for 'Lectures' and 'Short courses', with these being less valued by participants with the highest level of income (mean ranks of 95.28 and 92.63, respectively).

The results in Table 8 show that, once again, country differences were the most relevant, with significant differences for practically all analyzed learning supports and also for most of the assessment formats considered. While participants from Italy scored with 'E-books', 'Technical leaflets' and 'Educational games' as most useful (mean ranks of 159.09, 215.78 and 182.06, respectively), participants form Croatia rates attributed the highest mean scores to 'Interactive platforms', 'Videos' and "Specific programs or apps' (mean ranks of 153.60, 162.22 and 148.22, respectively). Age differences were significant for some of the learning supports, specifically 'E-books' and 'Educational games', which were less valued by older people (mean ranks of 105.81 and 95.52, respectively). The older participants also attributed lower usefulness to the assessment based on 'Tasks/reports' (MR = 103.27). Regarding sex, significant differences were found for some learning methodologies, such as 'Interactive platforms', 'Books/Paper manuals', and 'Educational games', with all these being more valued by female participants (mean ranks of 154.95, 162.52 and 140.69, respectively). The level of education showed significant differences only for 'E-books', with the highest level of usefulness assigned by participants with a university degree. Finally, significant differences were observed according to income for the 'Educational games', which were less valued by participants with the highest income level (MR = 89.49).

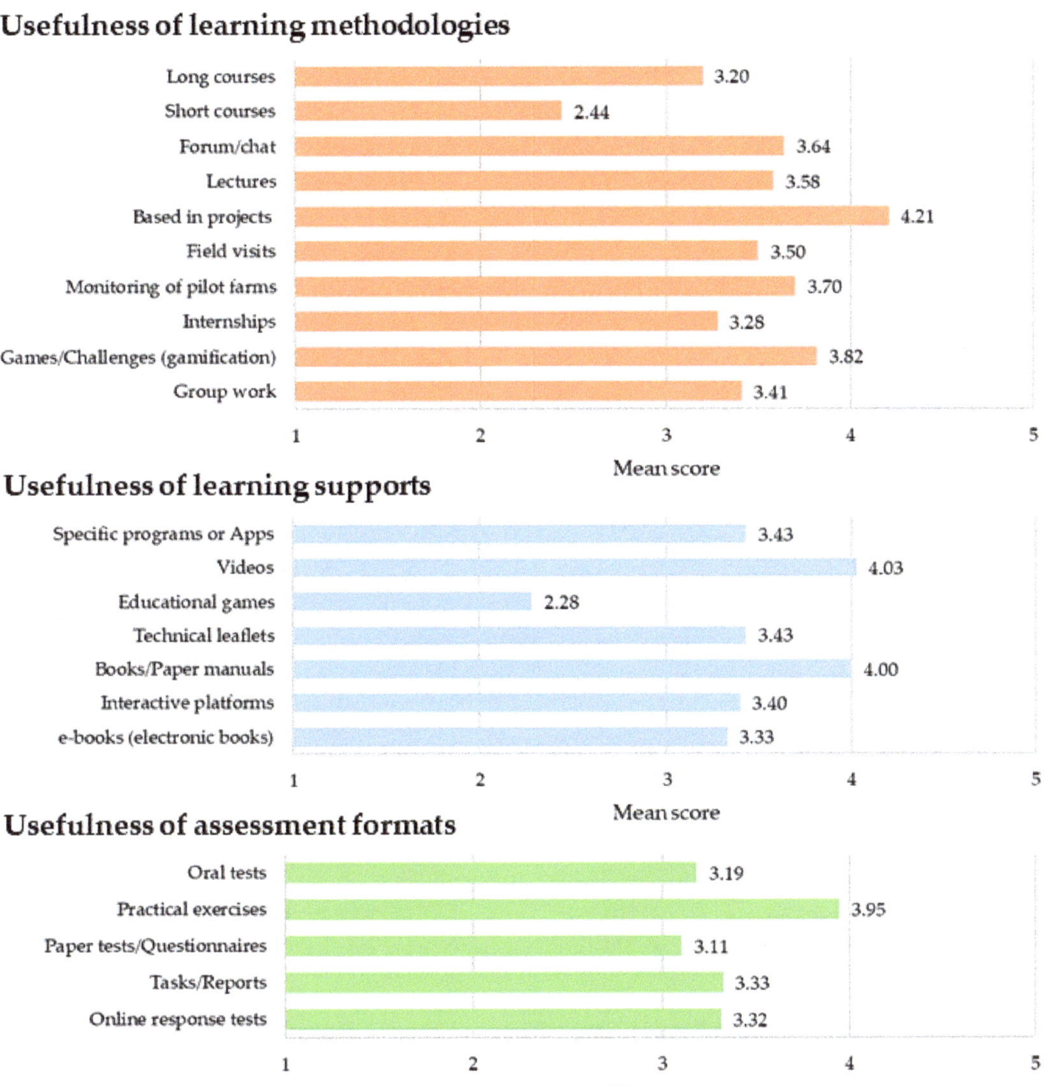


**Figure 7.** Rating the usefulness of learning methodologies, supports, and assessment formats.

**Table 7.** Group differences for the level of usefulness of different learning methodologies.

| Variable | Group | Mean Rank and Significance | | | | | | | | | |
|---|---|---|---|---|---|---|---|---|---|---|---|
| | | Group Work | Games/Challenges (Gamification) | Internships | Monitoring of Pilot Farms | Field Visits | Based in Projects | Lectures | Forum/Chat | Short Courses | Long Courses |
| Country | Portugal | 130.34 | 149.65 | 140.58 | 153.01 | 129.26 | 161.87 | 158.47 | 120.76 | 133.19 | 140.14 |
| | Estonia | 169.98 | 135.42 | 107.61 | 136.18 | 124.58 | 107.71 | 123.86 | 146.88 | 109.79 | 91.57 |
| | Norway | 135.88 | 160.39 | 121.42 | 133.21 | 113.16 | 123.02 | 106.43 | 128.16 | 111.80 | 125.31 |
| | Spain | 139.41 | 159.95 | 154.04 | 126.13 | 141.22 | 164.62 | 147.15 | 178.30 | 122.88 | 136.64 |
| | Italy | 40.94 | 193.38 | 105.25 | 87.66 | 83.72 | 137.84 | 99.25 | 46.41 | 166.19 | 69.33 |
| | Croatia | 137.75 | 84.08 | 168.13 | 150.50 | 171.23 | 151.19 | 166.02 | 135.58 | 156.46 | 135.92 |
| | Finland | 150.57 | 153.87 | 115.33 | 154.27 | 131.83 | 161.20 | 127.80 | 104.36 | 142.63 | 155.18 |
| | Sig. [1] | <0.001 | <0.001 | <0.001 | 0.064 | <0.001 | 0.001 | <0.00981 | <0.001 | 0.005 | <0.001 |
| Age | 18–30 y | 120.66 | 132.81 | 119.14 | 132.34 | 124.75 | 162.84 | 141.98 | 136.36 | 150.34 | 104.17 |
| | 31–59 y | 137.80 | 136.49 | 136.15 | 133.97 | 136.51 | 134.28 | 135.69 | 132.92 | 132.04 | 122.34 |
| | 60+ y | 124.09 | 138.62 | 121.00 | 137.57 | 107.65 | 135.35 | 117.56 | 115.17 | 98.05 | 131.21 |
| | Sig. [1] | 0.335 | 0.947 | 0.285 | 0.941 | 0.048 | 0.140 | 0.249 | 0.281 | 0.002 | 0.257 |
| Sex | Female | 149.67 | 152.89 | 149.19 | 146.60 | 148.92 | 152.69 | 142.67 | 142.88 | 154.18 | 133.63 |
| | Male | 132.20 | 136.37 | 130.57 | 134.55 | 127.46 | 137.34 | 133.81 | 129.86 | 122.71 | 122.40 |
| | Sig. [2] | 0.104 | 0.123 | 0.086 | 0.260 | 0.046 | 0.143 | 0.407 | 0.220 | 0.003 | 0.290 |
| Education | Secondary | 136.58 | 118.80 | 133.36 | 133.08 | 122.88 | 136.05 | 136.48 | 131.25 | 127.01 | 124.10 |
| | University | 131.61 | 148.03 | 130.39 | 134.56 | 132.71 | 138.37 | 130.00 | 128.44 | 126.20 | 121.48 |
| | Sig. [2] | 0.599 | 0.002 | 0.752 | 0.874 | 0.289 | 0.799 | 0.488 | 0.762 | 0.929 | 0.770 |
| Income | Low | 113.55 | 115.83 | 122.95 | 118.97 | 122.66 | 132.21 | 128.74 | 121.75 | 137.60 | 113.84 |
| | Medium | 129.04 | 117.24 | 119.49 | 120.21 | 122.77 | 126.79 | 129.62 | 121.24 | 117.68 | 112.14 |
| | High | 111.98 | 138.49 | 108.59 | 118.89 | 101.54 | 109.13 | 95.28 | 106.33 | 92.63 | 105.77 |
| | Sig. [1] | 0.177 | 0.082 | 0.430 | 0.989 | 0.089 | 0.095 | 0.003 | 0.299 | <0.001 | 0.752 |

[1] Kruskal–Wallis test (level of significance $p < 0.05$). [2] U-Mann–Whitney test (level of significance $p < 0.05$).

**Table 8.** Group differences for the level of usefulness of different learning supports and assessment formats.

| Variable | Group | Mean Rank and Significance | | | | | | | | | | | |
| --- | --- | --- | --- | --- | --- | --- | --- | --- | --- | --- | --- | --- | --- |
| | | Learning Supports | | | | | | | | Assessment Formats | | | |
| | | e-Books (Electronic Books) | Interactive Platforms | Books/ Paper Manuals | Technical Leaflets | Educational Games | Videos | Specific Programs or Apps | Online Response Tests | Tasks/Reports | Paper Tests/ Questionnaires | Practical Exercises | Oral Tests |
| Country | Portugal | 122.66 | 139.03 | 115.88 | 159.03 | 122.31 | 129.42 | 131.84 | 120.58 | 112.42 | 140.28 | 138.47 | 111.08 |
| | Estonia | 121.53 | 109.62 | 151.77 | 58.38 | 94.86 | 154.53 | 93.56 | 135.66 | 135.68 | 130.39 | 170.34 | 142.22 |
| | Norway | 119.61 | 124.98 | 154.83 | 135.92 | 106.82 | 130.34 | 115.81 | 158.88 | 132.85 | 146.41 | 124.52 | 125.88 |
| | Spain | 127.04 | 150.55 | 136.20 | 138.42 | 76.26 | 151.26 | 137.11 | 123.50 | 145.96 | 109.88 | 162.50 | 126.17 |
| | Italy | 159.09 | 143.84 | 132.88 | 215.78 | 182.06 | 89.69 | 120.81 | 131.16 | 115.56 | 76.00 | 85.22 | 68.19 |
| | Croatia | 157.51 | 153.60 | 136.07 | 144.71 | 133.68 | 162.22 | 148.22 | 132.27 | 135.17 | 144.81 | 134.37 | 142.80 |
| | Finland | 128.00 | 125.35 | 153.19 | 129.08 | 131.46 | 154.08 | 115.62 | 146.04 | 117.21 | 158.00 | 133.64 | 111.77 |
| | Sig. [1] | 0.064 | 0.083 | 0.233 | <0.001 | <0.001 | 0.018 | 0.013 | 0.197 | 0.453 | 0.005 | 0.001 | 0.006 |
| Age | 18–30 y | 116.69 | 128.23 | 131.86 | 138.37 | 136.45 | 125.79 | 122.73 | 107.83 | 115.48 | 125.61 | 143.19 | 142.30 |
| | 31–59 y | 136.94 | 136.97 | 139.61 | 134.66 | 116.11 | 142.78 | 126.36 | 139.26 | 135.27 | 133.99 | 139.44 | 121.13 |
| | 60+ y | 105.81 | 116.65 | 127.50 | 112.67 | 95.52 | 126.63 | 108.31 | 118.92 | 103.27 | 113.29 | 118.76 | 112.95 |
| | Sig. [1] | 0.020 | 0.238 | 0.547 | 0.154 | 0.037 | 0.255 | 0.326 | 0.048 | 0.016 | 0.215 | 0.187 | 0.204 |
| Sex | Female | 135.95 | 154.95 | 162.52 | 137.16 | 140.69 | 149.58 | 135.83 | 150.54 | 141.44 | 134.95 | 150.57 | 124.12 |
| | Male | 130.81 | 130.61 | 134.26 | 133.67 | 108.97 | 138.93 | 122.84 | 131.82 | 127.54 | 133.01 | 135.85 | 125.98 |
| | Sig. [2] | 0.635 | 0.029 | 0.009 | 0.746 | 0.001 | 0.320 | 0.211 | 0.081 | 0.182 | 0.854 | 0.165 | 0.855 |
| Education | Secondary | 115.90 | 124.12 | 133.86 | 135.57 | 115.94 | 132.64 | 118.72 | 116.40 | 122.04 | 129.86 | 137.85 | 125.32 |
| | University | 136.94 | 138.38 | 139.73 | 129.85 | 113.70 | 142.21 | 126.26 | 144.83 | 131.60 | 130.90 | 134.88 | 119.87 |
| | Sig. [2] | 0.024 | 0.131 | 0.530 | 0.543 | 0.798 | 0.304 | 0.378 | 0.002 | 0.298 | 0.911 | 0.750 | 0.544 |
| Income | Low | 115.60 | 122.94 | 122.94 | 131.51 | 123.13 | 127.14 | 116.16 | 105.50 | 109.22 | 122.65 | 124.17 | 111.72 |
| | Medium | 123.60 | 122.81 | 120.29 | 114.22 | 101.93 | 127.35 | 110.26 | 123.60 | 116.79 | 112.34 | 127.09 | 116.54 |
| | High | 107.32 | 108.10 | 129.31 | 114.09 | 89.49 | 115.22 | 112.34 | 130.85 | 118.11 | 120.65 | 113.17 | 96.90 |
| | Sig. [1] | 0.297 | 0.326 | 0.686 | 0.199 | 0.006 | 0.460 | 0.838 | 0.073 | 0.683 | 0.542 | 0.400 | 0.156 |

[1] Kruskal–Wallis test (level of significance $p < 0.05$). [2] U-Mann–Whitney test (level of significance $p < 0.05$).

## 4. Discussion

Education constitutes a privileged way to increase productivity and competitiveness in multiple business areas, including beekeeping. LL is relevant not only from the professional but also from the personal point of view, providing opportunities for self-development and continuous improvement. Allying LL with PT allows for a constant, or at least a regular, valorization of the individual and their skills and competencies, providing tools to become more resilient and successful in all areas of professional development. e-Learning takes the lead and will continue to play a prevailing role in the construction of educational management systems and related learning environments [19,23].

The integration of information technology (internet and other resources) and mobile devices used for learning (m-learning) with conventional education can have a significant impact on the improvement in LL capacity. It has been recommended that, particularly for rural environments, training programs for mobile education should address four main challenges related to the practical nature of the courses and specificity of learning environments, namely: scarce educational space and limited equipment; instructors and technicians with developed applied skills but without proper pedagogical support; the under-relevance attributed to parallel and additional experiences; unsatisfactory class management by the instructors and technicians. The agricultural sector and its related activities, such as beekeeping, are major contributors to the economies of many countries. Beekeeping, in particular, contributes through the great importance of bees as pollinators and regulators of biodiversity and ecosystems, and assumes an even greater role in global sustainability. Hence, a great challenge for the organizations teaching in this area might involve changes in the pedagogical methods adopted to address the needs and wishes of the students [24,25].

Despite the massive possibilities of distance learning methodologies having been acknowledged for many decades, it is also true that, until the year 2020, with the outbreak of COVID-19 pandemic, teaching methods continued to follow a mostly traditional approach based on in-person teaching inside a classroom. The pandemic brought an urgent need to shift rapidly from in-person learning systems to distance learning, supported by technology and digital content, causing an evolution not only in the technology itself but also in the didactic and pedagogical domains. Therefore, at present, professionals are more adapted to distance learning and innovative learning methodologies, as well as assessment formats [26–28].

Fischer et al. [23] describe a framework for reconsidering education, including novel components such as learning-on-demand or problem-based learning. The design of innovative learning approaches for the digital era entails meticulousness in designing learning experiences and evaluating them as a way to understand what effectively works, how it works, and why it works. The design of digital learning experiences is supported by multiple dimensions related to how learners interact with the digital tools they use, their learning environments, or services. These also relate to the pedagogical foundations leading to the established learning goals, the necessary activities to achieve those goals, and the chosen forms of assessment. Finally, it is necessary to investigate how learners interact with other peers and with instructors [29].

Distance learning tools for PT are particularly useful for active professionals, given their lack of time. Still, professionals feel a need to improve their knowledge, skills and competencies as a way to improve and expand their businesses and increase competitiveness, in addition to their natural desire to broaden their knowledge on certain topics [30–33].

Beekeeping is a complex activity once beekeepers manage upwards of 10,000 individual honeybees in a single colony. Honeybees are highly sensitive to environmental and seasonal changes and vulnerable to a range of diseases and pests. This makes beekeeping an activity that requires specialized skills and knowledge to ensure the health and productivity of honeybees [34].

In this work, beekeepers showed a preference for training needs on "Apiary health and pest control" and 'Colony management throughout the year". It can be explained by the high number of honeybee colonies lost every year and the beekeeper's will to increase the productivity of their apiaries. Gray et al. [35] showed that Spain was the European country with the highest rate of colonies lost in the winter of 2019/2020, with 36.5%, followed by Slovenia (28.9%) and Portugal (22.5%). Varroosis is the most prevalent worldwide disease of honey bees, and an important cause of beehive loss, with a high economic impact on beekeeping activity [36]. Increasing beekeepers' knowledge of these two issues is crucial to improving beehives' productivity and, consequently, beekeepers' income.

Jacques et al. [37] highlighted beekeeper background and apicultural practices as the major drivers of honey bee colony losses and reinforced the need for beekeeper training to promote the best beekeeping practices. The research suggests that access to beekeeping training could be an important mechanism influencing honey productivity and beekeeping incomes [38–40].

Regarding training activities, beekeepers prefer "in person" courses, followed by "in workplace/internship", to b-learning or e-learning courses. Beekeeping requires mostly practical training, which can explain beekeepers' preference for training that is carried out "in person", rather than b-learning or e-learning. However, the classical modes of teaching cause beekeepers to fall into a passive learning pattern and increase the gap between the practice and theory [41].

Schouten and Caldeira [40] recommend that beekeeping training focus on practical skills' development over classroom theory-based activities. Concerning the preferred tools for distance learning, beekeepers prefer knowledge-based projects, followed by gamification, as learning methodologies. The preferred learning materials were videos and books. Finally, the preferred assessment form was based on practical exercises. E-learning involves online instruction without any face-to-face contact, and beekeepers can learn at their own pace with online resources [42]. Training through e-learning can be engaging and interactive, using videos, presentations, chat, library, and assessments, with the goal of maximizing the learner's experience in the beekeeping learning process [41].

Beekeeping training can be delivered in a range of modes, in-person, e-learning or b-learning. Independent of beekeepers' preferences, training is important to improve their knowledge and skills. According to Schouten and Lloyd [43], the learning programs should be adjusted in developing countries, considering the strong necessity of beekeeping knowledge and the limited conditions required to enable the implementation of bee management in the colonies.

Education and learning are important means of supporting the knowledge needed to improve beekeeping management and create value-added hive products due to the new techniques and technology being adopted [44,45]. Even in more developed beekeeping structures, evolution, research, and innovation are only possible with LL, which is provided by different formal and informal modalities [46]. In a study conducted in Nagano, Japan, by Uchiyama et al. [47], it was found that tacit knowledge within the family promotes explicit knowledge in an ageing society, leading to a relatively large number of bee colonies and a perception of the necessary ecological conditions for sustainable beekeeping. In fact, beekeepers' environmental knowledge remains the backbone of the activity's sustainability [48].

## 5. Conclusions

The results of this study indicated valuable directions to implement proper professional training for actors in the beekeeping sector. The topics of highest interest include the health of apiaries and control of pests affecting the apiaries and bee colonies, or the management of the colonies throughout the year, with different specifications according to the season. The beekeepers seek new information mostly through family but also through professional training, and the preferred forms of training include in-person courses, workplace training or internships. The learning methodologies they consider most useful include project-based learning and learning through gamification and related tools. With respect to the learning supports, videos and paper books or manuals are particularly valued, and the assessment format rated as most valuable is practical exercises. Another inveistigated aspect was the effect of sociodemographic variables on the learning experiences and preferences of beekeeping actors, and in this respect, it was observed that the country was the most influential of the investigated factors.

The construction of courses adapted for mobile learning with adequate forms of assessment of the learning outcomes allows for the continuous updating of information, creation of knowledge, and development of skills that beekeepers consider essential for their activities. They want to take part in PT in topics they find crucial; therefore, the curriculum development needs to adapt to this reality. However, they find distance learning to be a useful means of training, but they recognize that complementing this with practical activities is necessary to achieve success, since these blended learning approaches bring together the best of the different approaches.

**Author Contributions:** Conceptualization, R.P.F.G.; methodology, R.P.F.G. and J.O.; software, R.P.F.G.; validation, R.P.F.G.; formal analysis, R.P.F.G.; investigation, R.P.F.G., C.A.C., H.E.C., B.D., R.R., R.K., L.T., S.B., E.B., I.S., M.O., C.C. and J.O.; resources, C.A.C.; data curation, R.P.F.G.; writing—original draft preparation, D.T.C., P.C., J.O., C.C. and R.P.F.G.; writing—review and editing, R.P.F.G.; visualization, R.P.F.G.; supervision, R.P.F.G.; project administration, C.A.C.; funding acquisition, C.A.C. All authors have read and agreed to the published version of the manuscript.

**Funding:** This research was funded by project beeB—Foster for beekeeping bridges through innovative and participative training (ref. ERASMUS+ 2019-1-PT01-KA202-060782). The APC was funded by FCT—Foundation for Science and Technology, I.P., within the scope of the project Ref. UIDB/00681/2020.

**Institutional Review Board Statement:** This research was implemented taking care to ensure all ethical standards and followed the guidelines of the Declaration of Helsinki.

**Informed Consent Statement:** Informed consent was obtained from all subjects involved in the study.

**Data Availability Statement:** Data are available from the first author (R.P.F.G.) upon reasonable request.

**Acknowledgments:** This work was developed under project beeB—Foster for beekeeping bridges through innovative and participative training (ref. ERASMUS+ 2019-1-PT01-KA202-060782). The authors would also like to acknowledge support from the FCT—Foundation for Science and Technology, I.P., within the scope of the project Ref. UIDB/00681/2020, as well as the CERNAS Research Centre and the Polytechnic Institute of Viseu.

**Conflicts of Interest:** The authors declare no conflict of interest.

## References

1. Abro, Z.; Kassie, M.; Tanga, C.; Beesigamukama, D.; Diiro, G. Socio-Economic and Environmental Implications of Replacing Conventional Poultry Feed with Insect-Based Feed in Kenya. *J. Clean. Prod.* **2020**, *265*, 121871. [CrossRef]
2. De Meio Reggiani, M.C.; Villar, L.B.; Vigier, H.P.; Brignole, N.B. An Evolutionary Approach for the Optimization of the Beekeeping Value Chain. *Comput. Electron. Agric.* **2022**, *194*, 106787. [CrossRef]
3. Malkamäki, A.; Toppinen, A.; Kanninen, M. Impacts of Land Use and Land Use Changes on the Resilience of Beekeeping in Uruguay. *For. Policy Econ.* **2016**, *70*, 113–123. [CrossRef]
4. Sillman, J.; Uusitalo, V.; Tapanen, T.; Salonen, A.; Soukka, R.; Kahiluoto, H. Contribution of Honeybees towards the Net Environmental Benefits of Food. *Sci. Total Environ.* **2021**, *756*, 143880. [CrossRef]
5. Armstrong, A.; Brown, L.; Davies, G.; Whyatt, J.D.; Potts, S.G. Honeybee Pollination Benefits Could Inform Solar Park Business Cases, Planning Decisions and Environmental Sustainability Targets. *Biol. Conserv.* **2021**, *263*, 109332. [CrossRef]
6. Fijen, T.P.M.; van Bodegraven, V.; Lucassen, F. Limited Honeybee Hive Placement Balances the Trade-off between Biodiversity Conservation and Crop Yield of Buckwheat Cultivation. *Basic Appl. Ecol.* **2022**, *65*, 28–38. [CrossRef]
7. Guiné, R.P.F.; Mesquita, S.; Oliveira, J.; Coelho, C.; Costa, D.T.; Correia, P.; Correia, H.E.; Dahle, B.; Oddie, M.; Raimets, R.; et al. Characterization of Beekeepers and Their Activities in Seven European Countries. *Agronomy* **2021**, *11*, 2398. [CrossRef]
8. El Agrebi, N.; Steinhauer, N.; Tosi, S.; Leinartz, L.; de Graaf, D.C.; Saegerman, C. Risk and Protective Indicators of Beekeeping Management Practices. *Sci. Total Environ.* **2021**, *799*, 149381. [CrossRef]
9. Sperandio, G.; Simonetto, A.; Carnesecchi, E.; Costa, C.; Hatjina, F.; Tosi, S.; Gilioli, G. Beekeeping and Honey Bee Colony Health: A Review and Conceptualization of Beekeeping Management Practices Implemented in Europe. *Sci. Total Environ.* **2019**, *696*, 133795. [CrossRef]

10. Abbott, L.F.; Nelson, S.B. Synaptic Plasticity: Taming the Beast. *Nat. Neurosci.* **2000**, *3*, 1178–1183. [CrossRef]

11. Gryshchuk, V.; Weber, C.; Loo, C.K.; Wermter, S. Go Ahead and Do Not Forget: Modular Lifelong Learning from Event-Based Data. *Neurocomputing* **2022**, *500*, 1063–1074. [CrossRef]

12. Parisi, G.I.; Kemker, R.; Part, J.L.; Kanan, C.; Wermter, S. Continual Lifelong Learning with Neural Networks: A Review. *Neural Netw.* **2019**, *113*, 54–71. [CrossRef] [PubMed]

13. Baghel, M.S.; Singh, B.; Dhuriya, Y.K.; Shukla, R.K.; Patro, N.; Khanna, V.K.; Patro, I.K.; Thakur, M.K. Postnatal Exposure to Poly (I:C) Impairs Learning and Memory through Changes in Synaptic Plasticity Gene Expression in Developing Rat Brain. *Neurobiol. Learn. Mem.* **2018**, *155*, 379–389. [CrossRef] [PubMed]

14. Goto, A. Synaptic Plasticity during Systems Memory Consolidation. *Neurosci. Res.* **2022**, *183*, 1–6. [CrossRef]

15. Evans, K.; Kersh, N. Lifelong Learning beyond Initial Schooling. In *International Encyclopedia of Education*, 4th ed.; Tierney, R.J., Rizvi, F., Ercikan, K., Eds.; Elsevier: Oxford, UK, 2023; pp. 520–529. ISBN 978-0-12-818629-9.

16. Irfan, M.; Jiangbin, Z.; Iqbal, M.; Masood, Z.; Arif, M.H.; Hassan, S.R. ul Brain Inspired Lifelong Learning Model Based on Neural Based Learning Classifier System for Underwater Data Classification. *Expert Syst. Appl.* **2021**, *186*, 115798. [CrossRef]

17. Colosimo, A.L.; Badia, G. Diaries of Lifelong Learners: Information Seeking Behaviors of Older Adults in Peer-Learning Study Groups at an Academic Institution. *Libr. Inf. Sci. Res.* **2021**, *43*, 101102. [CrossRef]

18. Huang, K.; Ma, X.; Song, R.; Rong, X.; Li, Y. Autonomous Cognition Development with Lifelong Learning: A Self-Organizing and Reflecting Cognitive Network. *Neurocomputing* **2021**, *421*, 66–83. [CrossRef]

19. Ekúndayò, O.T.; Tuluri, F. Learner Management Systems and Environments, Implications for Pedagogy and Applications to Resource Poor Environments. In *E-Learning Standards and Interoperability*; IGI Global: Hershey, PA, USA, 2011; Chapter 25; pp. 499–525. ISBN 9781616927899.

20. Segret, J.A.L.; García, M.C. The Management of the Human Resources and the Quality of the Services. In *Encyclopedia of Human Resources Information Systems: Challenges in e-HRM*; IGI Global: Hershey, PA, USA, 2009; Volume 2, pp. 632–639.

21. Bhavsar-Burke, I.; Dilly, C.K.; Oxentenko, A.S. How to Promote Professional Identity Development and Support Fellows-In-Training Through Teaching, Coaching, Mentorship, and Sponsorship. *Clin. Gastroenterol. Hepatol.* **2022**, *20*, 2166–2169. [CrossRef]

22. Garzón-Artacho, E.; Sola-Martínez, T.; Romero-Rodríguez, J.-M.; Gómez-García, G. Teachers' Perceptions of Digital Competence at the Lifelong Learning Stage. *Heliyon* **2021**, *7*, e07513. [CrossRef]

23. Fischer, G.; Lundin, J.; Lindberg, O.J. The Challenge for the Digital Age: Making Learning a Part of Life. *Int. J. Inf. Learn. Technol.* **2022**, *40*, 1–16. [CrossRef]

24. Guiné, R.P.F.; Costa, D.V.T.A.; Correia, P.M.R.; Costa, C.A.; Correia, H.E.; Castro, M.; Guerra, L.T.; Seeds, C.; Coll, C.; Radics, L.; et al. Designing Training in Organic Farming on a Multinational Basis. *Int. J. Inf. Learn. Technol.* **2016**, *33*, 99–114. [CrossRef]

25. Guiné, R.; Costa, D.; Correia, P.; Costa, C.; Correia, H.; Castro, M.; Guerra, L.; Seeds, C.; Coll, C.; Radics, L.; et al. Professional Training in Organic Food Production: A Cross-Country Experience. *Int. J. Inf. Learn. Technol.* **2017**, *34*, 259–273. [CrossRef]

26. Segbenya, M.; Bervell, B.; Minadzi, V.M.; Somuah, B.A. Modelling the Perspectives of Distance Education Students towards Online Learning during COVID-19 Pandemic. *Smart Learn. Environ.* **2022**, *9*, 13. [CrossRef]

27. Segbenya, M.; Anokye, F.A. Challenges and Coping Strategies among Distance Education Learners: Implication for Human Resources Managers. *Curr. Psychol.* **2022**, *29*, 1–15. [CrossRef] [PubMed]

28. Yorkovsky, Y.; Levenberg, I. Distance Learning in Science and Mathematics—Advantages and Disadvantages Based on Pre-Service Teachers' Experience. *Teach. Teach. Educ.* **2022**, *120*, 103883. [CrossRef]

29. Jahnke, I. Quality of Digital Learning Experiences—Effective, Efficient, and Appealing Designs? *Int. J. Inf. Learn. Technol.* **2022**, *40*, 17–30. [CrossRef]

30. Farida, I.; Setiawan, D. Business Strategies and Competitive Advantage: The Role of Performance and Innovation. *J. Open Innov. Technol. Mark. Complex.* **2022**, *8*, 163. [CrossRef]

31. Lee, K.; Choi, H.; Cho, Y.H. Becoming a Competent Self: A Developmental Process of Adult Distance Learning. *Internet High. Educ.* **2019**, *41*, 25–33. [CrossRef]

32. Mahdavinejad, M.; Ghasempourabadi, M.; Ghaedi, H.; Nikhoosh, N. Formal Architectural Education and Training Professional Technicians (Case Study: Iran). *Procedia Soc. Behav. Sci.* **2012**, *51*, 454–458. [CrossRef]

33. Shen, C.-C.; Yeh, C.-C.; Lin, C.-N. Using the Perspective of Business Information Technology Technicians to Explore How Information Technology Affects Business Competitive Advantage. *Technol. Forecast. Soc. Change* **2022**, *184*, 121973. [CrossRef]

34. Adams, E.C. How to Become a Beekeeper: Learning and Skill in Managing Honeybees. *Cult. Geogr.* **2018**, *25*, 31–47. [CrossRef]

35. Gray, A.; Adjlane, N.; Arab, A.; Ballis, A.; Brusbardis, V.; Bugeja Douglas, A.; Cadahía, L.; Charrière, J.-D.; Chlebo, R.; Coffey, M.F.; et al. Honey Bee Colony Loss Rates in 37 Countries Using the COLOSS Survey for Winter 2019–2020: The Combined Effects of Operation Size, Migration and Queen Replacement. *J. Apic. Res.* **2022**, *62*, 204–210. [CrossRef]

36. De Carolis, A.; Newmark, A.J.; Kim, J.; Cazier, J.; Hassler, E.; Pietropaoli, M.; Robinette, C.; Formato, G.; Song, J. Results of an International Survey for Risk Assessment of Honey Bee Health Concerning Varroa Management. *Appl. Sci.* **2023**, *13*, 62. [CrossRef]

37. Jacques, A.; Laurent, M.; Consortium, E.; Ribière-Chabert, M.; Saussac, M.; Bougeard, S.; Budge, G.E.; Hendrikx, P.; Chauzat, M.-P. A Pan-European Epidemiological Study Reveals Honey Bee Colony Survival Depends on Beekeeper Education and Disease Control. *PLoS ONE* **2017**, *12*, e0172591. [CrossRef] [PubMed]

38. Aksoy, A.; Demir, N.; Bilgiç, A. A Study on Identifying the Effectiveness of the Beekeeping Grants Provided by IPARD Program: Examples of Erzurum, Kars and Agri Provinces. *Custos E @Gronegócio Line* **2018**, *14*, 269–283.
39. Duah, H.K.; Segbefia, A.Y.; Adjaloo, M.K.; Fokuo, D. Income Sustainability and Poverty Reduction among Beekeeping Value Chain Actors in the Berekum Municipality, Ghana. *Int. J. Dev. Sustain.* **2017**, *6*, 667–684.
40. Schouten, C.N.; Caldeira, J. Improving the Effectiveness of Beekeeping Training: A Case Study of Beekeeping Instructors in Fiji. *Bee World* **2021**, *98*, 57–62. [CrossRef]
41. Momoh, J. Development of E-Learning Platform for Beekeepers. *SEEM Res. Dev. J.* **2013**, *2*, 63–72.
42. Gupta, S.B.; Gupta, M. Technology and E-Learning in Higher Education. *Int. J. Adv. Sci. Technol.* **2020**, *29*, 1320–1325.
43. Schouten, C.N.; Lloyd, D.J. Considerations and Factors Influencing the Success of Beekeeping Programs in Developing Countries. *Bee World* **2019**, *96*, 75–80. [CrossRef]
44. Vapa-Tankosić, J.; Miler-Jerković, V.; Jeremić, D.; Stanojević, S.; Radović, G. Investment in Research and Development and New Technological Adoption for the Sustainable Beekeeping Sector. *Sustainability* **2020**, *12*, 5825. [CrossRef]
45. Seifollahi, N. Investigating the Impact of knowledge Management Dimensions on Value Chain in Beekeeping Industry (Case Study: Ardebil Province). *Iran. J. Agric. Econ. Dev. Res.* **2018**, *49*, 797–804. [CrossRef]
46. Čavlin, M.; Prdić, N.; Ignjatijević, S.; Vapa Tankosić, J.; Lekić, N.; Kostić, S. Research on the Determination of the Factors Affecting Business Performance in Beekeeping Production. *Agriculture* **2023**, *13*, 686. [CrossRef]
47. Uchiyama, Y.; Matsuoka, H.; Kohsaka, R. Apiculture Knowledge Transmission in a Changing World: Can Family-Owned Knowledge Be Opened? *J. Ethn. Foods* **2017**, *4*, 262–267. [CrossRef]
48. Maderson, S. There's More Than One Way To Know A Bee: Beekeepers' Environmental Knowledge, and Its Potential Role in Governing for Sustainability. *Geoforum* **2023**, *139*, 103690. [CrossRef]

 *sustainability*

*Article*

# Hazards in Products of Plant Origin Reported in the Rapid Alert System for Food and Feed (RASFF) from 1998 to 2020

**Marcin Pigłowski [1,*] and Magdalena Niewczas-Dobrowolska [2]**

[1] Department of Quality Management, Faculty of Management and Quality Science, Gdynia Maritime University, Morska 81-87, 81-225 Gdynia, Poland
[2] Department of Quality Management, Institute of Quality Sciences and Product Management, College of Management and Quality Sciences, Cracow University of Economics, Rakowicka 27, 31-510 Kraków, Poland; niewczam@uek.krakow.pl
* Correspondence: m.piglowski@wznj.umg.edu.pl

**Abstract:** The elimination or reduction of hazards in plants is an important part of the "From field to fork" strategy adopted in the European Green Deal, where a sustainable model is pursued in the food system. In the European Union (EU), the Rapid Alert System for Food and Feed (RASFF) is in place to provide information on risks in the food chain. The largest number of notifications in this system concerns plants, followed by products of animal origin and other products. The goal of the study was to examine RASFF notifications for products of plant origin with respect to hazard, year, product, notifying country, origin country, notification type, notification basis, distribution status and actions taken in 1998–2020. Data were extracted from the RASFF notifications' pre-2021 public information database. A cluster analysis using joining and the two-way joining method was applied. The notifications mainly concerned aflatoxins in pistachios from Iran, ochratoxin A in raisins from Turkey, pesticide residues in peppers from Turkey, okra, curry, rice from India, tea from China and India, and pathogenic micro-organisms in sesame from India, and also basil, mint and betel from Thailand, Vietnam and Lao Republic. To ensure the safety of food of plant origin, it is necessary to adhere to good agricultural and manufacturing practices, involve producers in the control of farmers, ensure proper transport conditions (especially from Asian countries), ensure that legislative bodies set and update hazard limits, and ensure their subsequent control by the authorities of EU countries. Due to the broad period and scope of the studies that have been carried out and the significance of the European Union in the food chain, the research results can improve global sustainability efforts.

**Keywords:** food safety; food hazards; plants; RASFF; cluster analysis

**Citation:** Pigłowski, M.; Niewczas-Dobrowolska, M. Hazards in Products of Plant Origin Reported in the Rapid Alert System for Food and Feed (RASFF) from 1998 to 2020. *Sustainability* **2023**, *15*, 8091. https://doi.org/10.3390/su15108091

Academic Editor: Dimitris Skalkos

Received: 7 April 2023
Revised: 7 May 2023
Accepted: 12 May 2023
Published: 16 May 2023



## 1. Introduction

A sustainable global future should consider food security and food safety, taking public health into account to achieve long-term sustainability. According to the World Health Organization (WHO) definition, food security exists "when all people, at all times, have physical and economic access to sufficient, safe and nutritious food to meet their dietary needs and food preferences for an active and healthy life". This is closely linked to economic growth, social progress, political stability and peace. It should be noted that food safety can be recognised as a component of food security, as this refers to the fact that food is safe to eat and does not pose a risk to human health [1]. Food safety should include the sustainable development of the agri-food sector [1,2]. Thus, both sustainability and future food security require the consideration of food safety [1,3].

The most important challenge to food security and food safety is the growing human population [4]. However, it is important to point out that this mainly concerns developing countries. Considering sustainability in the context of food, it is noteworthy that in developing countries, attention is focused on food security, and in developed countries, on food safety [5]. Given this discrepancy, it therefore seems important to pay attention to the movement of food from developing countries to developed countries. In order to reduce the risk of foodborne disease hazards, developing countries that trade in food should have an integrated and inclusive development policy with regard to food security [6].

In the Sustainability Assessment of Food and Agriculture systems guidelines issued by the Food And Agriculture Organization of the United Nations (FAO), food safety is mentioned in the theme "Product quality & information" within the economic resilience dimension. In this document, food safety hazard is defined as "a biological, chemical or psychical agent in, or condition of, food with the potential to cause an adverse health effect" [7]. Among the biological agents, there are, for example, mycotoxins and pathogenic micro-organisms; chemical agents can include pesticide residues, and physical agents comprise foreign bodies [8].

According to the requirements for food safety included in the European law, food that is injurious to health is considered unsafe and should not be placed on the market [9]. Therefore, the Rapid Alert System for Food and Feed (RASFF) was established to provide information on risks in the food chain. During the period 1979–2020, the largest number of notifications in this system related to food of plant origin (more than 43%), followed by food of animal origin (30%), with the remaining notifications referring to other types of food, feed and food contact materials [8].

### 1.1. Characteristics of the RASFF

Currently, the legal basis for the operation of the RASFF is the Regulation (EC) No. 178/2002, laying down the general principles and requirements of food law, establishing the European Food Safety Authority, and setting up procedures in matters of food safety. This Regulation obliges each RASFF member to report to the European Commission with information on any serious health risks deriving from food or feed. The members of the system are the 27 countries of the European Union (EU), the European Commission, the European Food Safety Authority (EFSA), the European Free Trade Association Surveillance Authority (ESA), Norway, Liechtenstein, Iceland, and Switzerland [9,10].

Alert notifications are sent when food presenting a serious risk is already on the market, and also after the control at the external borders of the EU (in a broader sense, the European Economic Area (EEA)), if there is potential hazard, and when rapid action is required. The RASFF member who identifies the risk takes appropriate measures (e.g., a product withdrawal) and transmits the alert. In turn, other members of the system check whether the product in question is on their markets and, if so, also take appropriate measures. Information notifications are used when a risk in food or feed has been identified but other RASFF members do not need to take rapid action because the product has not reached their market or is no longer on their market, or the nature of the risk does not require rapid action. Border rejections may concern products that have been tested and rejected at the external borders of the EEA. Notifications of this type are sent to all other EEA border posts in order to introduce controls and prevent the rejected product from entering the EEA via another border post [9,10].

### 1.2. Products of Plant Origin in the RASFF

Among the product categories reported in the RASFF, the following can be considered as products of plant origin: cereals and bakery products, cocoa and cocoa preparations, coffee and tea, fruits and vegetables, herbs and spices and nuts, nut products and seeds (all product categories that appeared in the RASFF in the period 1979–2020 are shown in Table S1 in the Supplementary Material).

Notifications reported in the RASFF between 1979 and 2020 on products of plant origin are shown in Figure 1. During the period in question, 33,264 notifications were made regarding these products, representing more than 43% of the notifications in the system.

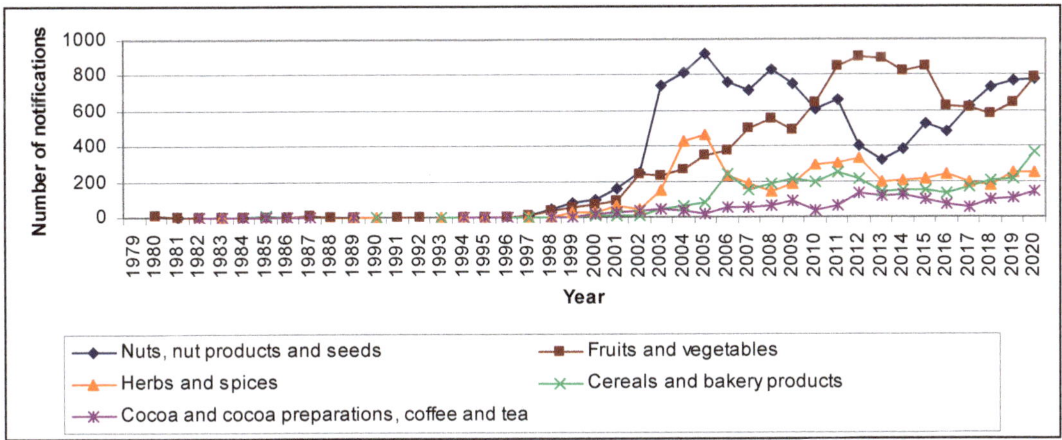



**Figure 1.** Number of notifications for product categories of plant origin in the RASFF in 1979–2020.

The largest number of notifications concerned nuts, nut products and seeds and fruits and vegetables (37% and 35%, respectively, of all notifications to plants in the period 1979–2020). Between 2009 and 2010, a 27% decrease in the number of notifications to nuts, nut products and seeds can be observed, and, in 2009, a 12% decrease in the number of notifications to fruits and vegetables can be seen. This may be related to the introduction of border rejections in the RASFF in 2008. However, in 2010, there was already an increase in the number of notifications for fruits and vegetables, and a slow growth for nuts, nut products and seeds, with around 800 notifications for both categories in 2020.

Annually, the RASFF reports approximately 2000 notifications on products of plant origin, accounting for a significant share of the notifications on all product categories, i.e., 4000–5000 per year (Figure 2).

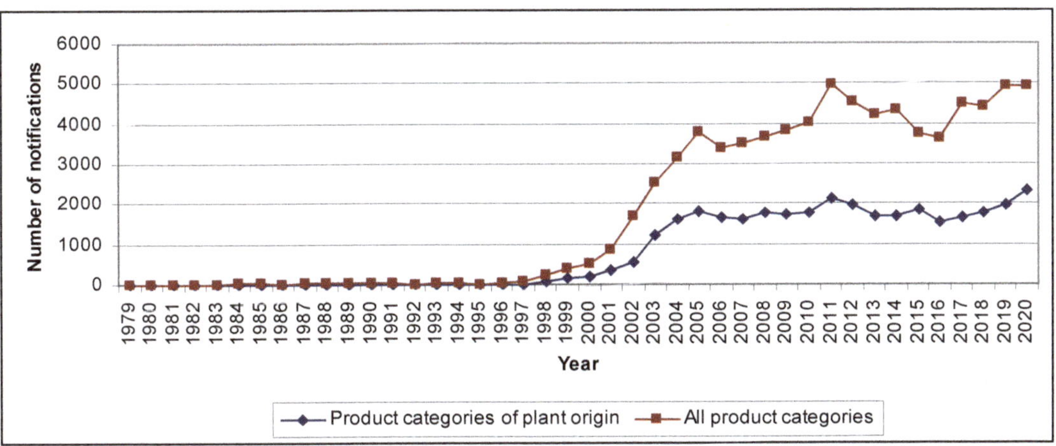



**Figure 2.** Number of notifications on product categories of plant origin and all product categories in the RASFF in 1979–2020.

### 1.3. Goal of the Study

The RASFF's annual reports contain general information about the notifications in this system and mostly only concern the year for which the report was issued. Furthermore, the report for 2021 was even more simplified, and no longer deals with the RASFF alone, but combines the functioning of the Alert and Cooperation Network (ACN), consisting of three networks: the RASFF, the Administrative Assistance and Cooperation (AAC) and the Agri-Food Fraud Network (FFN) [11].

Other studies relating to notifications on plants in the RASFF mostly concern short periods of a few years. Furthermore, they usually do not indicate individual products or hazards, but only product categories or hazard categories.

Thus, the goal of the study was to examine RASFF notifications for products of plant origin with respect to hazard, year, product, notifying country, origin country, notification type, notification basis, distribution status and action taken in 1998–2020 (23 years).

## 2. Materials and Methods

### 2.1. Data

2.1.1. Hazards Analysed

The data available in the RASFF notifications pre-2021 public information database cover 33,264 notifications (records) on products of plant origin reported in 1980–2020 [8]. This study was limited to the years 1998–2020, during which 33,163 notifications were made. They concerned 582 different hazards reported in 28 categories (Table S2 in the Supplementary Material). All these notifications were subjected to a general analysis (notification summaries in Section 3.1 and joining cluster analysis in Section 3.2).

Meanwhile, hazards with more than 200 notifications (Table 1) were examined in detail (two-way joining cluster analysis in Section 3.3). These were 22 hazards, for which 22,687 notifications were made between 1998 and 2020 (68% of the notifications on plants during this period). Most were notified in one category, but hazards such as colour, *Escherichia coli* and sulphite were reported in two categories, which resulted from the nature of the hazard and its classification by the supervisory authorities.

**Table 1.** The 22 most frequently reported hazards and other hazards regarding food of plant origin notified in the RASFF in 1998–2020.

| Hazard | Number | Hazard Category |
|---|---|---|
| Aflatoxins | 11,465 | Mycotoxins |
| Ochratoxin A | 900 | |
| Ethylene oxide | 463 | Pesticide residues |
| Chlorpyrifos | 437 | |
| Carbendazim | 328 | |
| Dimethoate | 326 | |
| Methomyl | 259 | |
| Acetamiprid | 254 | |
| Omethoate | 222 | |
| Triazophos | 209 | |
| Formetanate | 203 | |
| *Salmonella* | 2584 | Pathogenic micro-organisms |
| *Escherichia coli* | 201 | Microbial contaminants (other) (159), Pathogenic micro-organisms (42) |
| Moulds | 468 | Microbial contaminants (other) |
| Sudan | 790 | Composition |
| Iodine | 219 | |
| Sulphite | 790 | Food additives and flavourings (602), Allergens (188) |
| Colour | 536 | Food additives and flavourings (532), Composition (4) |
| Genetically modified plants | 708 | Genetically modified food or feed |
| Insects | 646 | Foreign bodies |
| Health certificate(s) | 472 | Adulteration/fraud |
| Milk | 207 | Allergens |
| All the above 22 hazards | 22,687 | |
| Other 560 hazards | 10,476 | |
| Total | 33,163 | |

### 2.1.2. Data Processing

The data were processed using Microsoft Excel (Microsoft Corporation, Redmond, DC, USA) applying pivot tables, a vertical search function, filtering and sorting. For product names, sometimes a common name, a Latin name (or in another language, but not in English) or even a proper name was recorded in the database. Different English names referring to the same products were also applied, or information regarding the part of the products (e.g., root, flakes, flower, sprouts), cultivation/breeding method (organic), country/area of origin, the state/degree of processing (e.g., canned, chopped, dried, frozen, milled, roasted, paste), kind/type (e.g., flour, kernels, pickles), the taste (e.g., sweet, bitter) or colour was included. All these entries needed to be standardised to distinguish the basic name, preferably the species name and name related to product category. If a product was classified in the inappropriate product category, this was also corrected.

In the case of products such as peppers and paprika, prunes and plums, the original names were retained (as the products were different), but in other cases the names have been changed to those used in British English (e.g., corn to maize, eggplants to aubergines) and sultanas were changed to raisins and noodles to pasta. In the case of ready-to-use or multi-ingredient products, the final name of the product was given (e.g., cake, bread). If the product consisted of multiple products, was a mixture of products, or could not be identified, the following phrases were used: "(other nuts)", "(other fruits)", "(other vegetables)", "(other herbs)", "(other spices)", "(other seeds)", "(other leaves)" or "(other product)".

Since 2011, information notifications have been divided into information for attention and information for follow-up. In order to standardise this type of notification throughout the research period, these names were shortened to their former name, i.e., information notifications. In the case of variables, such as notification basis, distribution status and action taken, the names of some values in the figures provided in the Supplementary Material were also shortened (by shortening or deleting certain words) to make them easier to handle (Table S3 in the Supplementary Material). In addition, in the case of variables, such as hazard category, notification basis, distribution status and action taken, the empty cells were filled with the phrase "(not specified)".

### 2.1.3. Comments on the RASFF Databases

The data used in this research (i.e., data from the RASFF notifications pre-2021 public information database) came from the archived European Commission website [8]. This is because the present official RASFF database (i.e., RASFF Portal) only contains data from 1.01.2020 onwards, and historical data are solely available to supervisory authorities. Furthermore, the source table exported from this database to a Microsoft Excel file does not contain information on the notification basis, the distribution status, or the action taken [12]. Obtaining information on the hazard category would require exporting data for each such category separately.

### *2.2. Methods*

To identify similarities in the notifications, a cluster analysis was applied using the joining and two-way joining methods. Data were first prepared in tables in Microsoft Excel. The use of these methods required the empty cells to be filled with the value 0. The data were then transferred to Statistica 13.3 (TIBCO Software Inc., Palo Alto, CA, USA).

### 2.2.1. The Joining Cluster Analysis

In order to apply the joining cluster analysis, eight data tables were constructed with all 582 hazards in the rows and the values of each variable in the columns, i.e., year, product, notifying country, origin country, notification type, notification basis, distribution status and action taken. Due to the statistical method used in the case of some hazards, the number of columns with products, origin countries and actions taken was limited to the first 30 with the highest number of notifications. In the joining cluster analysis, the following settings were used: linkage rule–Ward's method, distance measure–Euclidean measures and vertical icicle plots.

Charts showing the findings of the cluster analysis using the joining method are provided in Supplementary Material in Figure S1 (panels (a)–(h), separately for each mentioned variable).

### 2.2.2. The Two-Way Joining Cluster Analysis

Two-way joining cluster analysis was used when both cases and variables are expected to form clusters simultaneously. Difficulties in interpreting the results may arise because similarities between different clusters may lead to different subsets of variables, so the cluster structure is not homogeneous by nature. However, this method can be considered a powerful data exploration tool [13].

For each of the 22 most frequently reported hazards, seven tables were prepared. The years were placed in the rows and the values of the following variables were noted: product, notifying country, origin country, notification type, notification basis, distribution status and action taken. These were put into columns. Thus, a total of 154 data tables were constructed, with the number of columns for product, origin country and action taken limited to the 30 with the highest number of notifications.

Charts showing the findings of the cluster analysis using the two-way joining method are provided in the Supplementary Material in Figures S2–S23 (for the 22 hazards analysed) in panels (a)–(g), separately for each mentioned variable. These are contour/discrete charts and show the clusters by means of coloured squares (from green, through from yellow, orange and red to brown, where the clusters were highest). The dark green colour was faded (to white) as it would take up the largest part of each chart and would not provide any information.

### 3. Results

#### 3.1. General Results Related to All RASFF Notifications

The overall results considered all 33,163 notifications (based on 582 hazards) reported in the RASFF for products of plant origin in 1998–2020, but were limited to 10 values for particular variables.

#### 3.1.1. Product Categories and Products

Table 2 presents the product categories with 10 most-notified products of plant origin in the RASFF in 1998–2020.

**Table 2.** Product categories with the 10 most-notified products of plant origin in the RASFF in 1998–2020.

| Product Category (Notifications) | Product (Notifications) |
|---|---|
| Cereals and bakery products (3189) | Rice (984), Pasta (334), Maize (242), Biscuits (147), Wheat (109), Cake (91), Bread (86), Breakfast Cereals (80), Buckwheat (73), Linseed (71), Other (972) |
| Cocoa and cocoa preparations, coffee and tea (1490) | Tea (637), Chocolate (298), Coffee (166), Herbal Tea (110), Cocoa (89), Hibiscus (27), Jasmine (21), Senna (20), Fennel (15), Camomile (10), Other (97) |
| Fruits and vegetables (11,462) | Figs (1361), Peppers (1105), Beans (659), Raisins (472), Apricots (460), Betel (431), Okra (406), Mushrooms (342), Grapes (280), Chilli (266), Other (5680) |
| Herbs and spices (4601) | Chilli (779), Curry (547), Pepper (419), Paprika (276), Nutmeg (225), Mint (206), Basil (193), Peppers (168), Coriander (162), Ginger (158), Other (1468) |
| Nuts, nut products and seeds (12,421) | Pistachios (3824), Groundnuts (2281), Peanuts (1586), Sesame (1276), Hazelnuts (1023), Almonds (598), Melons (183), Brazil nuts (155), Rapeseed (150), Pine Nuts (148), Other (1197) |

As previously indicated in Figure 1, the highest number of notifications was reported for nuts (the three most frequently notified products were pistachios, groundnuts and peanuts) and also fruits and vegetables (figs, peppers and beans). Chilli, curry and pepper were mainly reported in the category "herbs and spices", rice, pasta and maize in the category "cereals and bakery products", and tea, chocolate and coffee were notified in the category covering cocoa, coffee and tea. It is also worth noting that peppers and chilli were reported under both the "fruits and vegetables" and "herbs and spices" categories.

### 3.1.2. Other Variables

Table 3 presents the values of the following variables: notifying country, origin country, notification type, notification basis, distribution status and action taken in relation to products of plant origin notified in the RASFF in 1998–2020. The number of these values was limited to the 10 types of notifications that were reported most frequently.

**Table 3.** The notifying country, origin country, notification type, notification basis, distribution status and action taken in notifications on products of plant origin in the RASFF in 1998–2020.

| Variable | Values (Notifications) |
| --- | --- |
| Notifying country | Germany (5735), United Kingdom (3960), Italy (3311), Netherlands (3240), Spain (2030), France (1799), Greece (1500), Poland (1266), Bulgaria (1187), Belgium (1144), other notifying countries (7991) |
| Origin country | Turkey (5137), India (3305), China (3190), Iran (2837), United States (1406), Thailand (1123), Egypt (1024), Netherlands (880), Germany (819), Italy (791), other country of origin (12,651) |
| Notification type | Border rejection (137,25), information (12,986), alert (6452) |
| Notification basis | Border control: consignment detained (17,822), official control on the market (8576), company's own check (2298), (not specified) (1405), border control-consignment released (1270), consumer complaint (851), border control: consignment under customs (633), food poisoning (204), official control in non-member country (45), official control following RASFF notification (44), monitoring of media (15) |
| Distribution status | No distribution (9000), product not (yet) placed on the distribution status market (7525), (not specified) (3533), distribution restricted to notifying country (3418), distribution to other member countries (3210), distribution possible (2677), information on distribution not (yet) available (788), product forwarded to destination (649), product (presumably) no longer on the market (648), product already consumed (623), other distribution status (1092) |
| Action taken | Re-dispatch (7550), destruction (4517), withdrawal from market (3749), official detention (2731), import not authorised (2423), recall from consumers (1767), (not specified) (1357), product recall or withdrawal (1230), return to consignor (783), informing recipient(s) (713), other action taken (6343) |

Products of plant origin were mainly notified by Germany, the United Kingdom, Italy, the Netherlands, and Spain, and originated from outside the European Union (Turkey, India, China, Iran and the United States). Consequently, the most common basis for notification was border control, followed by detention of the consignment and then border rejection. Information notifications and, to a much lesser extent, alerts were also reported. Notifications could also be based on official controls on the market or the company's own checks. Products were most often not distributed or not yet placed on the market, but distribution could also involve the notifying country as well as other member countries. Products were re-dispatched, destroyed or withdrawn from the market.

### 3.2. Results of Joining Cluster Analysis with all RASFF Notifications

In the joining cluster analysis, all 33,163 notifications (based on 582 hazards) were included. In tables prepared for this analysis, rows contained hazards and the columns contained the values of individual variables, i.e., year, product, notifying country, origin country, notification type, notification basis, distribution status and action taken. The number of products, origin countries and actions taken was limited to 30. The results of the joining cluster analysis are shown in the Supplementary Material in Figure S1 in panels (a)-(h) and summarised in Table 4. Next to the individual variable, the most distinct (separated) cluster was indicated first. The pairs of values of a given variable that were most similar to each other (with regard to the notified hazards) were linked by a long dash, but there were also single-element clusters.

**Table 4.** Results of the joining cluster analysis related to notifications regarding products of plant origin reported in the RASFF in 1998–2020.

| Variable (Figure in Supplemenary Material) | Clusters or Subclusters |
| --- | --- |
| Year (Figure S1a) | First: 2004–2005, 2006–2008, 2007–2010, 2003, 2009<br>Second: 2013–2014, 2015–2018, 2016–2017, 2011, 2012, 2019, 2020<br>Third: 1999–2000, 1998, 2001, 2002 |
| Product (Figure S1b) | First: figs–hazelnuts, peanuts, groundnuts, pistachios<br>Second: pepper–betel, sesame<br>Third: rice–apricots, chilli–almonds<br>Other products |
| Notifying country (Figure S1c) | First: Netherlands–Italy, Spain–France, United Kingdom, Germany, Greece<br>Second: Bulgaria–Belgium, Norway–Finland, Denmark–Czechia, Slovakia–Portugal, Slovenia–Luxemburg, Poland, Sweden, Austria<br>Other notifying countries |
| Origin country (Figure S1d) | First: Iran–Turkey<br>Second: United States–China, Brazil–Egypt, Argentina, India<br>Third: Sudan–Thailand<br>Other origin countries |
| Notification type (Figure S1e) | First: alert<br>Second: information–border rejection |
| Notification basis (Figure S1f) | First: border control: consignment detained<br>Second: official control on the market<br>Third: border control: consignment released–border control: consignment under customs, company's own check, (not specified)<br>Other notification basis– |
| Distribution status (Figure S1g) | First: product not (yet) placed on the market–(not specified), no distribution<br>Second: distribution restricted to notifying country–distribution on the market (possible), distribution to other member countries<br>Other distribution status |
| Action taken (Figure S1h) | First: re-dispatch<br>Second: import not authorised–official detention, withdrawal from market–destruction other action taken |

For the variable year, notifications can be divided into three sub-periods: 2003–2010 (a clear separate cluster), 2011–2020 and 1998–2002. In some cases, pairs of values were formed by consecutive years, meaning that similar hazards were reported at the turn of the year or even for two years (1999 and 2000, 2004 and 2005, 2013 and 2014, 2016 and 2017). Mostly, however, the clusters were formed by years not immediately following each other, meaning that there were fluctuations in the type of hazards reported.

In the case of the variable product, the first cluster was formed by different types of nuts, although the notifications for figs and hazelnuts were the most similar in terms of reported hazards. Notifications for pepper and betel, chilli and almonds, and rice and apricots were also similar.

Considering the notifying countries, the notifications reported by Western European countries, especially the Netherlands and Italy, as well as Spain and France, were the most similar (the United Kingdom was also included in this cluster). This may be indicative of the strong economic links between these countries. The second cluster included medium-sized countries, which were directly paired, e.g., Bulgaria and Belgium, Norway and Finland, Denmark and Czechia.

In the case of countries of origin, a distinct cluster was formed by Iran and Turkey. It is reasonable to assume that the number of notifications regarding the hazards originating from these countries was high. It is important to remark that, in the case of the variable notification basis, a one-element cluster "border control-consignment detained" was formed, with a significant linkage distance from the other values of this variable.

It is also worth noting that the second cluster of the variable origin country did not include EU countries. This means that most of the hazards regarding plant products came from non-EU countries. Considering the variable notification type, it can be seen that border rejections were more similar to information notifications than to alert notifications. Thanks to the border rejections, the border posts of the EU countries contributed, to a large extent, to the minimisation of hazards in products. This was also confirmed by the values of other variables. Indeed, considering the variable distribution status, the first cluster was formed by the values: "product not (yet) placed on the market" and "no distribution". However, in the case where a variable action was taken, the first one-element cluster was created by the value "re-dispatch", and in the second cluster, similar values were "import not authorised" and "official detention".

However, it is also important to note the other values of the individual variables, which can also be linked in a sequence concerning products of plant origin that are already on the EU market. In the case of the variable notification basis, the second cluster was formed by the value "official control on the market". When considering the variable distribution status, a similarity can be seen between the values: "distribution restricted to notifying country" and "distribution on the market (possible)" (second cluster). In turn, in the case of the variable action taken, a similarity can be observed between the values "withdrawal from market" and "destruction" (also second cluster).

### 3.3. Results of Two-Way Joining Cluster Analysis with Selected RASFF Notifications

The selected 22 hazards (reported under 22,687 notifications) indicated in Table 1 were considered for the two-way joining cluster analysis. The results of this analysis are presented in Figures S2–S23 in the Supplementary Material, where panels (a)–(g) show the similarity between year and product, notifying country, origin country, notification type, notification basis, distribution status and action taken, respectively. Based on the individual years of the variable year, the values of the other variables were indicated, with the product as the base variable, i.e., panel (a). If there was no coverage of the same years in the other variables, they were omitted. In some cases, the variation in cluster intensity (dependent on colours) in particular years caused the name of the product to be determined by the values of the other variables. This made it possible to focus only on the most distinct clusters that occurred simultaneously in the different variables.

In Sections 3.3.1–3.3.9, the hazards reported in particular categories are presented.

### 3.3.1. Mycotoxins (*Aflatoxins* and Ochratoxin A)

Notifications relating to mycotoxins (aflatoxins and ochratoxin A) are presented in Table 5. These notifications were reported most frequently and accounted for up to 55% of the notifications examined using two-way joining cluster analysis. They mainly concerned products from Asia, but aflatoxins in pistachios from Iran were the most prominent problem. This hazard was particularly prevalent between 2003 and 2006, and was reported by Germany and Spain using information notifications after border control. Consignments were detained and re-dispatched, resulting in products not being distributed.

Ochratoxin A in raisins from Turkey was notified in 2016–2019. These products were reported by Germany, the Netherlands, Poland and France, at both border and official controls at the market. They were withdrawn from the market, destroyed or dispatched.

**Table 5.** Results of the two-way joining cluster analysis related to notifications regarding mycotoxins in plants reported in the RASFF in 1998–2020.

| | Hazard/Variable | Value (Figure in Supplementary Material) |
|---|---|---|
| **Aflatoxins** | Year | 2003–2006 |
| | Product | Pistachios (2003–2006) (Figure S2a) |
| | Notifying country | Germany (2003–2006), Spain (2004, 2005) (Figure S2b) |
| | Origin country | Iran (2003–2006) (Figure S2c) |
| | Notification type | Information (2003–2006) (Figure S2d) |
| | Notification basis | Border control: consignment detained (2003–2006) (Figure S2e) |
| | Distribution status | (Not specified) (2003, 2004), no distribution (2005, 2006) (Figure S2f) |
| | Action taken | Re-dispatch (2003–2006) (Figure S2g) |
| **Ochratoxin A** | Year | 2006, 2016, 2018, 2019 (for some variables, there was not full coverage in years) |
| | Product | Raisins (2006, 2016, 2018, 2019) (Figure S3a) |
| | Notifying country | Czechia, Italy (2006), Germany (2016, 2018), Netherlands (2016, 2018, 2019), Poland (2018, 2019), France (2019) (Figure S3b) |
| | Origin country | Turkey (2018, 2019) (Figure S3c) |
| | Notification type | Information (2006), alert (2016, 2018, 2019), border rejections (2018, 2019) (Figure S3d) |
| | Notification basis | Official control on the market (2006, 2016, 2018, 2019), border control: consignment detained (2018, 2019) (Figure S3e) |
| | Distribution status | Distribution on the market (possible) (2006), product not (yet) placed on the market (2016, 2018, 2019) (Figure S3f) |
| | Action taken | Product recall or withdrawal (2006), re-dispatch (2006, 2018, 2019), destruction, informing recipient(s) (2016), re-dispatch, withdrawal from the market (2016, 2018, 2019), official detention, return to consignor (2019) (Figure S3g) |

### 3.3.2. Pesticide Residues (Ethylene oxide, Chlorpyrifos, Carbendazim, Dimethoate, Methomyl, Acetamiprid, Omethoate, Triazophos and Formetanate)

Notifications relating to pesticide residues are presented in Table 6. This was the largest group of reported hazards (9 different substances out of the 22 analysed hazards).

Products with these hazards usually originated from Asia. Of particular note is the presence of pesticide residues in peppers from Turkey, as notified by Bulgaria in several years. These included substances such as acetamiprid in 2020, chlorpyrifos in 2016, 2017 and 2019, formetanate in 2011, 2012, 2014 and 2017–2020 and methomyl in 2010, 2011 and 2018. The type of notification was border rejection based on border controls, after which the consignment was detained. Products were, therefore, not placed on the market or not distributed, and were most often destroyed thereafter.

Another country that frequently appeared in notifications relating to pesticide residues was India. France and the United Kingdom reported acetamiprid, dimethoate and triazophos in okra in 2012 and 2013 and triazophos in curry in the same years. Italy notified carbendazim in rice in 2014 and 2015. There were border rejections based on border controls, followed by detention of the consignment. The products were, therefore, not distributed and usually were destroyed. There was a more serious problem with ethylene oxide, which was notified by the Netherlands in sesame in 2020. Products with this hazard were reported as alerts after the companies' own checks, so distribution to other EU countries was possible. Actions such as informing consignors and recipients, recalls and withdrawals were then taken.

Acetamiprid was also notified in 2012 and 2013 by France and the United Kingdom in tea from China and India, and carbendazim was reported in 2010 by the United Kingdom in peppers from Thailand. Notifications related to dimethoate in beans and peas from Egypt and Kenya, respectively, were sent in 2013 by France. Many countries also reported the presence of omethoate in beans, aubergines, apples, okra and peppers from Thailand in 2006, 2008–2013 and 2019.

**Table 6.** Results of the two-way joining cluster analysis related to notifications on pesticide residues in plants reported in the RASFF in 1998–2020.

| | Hazard/Variable | Value (Figure in Supplementary Material) |
|---|---|---|
| **Ethylene oxide** | Year | 2020 (this year occurred for each value of each variable below) |
| | Product | Sesame (Figure S4a) |
| | Notifying country | Netherlands (Figure S4b) |
| | Origin country | India (Figure S4c) |
| | Notification type | Alert (Figure S4d) |
| | Notification basis | Company's own check (Figure S4e) |
| | Distribution status | Distribution to other member countries (Figure S4f) |
| | Action taken | Informing consignor, informing recipient(s), recall from consumers, withdrawal from the market (Figure S4g) |
| **Chlorpyrifos** | Year | 2016, 2017, 2019 (these years occurred for each value of each variable below) |
| | Product | Peppers (Figure S5a) |
| | Notifying country | Bulgaria (Figure S5b) |
| | Origin country | Turkey (Figure S5c) |
| | Notification type | Border rejections (Figure S5d) |
| | Notification basis | Border control: consignment detained (Figure S5e) |
| | Distribution status | Product not (yet) placed on the market (Figure S5f) |
| | Action taken | Destruction (Figure S5g) |
| **Carbendazim** | Year | 2010, 2014, 2015 |
| | Product | Peppers (2010), rice (2014, 2015) (Figure S6a) |
| | Notifying country | United Kingdom (2010), Italy (2014, 2015) (Figure S6b) |
| | Origin country | Thailand (2010), India (2014, 2015) (Figure S6c) |
| | Notification type | Information (2010), border rejection (2014, 2015) (Figure S6d) |
| | Notification basis | Official control on the market (2010), border control: consignment detained (2010, 2014, 2015), Border control: consignment under customs (2015) (Figure S6e) |
| | Distribution status | No distribution (2010), product not (yet) placed on the market (2014, 2015), product forwarded to distribution (2015) (Figure S6f) |
| | Action taken | Destruction (2010, 2014, 2015), withdrawal from market (2010, 2014), re-dispatch (2014, 2015) (Figure S6g) |
| **Dimethoate** | Year | 2012, 2013 |
| | Product | Okra (2012), beans, okra, peas (2013) (Figure S7a) |
| | Notifying country | United Kingdom (2012, 2013), France (2013) (Figure S7b) |
| | Origin country | India (2012), Egypt, India, Kenya (2013) (Figure S7c) |
| | Notification type | Border rejection (2012, 2013) (Figure S7d) |
| | Notification basis | Border control: consignment detained (2012, 2013) (Figure S7e) |
| | Distribution status | No distribution (2012), product not (yet) placed on the market (2013) (Figure S7f) |
| | Action taken | Destruction (2012, 2013) (Figure S7g) |
| **Methomyl** | Year | 2010, 2011, 2018 |
| | Product | Peppers (2010, 2011, 2018) (Figure S8a) |
| | Notifying country | Bulgaria (2010, 2011, 2018) (Figure S8b) |
| | Origin country | Turkey (2010, 2011, 2018) (Figure S8c) |
| | Notification type | Border rejections (2010, 2011, 2018) (Figure S8d) |
| | Notification basis | Border control: consignment detained (2010, 2011, 2018) (Figure S8e) |
| | Distribution status | No distribution (2010, 2011), product not (yet) placed on the market (2018) (Figure S8f) |
| | Action taken | Destruction (2010, 2011, 2018) (Figure S8g) |
| **Acetamiprid** | Year | 2012, 2013, 2020 |
| | Product | Tea (2012), tea, okra (2013), peppers (2020) (Figure S9a) |
| | Notifying country | France (2012), France, United Kingdom (2013), Bulgaria (2020) (Figure S9b) |
| | Origin country | China, India (2012, 2013), Turkey (2020) (Figure S9c) |
| | Notification type | Border rejection (2012, 2013, 2020) (Figure S9d) |
| | Notification basis | Border control: consignment detained (2012, 2013, 2020) (Figure S9e) |
| | Distribution status | No distribution (2012), product not (yet) placed on the market (2013, 2020) (Figure S9f) |
| | Action taken | Destruction (2012, 2020), import not authorised (2013) (Figure S9g) |

**Table 6.** *Cont.*

| | Hazard/Variable | Value (Figure in Supplementary Material) |
|---|---|---|
| **Omethoate** | Year | 2006, 2008–2013, 2019 (for some variables, there was not full coverage in years) |
| | Product | Beans (2006, 2008, 2011), aubergines (2009, 2010, 2012, 2013), apples (2010), okra (2013), peppers (2019) (Figure S10a) |
| | Notifying country | Norway (2006), Netherlands (2008, 2010, 2011), Finland (2009), Germany (2010, 2012), France (2013), United Kingdom (2013, 2019), Belgium, Bulgaria (2018) (Figure S10b) |
| | Origin country | Thailand (2006, 2008–2010) (Figure S10c) |
| | Notification type | Information (2006, 2008–2013, 2019), border rejection (2009–2013, 2019) (Figure S10d) |
| | Notification basis | Official detention (2006, 2008, 2010, 2011), border control: consignment detained (2009–2013, 2019) (Figure S10e) |
| | Distribution status | No distribution (2009–2012), product already consumed (2012), product not (yet) placed on the market (2013, 2019) (Figure S10f) |
| | Action taken | Withdrawal from the market (2009, 2011), destruction (2009, 2010, 2013, 2019), informing authorities (2012, 2013) (Figure S10g) |
| **Triazophos** | Year | 2012, 2013 |
| | Product | Curry (2012), okra (2012, 2013) (Figure S11a) |
| | Notifying country | France (2012), United Kingdom (2012, 2013) (Figure S11b) |
| | Origin country | India (2012, 2013) (Figure S11c) |
| | Notification type | Border rejection (2012, 2013) (Figure S11d) |
| | Notification basis | Border control: consignment detained (2012, 2013) (Figure S11e) |
| | Distribution status | No distribution (2012, 2013) (Figure S11f) |
| | Action taken | Destruction (2012, 2013) (Figure S11g) |
| **Formetanate** | Year | 2011, 2012, 2014, 2017–2020 |
| | Product | Peppers (2011, 2012, 2014, 2017–2020) (Figure S12a) |
| | Notifying country | Bulgaria (2011, 2012, 2014, 2017–2020) (Figure S12b) |
| | Origin country | Turkey (2011, 2012, 2014, 2017–2020) (Figure S12c) |
| | Notification type | Border rejection (2011, 2012, 2014, 2017–2020) (Figure S12d) |
| | Notification basis | Border control: consignment detained (2011, 2012, 2014, 2017–2020) (Figure S12e) |
| | Distribution status | No distribution (2011, 2012), product not (yet) placed on the market (2014, 2017–2020) (Figure S12f) |
| | Action taken | Re-dispatch or destruction (2011), placed under customs seals (2012), destruction (2017–2020) (Figure S12g) |

### 3.3.3. Pathogenic Micro-Organisms and Microbial Contaminants (*Salmonella, Escherichia coli* and Moulds)

Notifications regarding pathogenic micro-organisms are presented in Table 7. Hazards related to *Salmonella* presence have been reported in recent years (2015, 2018 and 2019) by the United Kingdom, Greece and Germany in sesame from India, Sudan and Brazil. Notifications were reported as border rejections on the basis of controls, after which shipments were detained. Consequently, the products were not placed on the market and were destroyed, re-dispatched or physically/chemically treated.

*Escherichia coli* was reported by Norway and the United Kingdom in 2005, 2012, 2013, 2016 and 2020 in basil, mint and betel from Asian countries, i.e., Thailand, Vietnam and the Lao Republic. These were information notifications sent after official controls on the market or border controls, after which consignment was detained. Distribution was limited to the notifying country or the product was removed from the market. The trade of these products was prohibited, and they were also withdrawn from the market and destroyed.

Mould has been reported over a wide range of time (2007, 2008, 2011, 2012, 2017 and 2018), mainly by Poland in nuts (peanuts, hazelnuts, groundnuts), raisins and beans from Turkey and China. These were information notifications or border rejections, after which the shipments were detained. Consequently, the products were not distributed or had not yet been placed on the market, and were most often re-dispatched.

**Table 7.** Results of the two-way joining cluster analysis related to notifications regarding pathogenic micro-organisms and microbial contaminants in plants reported in the RASFF in 1998–2020.

| | Hazard/Variable | Value (Figure in Supplementary Material) |
|---|---|---|
| *Salmonella* | Year | 2015, 2018,2019 |
| | Product | Sesame (2015, 2018–2020) (Figure S13a) |
| | Notifying country | United Kingdom (2015), Greece (2018, 2019), Germany (2019, 2020) (Figure S13b) |
| | Origin country | India (2015), Sudan (2018, 2019), Brazil (2019, 2020) (Figure S13c) |
| | Notification type | Border rejection (2015, 2018–2020) (Figure S13d) |
| | Notification basis | Border control: consignment detained (2015, 2018–2020) (Figure S13e) |
| | Distribution status | Product not (yet) placed on the market (2015, 2018–2020) (Figure S13f) |
| | Action taken | Destruction (2015), re-dispatch (2015, 2018, 2019), physical/chemical treatment (2019, 2020), official detention (2020) (Figure S13g) |
| *Escherichia coli* | Year | 2005, 2012, 2013, 2016, 2020 |
| | Product | Basil (2005, 2012, 2013, 2016), mint (2005), betel (2020) (Figure S14a) |
| | Notifying country | Norway (2005, 2012, 2013), United Kingdom (2016, 2020) (Figure S14b) |
| | Origin country | Thailand (2005), Vietnam (2012, 2013, 2020), Lao Republic (2016) (Figure S14c) |
| | Notification type | Information (2005, 2012, 2013, 2016, 2020) (Figure S14d) |
| | Notification basis | Official control on the market (2005, 2012, 2013), border control: consignment detained (2016, 2020) (Figure S14e) |
| | Distribution status | Distribution restricted to notifying country (2005, 2012, 2013), product (presumably) no longer on the market (2016) (Figure S14f) |
| | Action taken | Prohibition to trade (2005), withdrawal from the market (2012, 2013, 2016, 2020), destruction (2016) (Figure S14g) |
| *Moulds* | Year | 2007, 2008, 2011, 2012, 2017, 2018 (for some variables there was not full coverage in years) |
| | Product | Peanuts (2007), beans (2008, 2011), hazelnuts, raisins (2012), groundnuts (2017, 2018) (Figure S15a) |
| | Notifying country | Poland (2007, 2008, 2011) (Figure S15b) |
| | Origin country | China (2007, 2008, 2011), Turkey (2012) (Figure S15c) |
| | Notification type | Information (2007, 2012), border rejection (2008, 2011, 2012, 2017, 2018) (Figure S15d) |
| | Notification basis | Border control: consignment detained (2007, 2008, 2011, 2012, 2017, 2018) (Figure S15e) |
| | Distribution status | No distribution (2007, 2008, 2011, 2012), product not (yet) placed on the market (2017, 2018) (Figure S15f) |
| | Action taken | Re-dispatch (2007, 2008, 2012), return to consignor (2011), withdrawal from the market (2012) (Figure S15g) |

### 3.3.4. Composition (Sudan and Iodine)

Problems regarding composition (Table 8) were reported in products originating from Asia and Europe.

**Table 8.** Results of the two-way joining cluster analysis related to notifications on composition in plants reported in the RASFF in 1998–2020.

| | Hazard/Variable | Value (Figure in Supplementary Material) |
|---|---|---|
| *Sudan* | Year | 2004, 2005 |
| | Product | (Other spices) (2004, 2005), chilli (2005) (Figure S16a) |
| | Notifying country | Germany (2004, 2005) (Figure S16b) |
| | Origin country | Germany, Italy, Turkey (2004), India (2004, 2005) (Figure S16c) |
| | Notification type | Information, alert (2004, 2005) (Figure S16d) |
| | Notification basis | Official control on the market (2004, 2005) (Figure S16e) |
| | Distribution status | (Not specified) (2004), distribution on the market (possible) (2004, 2005) (Figure S16f) |
| | Action taken | Destruction (2004), product recall or withdrawal (2004, 2005) (Figure S16g) |
| *Iodine* | Year | 2004, 2005, 2008–2010, 2014, 2018, 2019 (for some variables, there was not full coverage in years) |
| | Product | Seaweed (2004, 2005, 2008–2010, 2014, 2018, 2019), algae (2004, 2005) (Figure S17a) |
| | Notifying country | Germany (2004, 2005, 2008–2010, 2014) (Figure S17b) |
| | Origin country | South Korea (2004, 2005, 2009, 2014, 2018), Netherlands (2004, 2009), China (2004, 2005, 2008, 2010, 2018, 2019) (Figure S17c) |
| | Notification type | Alert (2004, 2005, 2009, 2010, 2014, 2018, 2019), information (2008) (Figure S17d) |
| | Notification basis | Official control on the market (2004, 2005, 2008–2010, 2014, 2018, 2019) (Figure S17e) |
| | Distribution status | (Not specified) (2004), distribution on the market (possible) (2004, 2005, 2008–2010), distribution restricted to notifying country (2008), distribution to other member countries (2018) (Figure S17f) |
| | Action taken | Product recall or withdrawal (2004, 2005), destruction (2008), withdrawal from the market (2009, 2010, 2014, 2018) (Figure S17g) |

Sudan dye was notified by Germany in earlier years (2004 and 2005) in chilli and other spices from Germany, Italy, Turkey and India. These notifications were in the form of information or alert notifications based on official controls on the market, and the reported products were destroyed or withdrawn.

Iodine was reported mainly by Germany in 2004, 2005, 2008–2010, 2014, 2018 and 2019. Algae was notified only in the earlier years (2004 and 2005), while seaweed was submitted in all the mentioned years. The reported products were from South Korea, China and the Netherlands. These were mainly alert notifications and, to a lesser extent, information notifications, sent on the basis of official controls on the market. Distribution status varied widely, with products being destroyed or withdrawn from the market.

### 3.3.5. Food Additives and Flavourings (Sulphite and Colour)

Sulphites and colours (Table 9) were mainly notified in food additives and flavourings category; however, sulphites were also reported as allergens, and colours were also notified within each composition category.

**Table 9.** Results of the two-way joining cluster analysis related to notifications regarding food additives and flavourings in plants reported in the RASFF in 1998–2020.

| | Hazard/Variable | Value (Figure in Supplementary Material) |
|---|---|---|
| **Sulphite** | Year | 2003, 2005, 2014–2018 (for some variables, there was not full coverage in years) |
| | Product | Apricots (2003, 2005, 2014–2018) (Figure S18a) |
| | Notifying country | Spain (2003), Cyprus (2005) (Figure S18b) |
| | Origin country | Turkey (2003, 2005, 2014–2018) (Figure S18c) |
| | Notification type | Information (2003, 2005), alert (2005), border rejection (2014–2018) (Figure S18d) |
| | Notification basis | Border control: consignment detained (2003, 2014–2018), official control on the market (2005) (Figure S18e) |
| | Distribution status | (Not specified) (2003), distribution on the market (possible) (2005), product not (yet) placed on the market (2014–2018) (Figure S18f) |
| | Action taken | Re-dispatch (2003), product recall or withdrawal (2005), import not authorised (2017), recall from consumers (2018) (Figure S18g) |
| **Colour** | Year | 2020 (this year occurred for each value of each variable below) |
| | Product | Breakfast cereals (Figure S19a) |
| | Notifying country | United Kingdom (Figure S19b) |
| | Origin country | United States (Figure S19c) |
| | Notification type | Border rejection (Figure S19d) |
| | Notification basis | Border control: consignment detained (Figure S19e) |
| | Distribution status | Product not (yet) placed on the market (Figure S19f) |
| | Action taken | Official detention (Figure S19g) |

Sulphites in apricots from Turkey were reported both in earlier years (2003 and 2005 as information or alert notifications) and more recently (2014–2018 as border rejections). These notifications were sent by Spain and Cyprus on the basis of an official control on the market or a border control, after which the consignment was detained. The notified products were dispatched and, if found on the market, were withdrawn or recalled from consumers.

Hazards regarding colour were reported by the United Kingdom in 2020 on breakfast cereals originating from the United States. These were border rejections based on border controls, after which the consignment was detained. The products were not (yet) placed on the market, because they were officially detained.

### 3.3.6. Genetically Modified Food

Alerts regarding genetically modified products were raised in 2006 by Austria regarding linseed originating from the United States. In turn, Germany reported this hazard in 2009 in rice from Canada as an information notification. These products were reported on the basis of official control and were withdrawn from the market (Table 10).

**Table 10.** Results of the two-way joining cluster analysis related to notifications regarding genetically modified plants reported in the RASFF in 1998–2020.

| | Hazard/Variable | Value (Figure in Supplementary Material) |
|---|---|---|
| **Genetically modified** | Year | 2006, 2009 |
| | Product | Linseed (2006), rice (2009) (Figure S20a) |
| | Notifying country | Austria (2006), Germany (2009) (Figure S20b) |
| | Origin country | United States (2006), Canada (2009) (Figure S20c) |
| | Notification type | Alert (2006), information (2009) (Figure S20d) |
| | Notification basis | Official control on the market (2006, 2009) (Figure S20e) |
| | Distribution status | Distribution on the market (possible) (2006, 2009) (Figure S20f) |
| | Action taken | Product recall or withdrawal (2006), withdrawal from the market (2006, 2009) (Figure S20g) |

### 3.3.7. Foreign Bodies (Insects)

Insects (as foreign bodies) were reported mainly in 2006–2009, 2011, 2012 and 2017 (Table 11).

**Table 11.** Results of the two-way joining cluster analysis related to notifications on foreign bodies (insects) in plants reported in the RASFF in 1998–2020.

| | Hazard/Variable | Value (Figure in Supplementary Material) |
|---|---|---|
| **Insects** | Year | 2006–2009, 2011, 2012, 2017 |
| | Product | Rice (2006, 2011), dates (2007, 2008, 2014), figs (2007, 2008, 2011), almonds (2009, 2011), tea (2009), chocolate (2011), rapeseed (2012), peanuts (2007) (Figure S21a) |
| | Notifying country | Poland (2006–2009, 2011, 2012, 2014), Slovenia (2008), Spain (2009), Italy (2011, 2012), Czechia (2012) (Figure S21b) |
| | Origin country | Turkey (2006–2008), China (2007, 2009), Italy (2007, 2008), United States (2009), Ukraine (2011, 2012), India, Tunisia (2014) (Figure S21c) |
| | Notification type | Information (2006–2008, 2011, 2012), border rejection (2008, 2009, 2011, 2012, 2014) (Figure S21d) |
| | Notification basis | Border control: consignment detained (2006–2009, 2011, 2012, 2014), official control on the market (2006–2008), consumer complaint (2008, 2011) (Figure S21e) |
| | Distribution status | No distribution (2006–2009, 2011, 2012), distribution on the market (possible) (2008, 2009), information on the product not (yet) available (2011), product not (yet) placed on the market (2014) (Figure S21f) |
| | Action taken | Re-dispatch (2006–2009, 2011, 2012, 2017), withdrawal from the market (2007, 2008, 2011, 2012) (Figure S21g) |

These notifications concerned products such as rice, dates, figs, almonds, tea, chocolate, rapeseed and peanuts originated from Asian countries (Turkey, China, India), European countries (Italy and Ukraine), and also the United Stated and Tunisia. They were sent by Poland, Slovenia, Spain, Italy and Czechia as information notifications or border rejections. The notification basis was official control on the market or border control, after which the consignment was detained, as well as consumer complaints. The distribution status of notified products was very diverse, and they were withdrawn from the market or re-dispatched.

### 3.3.8. Adulteration/Fraud (Health Certificate(s))

Problems with health certificates were the cause of notifications within the adulteration/fraud category (Table 12). The notifications concerned products such as nutmeg (in 2016) and chilli, sesame and pistachios (in 2017) from India, reported by the United Kingdom. These were border rejections on the basis of border controls, after which the consignment was detained. Consequently, the products were not placed on the market and were destroyed.

**Table 12.** Results of the two-way joining cluster analysis related to notifications regarding adulteration/fraud (health certificate(s)) in plants reported in the RASFF in 1998–2020.

| | Hazard/Variable | Value (Figure in Supplementary Material) |
|---|---|---|
| Health certificate(s) | Year | 2016, 2017 |
| | Product | Nutmeg (2016), chilli, sesame, pistachios (2017) (Figure S22a) |
| | Notifying country | United Kingdom (2016, 2017) (Figure S22b) |
| | Origin country | India (2016, 2017) (Figure S22c) |
| | Notification type | Border rejection (2016, 2017) (Figure S22d) |
| | Notification basis | Border control: consignment detained (2016, 2017) (Figure S22e) |
| | Distribution status | Product not (yet) placed on the market (2016, 2017) (Figure S22f) |
| | Action taken | Destruction (2016, 2017) (Figure S22g) |

### 3.3.9. Allergens (Milk)

Milk as an allergen (Table 13) in chocolate originating from Germany was reported primarily by Austria in 2009, using alert notifications. These notifications were based on the official controls on the market, and the action taken was to issue a public warning.

**Table 13.** Results of the two-way joining cluster analysis related to notifications regarding allergens (milk) in plants reported in the RASFF in 1998–2020.

| | Hazard/Variable | Value (Figure in Supplementary Material) |
|---|---|---|
| Milk | Year | 2009 (this year occurred for each value of each variable below) |
| | Product | Chocolate (Figure S23a) |
| | Notifying country | Austria (Figure S23b) |
| | Origin country | Germany (Figure S23c) |
| | Notification type | Alert (Figure S23d) |
| | Notification basis | Official control on the market (Figure S23e) |
| | Distribution status | Distribution on the market (possible) (Figure S23f) |
| | Action taken | Public warning: press release (Figure S23g) |

### 3.4. Limitations of Using RASFF Data

The research used data from the archived RASFF database, covering notifications up to 2020 at the time of data extraction [8]. At present, 2021 is also available in this database. A study that also covers the year 2022 would require the data from this database to be combined with the data from the database currently available on the European Commission website [10]. However, this would be difficult due to their different structure, especially once exported to an Excel file. It is also unknown if and when the Commission will officially make the historical data available. At present, they are only available to the supervisory authorities of the member countries. It is also worth mentioning that the current database is much less accessible to the user than the one made officially available a few years ago.

The actual number of notifications placed in the RASFF database was about 20% less than the number of records, as one notification could include several records (concerning, for example, the different countries of origin of the notified product). However, combining the records into a single notification would lead to the loss of a large amount of data, as it would require the adoption of the principle that only the value from one (e.g., the first) record of a notification can be taken into account. Indeed, only one value could occur in each notification within a given variable for further analysis. However, it should be noted that the inclusion of all records allowed for proportionality, and so should not significantly affect the final results.

In the earlier years of the RASFF functioning (1980s and 1990s), missing data could be observed for the variables of hazard category, notification basis, distribution status and action taken (empty cells were filled with the phrase "(not specified)"). It should be added, however, that, due to the small number of notifications in that period, these years were excluded from the study. A major difficulty was the wide variety of product names, as these were given with their characteristics or states or under different English names. The inability to clearly identify the product, or to only identify the few notifications regarding little-known products, required the creation of new names: "(other fruits)", "(other vegetables)", "(other herbs)", "(other spices)", "(other nuts)", "(other seeds)", "(other leaves)" and "(other product)". Differentiated products were thus concentrated under the same group name. However, this applied to only 3% of the total examined population, and was dispersed across the five studied product categories.

In the source tables prepared for the cluster analysis in Statistica 13.3, a maximum of approximately thirty columns (for the joining method) and, similarly, a maximum of approximately thirty columns and thirty rows (for the two-way joining method) could be included. A larger number of columns and/or rows could significantly impair the readability of the charts generated based on these. Therefore, sorting was carried out from the largest to the smallest sum of values (up to the aforementioned number of about thirty), and the others were omitted. However, this allowed for us to focus on the most significant clusters. In turn, the use of Ward's method as a linkage rule in joining cluster analysis enabled a good separation of clusters, but caused them to be flattened (this is, however, a characteristic of this method), which sometimes made it difficult to read the charts.

Difficulties also arose from the use of the two-way joining cluster analysis method. Although each variable (i.e., product, notifying country, origin country, notification type, notification basis, distribution status and action taken) consecutively referred to the same variable, i.e., year, it was sometimes possible to see values concentrating (clustering) within one variable and dispersing within another variable. This caused difficulties in interpretation, as, for some variables, there was not full clustering coverage within the same years. In addition, when generating the charts, the Statistica program did not accurately map the colours from the legend to the colours on the chart and omitted every second mark (on each axis), resulting in the need to manually modify each chart. Furthermore, the clusters were not arranged according to consecutive years, but according to the number of notifications in different years (so there was no continuity over time in the charts). As a final difficulty, the charts were automatically rescaled in such a way that they did not take up all the available space (much of the space was left blank). This caused the graphical and textual elements of the charts to be reduced in size, and thus compromised its readability.

## 4. Discussion

### *4.1. The Annual RASFF Reports*

Notifications regarding hazards in products of plant origin occurred each year among the so-called "Top 10" included in the annual RASFF reports for 2010–2020. They covered information on hazard, product category, origin country and notifying country (Table 14). In reports for earlier years, information on the "Top 10" was not provided. It is also worth mentioning that, in the RASFF annual reports, information on the notification basis, distribution status and action taken is not given in "Top 10", but only within a selected case study for a particular product in a given year.

Of particular note is the indication of aflatoxins in nuts (mainly from China, Iran and Turkey, and notified by Germany, Italy, the Netherlands and the United Kingdom) in each of the 2010–2020 RASFF annual reports. In turn, according to the results of the two-way joining cluster analysis presented in Section 3.3.1, this hazard was reported on nuts from Iran between 2003 and 2006. There may be two reasons for this difference: firstly, the "Top 10" summaries were not included in earlier RASFF annual reports (i.e., for years prior to 2010), and secondly, the number of notifications for nuts between 2003 and 2006 was so high that the cluster analysis showed it to be the highest concentration, omitting the subsequent years of the analysed period. It should be noted, however, that this hazard is an ongoing, significant problem signalled in the annual reports, despite its noticeable reduction. Importantly, aflatoxins were also reported almost every year in fruits and vegetables from Turkey, and in 2018 and 2019 this was also related to ochratoxin A (which coincides with the results of the cluster analysis).

**Table 14.** Hazards in products of plant origin in the annual RASFF reports for 2010–2020.

| Year | Hazard | Product Category | Origin Country * | Notifying Country * | Reference |
|---|---|---|---|---|---|
| 2010 | Aflatoxins | Fruits and vegetables | Turkey | NDA | [14] |
| | | Herbs and spices | India | United Kingdom | |
| | | Nuts, nut products and seeds | Argentina, China, Iran, Turkey, United States | Germany, Greece, Italy, The Netherlands, Spain, United Kingdom | |
| | Unauthorized genetically modified | Cereals and bakery products | China | NDA | |
| 2011 | Aflatoxins | Fruits and vegetables | Turkey | NDA | [15] |
| | | Herbs and spices | India | NDA | |
| | | Nuts, nut products and seeds | China, Turkey, Iran | Germany, The Netherlands, United Kingdom | |
| | *Salmonella* | Fruits and vegetables | Bangladesh | United Kingdom | |
| | Living and died mites | Nuts, nut products and seeds | Ukraine | Poland | |
| 2012 | Aflatoxins | Fruits and vegetables | Turkey | France | [16] |
| | | Nuts, nut products and seeds | China | Germany, The Netherlands, United Kingdom | |
| | Monocrotophos | Fruits and vegetables | India | NDA | |
| | *Salmonella* | Fruits and vegetables | Bangladesh | United Kingdom | |
| 2013 | Aflatoxins | Fruits and vegetables | Turkey | NDA | [17] |
| | | Nuts, nut products and seeds | China, Turkey | Germany, Italy, The Netherlands | |
| 2014 | Aflatoxins | Fruits and vegetables | Turkey | NDA | [18] |
| | | Nuts, nut products and seeds | China, Iran, Turkey | Germany, Italy, The Netherlands, United Kingdom | |
| | Dichlorvos | Fruits and vegetables | Nigeria | United Kingdom | |
| 2015 | Aflatoxins | Fruits and vegetables | Turkey | NDA | [19] |
| | | Nuts, nut products and seeds | China, Iran, Turkey, United States | Belgium, Germany, Italy, The Netherlands, Spain, United Kingdom | |
| | *Salmonella* | Fruits and vegetables | India | United Kingdom | |
| | | Nuts, nut products and seeds | India | NDA | |
| 2016 | Aflatoxins | Fruits and vegetables | Turkey | NDA | [20] |
| | | Herbs and spices | India | NDA | |
| | | Nuts, nut products and seeds | China, Egypt, Iran, Turkey, United States | Germany, Italy, The Netherlands, United Kingdom | |
| | Pesticide residues | Fruits and vegetables | Turkey | Bulgaria, The Netherlands | |
| | *Salmonella* | Fruits and vegetables | India | United Kingdom | |
| 2017 | Aflatoxins | Fruits and vegetables | Turkey | NDA | [21] |
| | | Nuts, nut products and seeds | China, Iran, Turkey | Germany, Italy, The Netherlands, Spain | |
| | Absence of health certificate(s) | Nuts, nut products and seeds | NDA | United Kingdom | |
| | Pesticide residues | Fruits and vegetables | Turkey | NDA | |
| 2018 | Aflatoxins | Nuts, nut products and seeds | Argentina, China, Egypt, Turkey, United States | Germany, Italy, The Netherlands, Spain, United Kingdom | [22] |
| | Ochratoxin A | Fruits and vegetables | NDA | Turkey | |
| | *Salmonella* | Nuts, nut products and seeds | Sudan | Greece | |

**Table 14.** *Cont.*

| Year | Hazard | Product Category | Origin Country * | Notifying Country * | Reference |
|---|---|---|---|---|---|
| 2019 | Aflatoxins | Fruits and vegetables | Turkey | NDA | [23] |
| | | Nuts, nut products and seeds | Argentina, Turkey, United States | Germany, Italy, The Netherlands, Spain | |
| | Ochratoxin A | Fruits and vegetables | NDA | Turkey | |
| | *Salmonella* | Herbs and spices | Brazil | NDA | |
| | | Nuts, nut products and seeds | Sudan | Greece | |
| 2020 | Aflatoxins | Fruits and vegetables | Turkey | NDA | [24] |
| | | Nuts, nut products and seeds | Argentina, Iran, Turkey, United States | Germany, The Netherlands | |
| | Ethylene oxide | Nuts, nut products and seeds | India | Germany, The Netherlands | |
| | Pesticide residues | Fruits and vegetables | Turkey | Bulgaria | |
| | *Salmonella* | Herbs and spices | Brazil | Germany | |

* NDA—No Data Available.

Mycotoxins (including afltotocins) are highly carcinogenic and mutagenic, and are therefore an important issue in food production [25,26]. Cereals, spices and nuts can be infected with mycotoxins [25]. It is estimated that 25% of the world's cereal production is contaminated by these compounds [25,27]. Therefore, they cause significant losses in agriculture [28], especially in developing countries [29]. Economic losses associated with mycotoxin contamination include the costs of prevention, storage of infected wastes and quality control, and are calculated at billions of euros per year [30].

Some findings presented in annual RASFF reports overlap with results from the aforementioned cluster analysis. These were pesticide residues in fruits and vegetables from Turkey (in 2016, 2017 and 2020), *Salmonella* in fruits and vegetables from India (in 2015), nuts and seeds from Sudan (in 2018 and 2109), herbs and spices from Brazil (in 2019), the absence of health certificate(s) for nuts and seeds (in 2017), and ethylene oxide in seeds from India (in 2020).

The problem with the presence of pesticide residues in fruits and vegetables has been recognised in Turkey. Therefore, in this country a pesticide monitoring programme is required due to health and environmental concerns [31]. Turkey is a world leader in fresh produce, so it is believed that a surveillance system is needed to ensure food safety [32]. It should also be mentioned that the fruit and vegetable sector is very important for Turkey because of trade with the European Union [33]. In the context of pesticide residues, it is also worth noting the presence of ethylene oxide in sesame seeds from India. This problem continued and further diversified in 2021, leading to the largest food recall in the EU's history [9]. *Salmonella* is also frequently indicated in RASFF annual reports. Due to the environmental changes in the food chain, reducing the presence of this bacterium is more difficult than laboratory tests might suggest [34]. Therefore, it is important to understand how *Salmonella* can adopt, avoid and/or suppress plant defences in order to take appropriate strategies [35].

*4.2. RASFF Notifications in Studies by Various Authors*

Various authors often referred to RASFF notifications for plants in their studies, but provided very little information. Therefore, after reviewing the studies on notifications in this system, only those with the following variables were selected: year(s), hazard or hazard category and product or product category. They were sorted by year(s) of notification and then in alphabetical order. The product names were, in fact, given by authors with varying degrees of detail and, in addition, sometimes modified. If the authors also provided the country of origin, this is indicated in brackets after the name of the product category or product (Table 15).

**Table 15.** Hazards in products of plant origin by studies of various authors on RASFF notifications.

| Year(s) | Hazard or Hazard Category | Product or Product Category | Reference |
|---|---|---|---|
| 1979–2020 | Food additives and flavourings, pathogenic micro-organisms, pesticide residues | Fruits and vegetables | [36] |
| 1979–2020 | Mycotoxins | Herbs and spices | [36] |
| 1979–2020 | Mycotoxins, pathogenic micro-organisms | Nut products and seeds | [36] |
| 1999–2020 | *Bacillus cereus, Clostridium* spp., *Listeria monocytogenes, Salmonella* spp. | Mushrooms | [37] |
| 2000–2010 | Noroviruses | Berries, Tomatoes | [38] |
| 2000–2015 | Health certificate(s), illegal importation, tampering | Cereals and bakery products, fruits and vegetables, nuts, nut products and seeds | [39] |
| 2001–2010 | Aflatoxins | Fruits, nuts (from Argentina, Brazil, China, Egypt, Ghana, India, Iran, Turkey and United States) | [40] |
| 2001–2013 | Carbendazim | Aubergines, beans, broccoli, celery, chamomile, grapes, mint, okra, papaya | [41] |
| 2001–2015 | *Listeria monocytogenes* | Fruits and vegetables (from Germany) | [42] |
| 2002–2014 | Aflatoxins | Groundnuts, hazelnuts, pistachios, figs, herbs and spices | [43] |
| 2002–2018 | Genetically modified | Linseed, maize, papaya, rice | [44] |
| 2002–2019 | Pesticides | Gherkins (from Turkey) | [31] |
| 2002–2019 | Aflatoxins | Figs, hazelnuts, pistachios (from China, Iran, Turkey, United States) | [45] |
| 2002–2020 | Pesticide residues | Fruits, vegetables, nuts | [46] |
| 2002–2020 | Pesticide residues | Apples, pomegranates, peppers (from Turkey), rice (from India), tea (from China) | [47] |
| 2003 | Aflatoxins | Maize | [48] |
| 2003–2005 | Sudan | Chilli, paprika, turmeric-derived spicy products, palm oil | [49] |
| 2003–2006 | Aflatoxins | Peanuts, tree nuts | [50] |
| 2003–2007 | Aflatoxins | Pistachios (from Iran) | [51] |
| 2003–2007 | *Escherichia coli* | Spice and condiments | [51] |
| 2003–2007 | Genetically modified | Rice (from China, United States) | [51] |
| 2003–2007 | Noroviruses | Raspberries | [51] |
| 2003–2007 | Ochratoxin A | Cereals, figs, pepper, raisins/sultanas, vegetables | [51] |
| 2003–2007 | Pesticides | Fruits and vegetables | [51] |
| 2003–2007 | Sudan 4 | Palm oil (from African countries) | [51] |
| 2003–2009 | Sudan | Palm oil (from African countries) | [52] |
| 2004–2007 | *Bacillus cereus, Escherichia coli, Listeria monocytogenes, Salmonella* | Mushrooms | [53] |
| 2004–2008 | Dimethoate, insect, mould, rodent excrements, *Salmonella*, sulphite | Edible flowers (from Albania, Egypt, Sri Lanka and Thailand) | [54] |
| 2004–2009 | *Salmonella* | Rucola (from Italy) | [55] |
| 2004–2013 | Genetically modified | Papaya (from China, Thailand, Vietnam, United States) | [56] |
| 2004–2014 | Aflatoxin B1, ochratoxin A | Chilli, nutmeg, paprika, pepper | [57] |
| 2004–2014 | *Salmonella* spp. | Basil, coriander, black pepper, peppermint | [57] |
| 2004–2014 | *Bacillus* spp. | Chilli, curry | [57] |
| 2004–2014 | Aflatoxins, pesticide residues, Sudan | Herbs and spices | [58] |
| 2004–2018 | Dimethoate, insects, mould, rodent excrements, *Salmonella*, sulphite | Edible flowers | [59] |
| 2005 | Aflatoxins, Ochratoxin A | Fruits and vegetables, herbs and spices, nuts and nut products (pistachios from Iran) | [60] |
| 2005 | Aflatoxins | Pistachios | [61] |
| 2005–2006 | Microbiological contamination | Herbs and spices | [62] |
| 2005–2014 | Pathogenic micro-organisms | Almonds, coconuts, hazelnuts, pine nuts, pistachios | [63] |
| 2005–2015 | Chemical contaminants, foreign bodies, mycotoxins, pesticide residues, unauthorized additives and adulteration | Fruits and vegetables (from Turkey, India and Thailand) | [64] |

**Table 15.** *Cont.*

| Year(s) | Hazard or Hazard Category | Product or Product Category | Reference |
|---|---|---|---|
| 2005–2020 | *Salmonella* | Parsley | [65] |
| 2006 | Genetically modified | Rice (from China) | [40] |
| 2006–2015 | Aflatoxins | Paprika | [66] |
| 2007 | *Salmonella* | Alfalfa (from Pakistan) | [67] |
| 2008 and before | Aflatoxins | Pistachios | [68] |
| 2008–2011 | Additives, bacterial pathogens, chemical hazards, heavy metals, hygiene hazard/insufficient quality, mycotoxins, pesticide residues, physical hazard, viruses | Fruits and vegetables | [69] |
| 2008–2011 | bacterial pathogens, hygiene hazard/insufficient quality, mycotoxins, pesticide residues | Herbs and spices | [69] |
| 2008–2011 | Bacterial pathogens, hygiene hazard/insufficient quality, mycotoxins, genetically modified | Nuts, nut products and seeds | [69] |
| 2009 | Norovirus | Raspberries | [70] |
| 2009 and before | Aflatoxins, ochratoxin A | Cereals | [71] |
| 2009–2012 | Norovirus | Raspberries, strawberries | [72] |
| 2010–2011 | Aflatoxins | Nuts, nut products and seeds | [73] |
| 2010–2011 | Genetically modified | Rice (from China) | [74] |
| 2010–2012 | Norovirus, hepatovirus A | Dates (from Algeria), lettuce (from France, Germany), raspberries (from Chile, China, Poland, Serbia) | [75] |
| 2010–2014 | Norovirus, hepatovirus A | Dates (from Algeria), lettuce (from France), raspberries (from Chile, China, Poland, Serbia) | [76] |
| 2011 | Aflatoxins | Groundnuts | [77] |
| 2011 | *Salmonella, Escherichia coli* | Betel (from Bangladesh, India and Thailand) | [40] |
| 2011 | Aflatoxins, Ochratoxin A | herbs and spices, fruits and vegetables, nuts, nut products and seeds | [67] |
| 2011 | Norovirus | Raspberries | [78] |
| 2011 and before | Microbiological hazards | Fruits and vegetables, herbs and spices | [70] |
| 2011–2012 | Aflatoxins, ochratoxin A | Cereals and bakery products | [30] |
| 2011–2013 | Norovirus, hepatovirus A | Raspberries, strawberries | [79] |
| 2011–2014 | Allergens | Cereals and bakery products, cocoa, cocoa preparations, coffee and tea, fruits and vegetables, herbs and spices, nuts, nut products and seeds | [80] |
| 2011–2017 | Allergens | Cereals and bakery products | [81] |
| 2012 | Aflatoxins | Hazelnuts, figs, pistachios | [33] |
| 2012 | Pesticide residues | Pepper | [33] |
| 2012 | Genetically modified | Rice (from China) | [82] |
| 2012 | Allergens | Wheat | [83] |
| 2012 and before | Pesticide residues | Tea | [84] |
| 2012–2015 | Aflatoxins | Maize (from Bulgaria, Croatia, Greece, Hungary, Italy, Serbia, Slovakia, Poland, Romania) | [26] |
| 2012–2017 | *Salmonella* | Herbs and spices, nuts, nut products and seeds | [85] |
| 2012–2021 | Pyrrolizidine alkaloids | Spices and aromatic herbs, tea | [86] |
| 2013 | Mycotoxins, pathogenic micro-organisms, pesticide residues | Fruits and vegetables | [87] |
| 2013–2014 | Carbendazim | Mint | [88] |
| 2014 | Norovirus | Raspberries, strawberries | [89] |
| 2014 | Aflatoxins | Nuts and nut products | [90] |
| 2014–2018 | Chlorpyrifos | Herbs and spices | [91] |
| 2015 and before | Aflatoxins | Chilli | [92] |
| 2015–2018 | Norovirus, Hepatovirus A | Strawberries | [93] |

**Table 15.** *Cont.*

| Year(s) | Hazard or Hazard Category | Product or Product Category | Reference |
|---|---|---|---|
| 2015–2020 | Pesticide residues | Fruits and vegetables | [94] |
| 2016 | Aflatoxins | Nuts | [95] |
| 2016 | Mycotoxins | Herbs | [96] |
| 2016 | Mycotoxins | Herbs and spices, nuts, nut products and seeds | [97] |
| 2017 | Aflatoxins | Fruits and vegetables, herbs and spices, Nuts, nut products and seeds (from India) | [98] |
| 2017 | Aflatoxins | Nuts, nut products and seeds | [99] |
| 2017 and before | Pesticide residues | Chilli, paprika | [100] |
| 2017 and before | Additives | Chilli, curcuma, curry, palm oil, pepper | [101] |
| 2017 and before | Sudan | Herbs and spices | [102] |
| 2017–2021 | Aflatoxins, ochratoxin A, insects, missing documents, pesticides, sulphites | Figs (from Turkey and Spain) | [103] |
| 2018 | Enteric viruses | Berries | [104] |
| 2019 | Aflatoxins | Nuts | [105] |
| 2019 | Aflatoxins | Nuts | [106] |
| 2019 | Ochratoxin A | Figs, raisins | [106] |
| 2019 | Chlorpyrifos | Fruits and vegetables | [106] |
| 2019 and before | Mycotoxins | Maize, rice, wheat | [107] |
| 2020–2022 | Aflatoxins, ochratoxin A | Figs (from Turkey) | [108] |

The studies carried out by these authors confirm the results presented in Section 3 (Results) regarding the three most frequently reported hazard categories in the RASFF. It was noted that the notifications mainly concerned mycotoxins (aflatoxins and ochratoxin A) in fruits and vegetables, herbs and spices, and nuts. Another hazard category was pesticide residues (including, e.g., carbendazim, chlorpyrifos, dimethoate) notified in fruits and vegetables and herbs and spices. A third clearly noticeable hazard category was pathogenic micro-organisms (including *Escherichia coli*, *Listeria monocytogenes* and *Salmonella* spp.), which were similarly reported in fruits and vegetables and herbs and spices and, to a lesser extent, also in nuts.

Attention was also paid to RASFF notifications of other hazards (most of which were presented in the Section 3): additives including Sudan dye in herbs and spices and palm oil, genetically modified rice, foreign bodies in fruits and vegetables, lack of health certificates for fruits and vegetables, herbs and spices, and nuts or allergens in cereals and bakery products. However, the authors also highlighted hazards reported in the RASFF in other products: pesticides (dimethoate), foreign bodies, *Salmonella* spp. and sulphites in edible flowers, pathogenic micro-organisms (including *Bacillus cereus*, *Clostridium* spp., *Escherichia coli*, *Listeria monocytogenes*, *Salmonella* spp.) in mushrooms, and norovirus and hepatovirus A in strawberries and raspberries.

When the origin of the notified products was indicated, they were mainly Asian countries (Turkey, India, China, Thailand), the United States, and African and South American countries.

## 5. Conclusions

The three most commonly encountered hazards in foods of plant origin, i.e., mycotoxins, pesticide residues and pathogenic micro-organisms (including microbial contamination) related to 80% of notifications in the European Rapid Alert System for Food and Feed (RASFF) in 1998–2020. Particular attention should be paid to the hazards that have occurred in recent years: pesticide residues in peppers, moulds in groundnuts, ochratoxin A in raisins and sulphite in apricots from Turkey, ethylene oxide in sesame and problems with health certificate(s) for chilli, nutmeg, pistachios and sesame from India, iodine in seaweed from China and South Korea, *Salmonella* in sesame from Brazil, India and Sudan, *Escherichia coli* in basil from Lao Republic and betel from Thailand, and colour in breakfast cereals from the United States.

The notified products were, therefore, mainly from non-EU countries (particularly from Asia), i.e., Turkey, followed by India, China and Iran, and also from the United States. Given their proximity to the EU common market, hazards in products from Turkey (which shares a land border with Bulgaria and Greece) are of particular concern. These products were reported on the basis of border rejections, information notifications and, to a lesser extent, alerts. Notifications were based on border control, after which the consignments were detained, or official controls placed on the market; consequently, products were re-dispatched, withdrawn or destroyed.

Measures leading to the elimination of unsafe food products of plant origin from the European Union common market were necessary, but resulted in high costs and image losses for farmers, producers and other economic operators. Therefore, farmers need to pay particular attention to the use of methods such as Good Agricultural Practice (GAP), Good Hygiene Practice (GHP) and Good Manufacturing Practice (GMP), because, through these methods, hazards in food products of plant origin can be largely prevented or eliminated. It is also important that pesticides used by farmers to reduce or suppress the presence of pathogenic micro-organisms and the effects of their activities should be applied in an appropriate and proportionate manner, and with withdrawal periods. Producers (processors) should be more involved in the control of fresh produce delivered by farmers. Transporters should pay attention to maintaining the right parameters (temperature and humidity), especially in sea transport from distant Asian countries. It is also important for hazard limits to be set and updated by legislative bodies, and subsequently controlled by the authorities of the EU countries.

The "From field to fork" strategy adopted in the European Green Deal emphasises the need to build a sustainable model in the food system, and the elimination or reduction of hazards in plants is an important part of this strategy. Therefore, the research carried out, covering a wide time period and range of hazards found in food products of plant origin, can contribute to improvements in sustainability efforts.

**Supplementary Materials:** The following supporting information can be downloaded at: https://www.mdpi.com/article/10.3390/su15108091/s1, Table S1: Number of notifications in the RASFF under product categories for the period 1979–2020; Table S2: Hazard categories and hazards notified in products of plant origin in the RASFF in 1998–2020; Table S3: Shortened values names of variables notification basis, distribution status and action taken used in figures in Supplementary Materials; Figure S1: Results of joining cluster analysis; (a) year; (b) product; (c) notifying country; (d) origin country; (e) notification type; (f) notification basis; (g) distribution status; (h) action taken; Figure S2: Results of two-way joining cluster analysis for aflatoxins; (a) product; (b) notifying country; (c) origin country; (d) notification type; (e) notification basis; (f) distribution status; (g) action taken; Figure S3: Results of two-way joining cluster analysis for ochratoxin A; (a) product; (b) notifying country; (c) origin country; (d) notification type; (e) notification basis; (f) distribution status; (g) action taken; Figure S4: Results of two-way joining cluster analysis for ethylene oxide; (a) product; (b) notifying country; (c) origin country; (d) notification type; (e) notification basis; (f) distribution status; (g) action taken; Figure S5: Results of two-way joining cluster analysis for chlorpyrifos; (a) product; (b) notifying country; (c) origin country; (d) notification type; (e) notification basis; (f) distribution status; (g) action taken; Figure S6: Results of two-way joining cluster analysis for carbendazim; (a) product; (b) notifying country; (c) origin country; (d) notification type; (e) notification basis; (f) distribution status; (g) action taken; Figure S7: Results of two-way joining cluster analysis for dimethoate; (a) product; (b) notifying country; (c) origin country; (d) notification type; (e) notification basis; (f) distribution status; (g) action taken; Figure S8: Results of two-way joining cluster analysis for methomyl; (a) product; (b) notifying country; (c) origin country; (d) notification type; (e) notification basis; (f) distribution status; (g) action taken; Figure S9: Results of two-way joining cluster analysis for acetamiprid; (a) product; (b) notifying country; (c) origin country; (d) notification type; (e) notification basis; (f) distribution status; (g) action taken; Figure S10: Results of two-way joining cluster analysis for omethoate; (a) product; (b) notifying country; (c) origin country; (d) notification type; (e) notification basis; (f) distribution status; (g) action taken; Figure S11: Results of two-way joining cluster analysis for triazophos; (a) product; (b) notifying country; (c) origin country; (d) notification type; (e) notification basis; (f) distribution status; (g) action taken; Figure S12: Results of two-way joining cluster analysis for formetanate; (a) product; (b) notifying country; (c) origin country; (d) notification type; (e) notification basis; (f) distribution status; (g) action taken; Figure S13: Results of two-way joining cluster analysis for *Salmonella*; (a) product; (b) notifying country; (c) origin country; (d) notification type; (e) notification basis; (f) distribution status; (g) action taken; Figure S14: Results of two-way joining cluster analysis for *Escherichia coli*; (a) product; (b) notifying country; (c) origin country; (d) notification type; (e) notification basis; (f) distribution status; (g) action taken; Figure S15: Results of two-way joining cluster analysis for moulds; (a) product; (b) notifying country; (c) origin country; (d) notification type; (e) notification basis; (f) distribution status; (g) action taken; Figure S16: Results of two-way joining cluster analysis for Sudan; (a) product; (b) notifying country; (c) origin country; (d) notification type; (e) notification basis; (f) distribution status; (g) action taken; Figure S17: Results of two-way joining cluster analysis for iodine; (a) product; (b) notifying country; (c) origin country; (d) notification type; (e) notification basis; (f) distribution status; (g) action taken; Figure S18: Results of two-way joining cluster analysis for sulphite; (a) product; (b) notifying country; (c) origin country; (d) notification type; (e) notification basis; (f) distribution status; (g) action taken; Figure S19: Results of two-way joining cluster analysis for colour; (a) product; (b) notifying country; (c) origin country; (d) notification type; (e) notification basis; (f) distribution status; (g) action taken; Figure S20: Results of two-way joining cluster analysis for genetically modified plants; (a) product; (b) notifying country; (c) origin country; (d) notification type; (e) notification basis; (f) distribution status; (g) action taken; Figure S21: Results of two-way joining cluster analysis for insects; (a) product; (b) notifying country; (c) origin country; (d) notification type; (e) notification basis; (f) distribution status; (g) action taken; Figure S22: Results of two-way joining cluster analysis for health certificate(s); (a) product; (b) notifying country; (c) origin country; (d) notification type; (e) notification basis; (f) distribution status; (g) action taken; Figure S23: Results of two-way joining cluster analysis for milk; (a) product; (b) notifying country; (c) origin country; (d) notification type; (e) notification basis; (f) distribution status; (g) action taken.

**Author Contributions:** Conceptualization, methodology, formal analysis, investigation, data curation, writing—original draft preparation, M.P.; writing—review and editing, M.N.-D. All authors have read and agreed to the published version of the manuscript.

**Funding:** The APC was funded by Gdynia Maritime University, Department of Quality Management, team research project "Systemic quality, environment and safety management in the product life cycle" number WZNJ/2023/PZ/04 and Cracow University of Economics, Department of Quality Management.

**Institutional Review Board Statement:** Not applicable.

**Informed Consent Statement:** Not applicable.

**Data Availability Statement:** Not applicable.

**Conflicts of Interest:** The authors declare no conflict of interest.

# References

1. Ene, C. Food Security and Food Safety: Meanings and Connections. *Econ. Insights Trends Chall.* **2020**, *IX*, 59–68.
2. Răbonţu, C. Food Safety and its Role in the Evolution of Food Trade. *An. Univ. "Constantin Brâncuşi" Din Târgu Jiu Ser. Econ.* **2010**, *2*, 161–172.
3. Vågsholm, I.; Arzoomand, N.S.; Boqvist, S. Food Security, Safety, and Sustainability—Getting the Trade-Offs Right. *Front. Sustain. Food Syst.* **2020**, *4*, 16. [CrossRef]
4. Garcia, S.N.; Osburn, B.I.; Jay-Russell, M.T. One Health for Food Safety, Food Security, and Sustainable Food Production. *Front. Sustain. Food Syst.* **2020**, *4*, 1. [CrossRef]
5. Aiking, H.; de Boer, J. Food sustainability: Diverging interpretations. *Br. Food J.* **2004**, *106*, 359–365. [CrossRef]
6. Pozza, L.; Field, D.J. The science of Soil Security and Food Security. *Soil Secur.* **2020**, *1*, 100002. [CrossRef]
7. FAO (Food and Agriculture Organization of the United Nations). *SAFA Sustainability Assessment of Food and Agriculture Systems Guidelines*, 3rd ed.; FAO: Rome, Italy, 2014.
8. RASFF Notifications Pre-2021 Public Information. Available online: https://data.europa.eu/data/datasets/restored_rasff?locale=en (accessed on 25 March 2022).
9. European Parliament; Council of the European Union. *Regulation (EC) 178/2002 of the European Parliament and of the Council of 28 January 2002 Laying Down the General Principles and Requirements of Food Law, Establishing the European Food Safety Authority and Laying Down Procedures in Matters of Food Safety*; OJ L 31, 1.02.2002; European Parliament and Council of the European Union: Brussels, Belgium, 2002; pp. 1–24.

10. RASFF—Food and Feed Safety Alerts. Available online: https://ec.europa.eu/food/safety/rasff-food-and-feed-safety-alerts_en (accessed on 3 November 2022).
11. European Union. *2021 Annual Report Alert and Cooperation Network*; Publications Office of the European Union: Luxembourg, 2022.
12. RASFF Window. Available online: https://webgate.ec.europa.eu/rasff-window/screen/search (accessed on 2 November 2022).
13. TIBCO Statistica®User's Guide. Available online: https://docs.tibco.com/pub/stat/14.0.0/doc/html/UsersGuide/ (accessed on 3 November 2022).
14. European Communities. *The Rapid Alert System for Food and Feed (RASFF) Annual Report 2010*; Office for Official Publications of the European Communities: Luxembourg, 2011.
15. European Communities. *RASFF. The Rapid Alert System for Food and Feed 2011 Annual Report*; Office for Official Publications of the European Communities: Luxembourg, 2012.
16. European Union. *RASFF—The Rapid Alert System for Food and Feed—2012 Annual Report*; Publications Office of the European Union: Luxembourg, 2013.
17. European Union. *RASFF—The Rapid Alert System for Food and Feed—2013 Annual Report*; Publications Office of the European Union: Luxembourg, 2014.
18. European Union. *RASFF for Safer Food—The Rapid Alert System for Food and Feed—2014 Annual Report*; Publications Office of the European Union: Luxembourg, 2015.
19. European Union. *RASFF—The Rapid Alert System for Food and Feed—2015 Annual Report*; Publications Office of the European Union: Luxembourg, 2016.
20. European Union. *RASFF—The Rapid Alert System for Food and Feed—2016 Annual Report*; Publications Office of the European Union: Luxembourg, 2017.
21. European Union. *RASFF—The Rapid Alert System for Food and Feed—2017 Annual Report*; Publications Office of the European Union: Luxembourg, 2018.
22. European Union. *RASFF—The Rapid Alert System for Food and Feed—2018 Annual Report*; Publications Office of the European Union: Luxembourg, 2019.
23. European Union. *RASFF—The Rapid Alert System for Food and Feed—Annual Report 2019*; Publications Office of the European Union: Luxembourg, 2020.
24. European Union. *RASFF—The Rapid Alert System for Food and Feed—Annual Report 2020*; Publications Office of the European Union: Luxembourg, 2021.
25. Iqbal, S.Z. Mycotoxins in food, recent development in food analysis and future challenges: A review. *Curr. Opin. Food Sci.* **2021**, *42*, 237–247. [CrossRef]
26. Tian, F.; Woo, S.Y.; Lee, S.Y.; Park, S.B.; Im, J.H.; Chun, H.S. Plant-based natural flavonoids show strong inhibition of aflatoxin production and related gene expressions correlated with chemical structure. *Food Microbiol.* **2023**, *109*, 104141. [CrossRef]
27. Leite, M.; Freitas, A.; Silva, A.S.; Barbosa, J.; Ramos, F. Maize food chain and mycotoxins: A review on occurrence studies. *Trends Food Sci. Tech.* **2021**, *115*, 307–331. [CrossRef]
28. Jallow, A.; Xie, H.; Tang, X.; Qi, Z.; Li, P. Worldwide aflatoxin contamination of agricultural products and foods: From occurrence to control. *Compr. Rev. Food Saf.* **2021**, *20*, 2332–2381. [CrossRef]
29. Söylemez, T.; Yamaç, M.; Yıldız, Z. Statistical optimization of cultural variables for enzymatic degradation of aflatoxin B1 by *Panus neostrigosus*. *Toxicon* **2020**, *186*, 141–150. [CrossRef]
30. Oliveira, P.M.; Zannini, E.; Arendt, E.K. Cereal fungal infection, mycotoxins, and lactic acid bacteria mediated bioprotection: From crop farming to cereal products. *Food Microbiol.* **2014**, *37*, 78–95. [CrossRef]
31. Golge, O.; Cinpolat, S.; Kabak, B. Quantification of pesticide residues in gherkins by liquid and gas chromatography coupled to tandem mass spectrometry. *J. Food Compos. Anal.* **2021**, *96*, 103755. [CrossRef]
32. Gunel, E.; Kilic, G.P.; Bulut, E.; Durul, B.; Acar, S.; Alpas, H.; Soyer, Y. *Salmonella* surveillance on fresh produce in retail in Turkey. *Int. J. Food Microbiol.* **2015**, *199*, 72–77. [CrossRef]
33. Engelbert, T.; Bektasoglu, B.; Brockmeier, M. Moving toward the EU or the Middle East? An assessment of alternative Turkish foreign policies utilizing the GTAP framework. *Food Policy* **2014**, *47*, 46–61. [CrossRef]
34. Ortiz-Solà, J.; Colás-Medà, P.; Nicolau-Lapeña, I.; Alegre, I.; Abadias, M.; Vinas, I. Pathogenic potential of the surviving Salmonella Enteritidis on strawberries after disinfection treatments based on ultraviolet-C light and peracetic acid. *Int. J. Food Microbiol.* **2022**, *364*, 109536. [CrossRef]
35. Zarkani, A.A.; Schikora, A. Mechanisms adopted by Salmonella to colonize plant hosts. *Food Microbiol.* **2021**, *99*, 103833. [CrossRef]
36. Nogales, A.; Mora-Cantallops, M.; Morón, R.; García-Tejedor, A.J. Network analysis for food safety: Quantitative and structural study of data gathered through the RASFF system in the European Union. *Food Control* **2023**, *145*, 109422. [CrossRef]
37. Kragh, M.L.; Obari, L.; Caindec, A.M.; Jensen, H.A.; Hansen, L.T. Survival of *Listeria monocytogenes*, *Bacillus cereus* and *Salmonella* Typhimurium on sliced mushrooms during drying in a household food dehydrator. *Food Control* **2022**, *134*, 108715. [CrossRef]
38. Le Guyader, F.S.; Atmar, R.; Le Pendu, J. Transmission of viruses through shellfish: When specific ligands come into play. *Curr. Opin. Virol.* **2012**, *2*, 103–110. [CrossRef]

39. Bouzembrak, Y.; Steen, B.; Neslo, R.; Linge, J.; Mojtahed, V.; Marvin, H.J.P. Development of food fraud media monitoring system based on text mining. *Food Control* **2018**, *93*, 283–296. [CrossRef]

40. Jezsó, V. Managing Risks in Imports of Non-animal Origin: EU Emergency Measures. In *Risk Regulation in Non-Animal Food Imports*; Montanari, F., Jezsó, V., Donati, C., Eds.; Springer: Cham, Switzerland, 2015; pp. 57–95.

41. Rubio, L.; Ortiz, M.C.; Sarabia, L.A. Identification and quantification of carbamate pesticides in dried lime tree flowers by means of excitation-emission molecular fluorescence and parallel factor analysis when quenching effect exists. *Anal. Chim. Acta* **2014**, *820*, 9–22. [CrossRef] [PubMed]

42. Lüth, S.; Boone, I.; Kleta, S.; Dahouk, S. Analysis of RASFF notifications on food products contaminated with *Listeria monocytogenes* reveals options for improvement in the rapid alert system for food and feed. *Food Control* **2019**, *96*, 479–487. [CrossRef]

43. Kabak, B. Aflatoxins in hazelnuts and dried figs: Occurrence and exposure assessment. *Food Chem.* **2016**, *211*, 8–16. [CrossRef] [PubMed]

44. Rostoks, N.; Grantina-Ieviņa, L.; Ieviņa, B.; Evelone, V.; Valciņa, O.; Aleksejeva, I. Genetically modified seeds and plant propagating material in Europe: Potential routes of entrance and current status. *Heliyon* **2019**, *5*, e01242. [CrossRef]

45. Kabak, B. Aflatoxins in foodstuffs: Occurrence and risk assessment in Turkey. *J. Food Compos. Anal.* **2021**, *96*, 103734. [CrossRef]

46. Kuchheuser, P.; Birringer, M. Pesticide residues in food in the European Union: Analysis of notifications in the European Rapid Alert System for Food and Feed from 2002 to 2020. *Food Control* **2022**, *133 Pt A*, 108575. [CrossRef]

47. Kuchheuser, P.; Birringer, M. Evaluation of specific import provisions for food products from third countries based on an analysis of RASFF notifications on pesticide residues. *Food Control* **2022**, *133 Pt B*, 108581. [CrossRef]

48. Baranyi, N.; Kocsubé, S.; Varga, J. Aflatoxins: Climate change and biodegradation. *Curr. Opin. Food Sci.* **2015**, *5*, 60–66. [CrossRef]

49. Reile, C.G.; Rodriguez, M.S.; De Sousa Fernandes, D.D.; de Araújo Gomes, A.; Gonçalves Dias Diniz, P.H.; Di Anibal, C.V. Qualitative and quantitative analysis based on digital images to determine the adulteration of ketchup samples with Sudan I dye. *Food Chem.* **2020**, *328*, 127101. [CrossRef]

50. Van Der Fels-Klerx, H.J.; Dekkers, S.; Kandhai, M.C.; Jeurissen, S.M.F.; Booij, C.J.H.; De Heer, C. Indicators for early identification of re-emerging mycotoxins. *NJAS-Wagen J. Life Sci.* **2010**, *57*, 133–139. [CrossRef]

51. Kleter, G.A.; Prandini, A.; Filippi, L.; Marvin, H.J.P. Identification of potentially emerging food safety issues by analysis of reports published by the European Community's Rapid Alert System for Food and Feed (RASFF) during a four-year period. *Food Chem. Toxicol.* **2009**, *47*, 932–950. [CrossRef]

52. Rebane, R.; Leito, I.; Yurchenko, S.; Herodes, K. A review of analytical techniques for determination of Sudan I-IV dyes in food matrixes. *J. Chromatogr. A* **2010**, *1217*, 2747–2757. [CrossRef]

53. Venturini, M.E.; Reyes, J.E.; Rivera, C.S.; Oria, R.; Blanco, D. Microbiological quality and safety of fresh cultivated and wild mushrooms commercialized in Spain. *Food Microbiol.* **2011**, *28*, 1492–1498. [CrossRef]

54. Fernandes, L.; Casal, S.; Pereira, J.; Saraiva, J.; Ramalhosa, E. Edible flowers: A review of the nutritional, antioxidant, antimicrobial properties and effects on human health. *J. Food Compos. Anal.* **2017**, *60*, 38–50. [CrossRef]

55. Volpe, G.; Delibato, E.; Fabiani, L.; Pucci, E.; Piermarini, S.; D'Angelo, A.; Capuano, F.; De Medici, D.; Palleschi, G. Development and evaluation of an ELIME assay to reveal the presence of *Salmonella* in irrigation water: Comparison with Real-Time PCR and the Standard Culture Method. *Talanta* **2016**, *149*, 202–210. [CrossRef]

56. Prins, T.W.; Scholtens, I.M.J.; Bak, A.W.; Van Dijk, J.P.; Voorhuijzen, M.M.; Laurensse, E.J.; Kok, E.J. A case study to determine the geographical origin of unknown GM papaya in routine food sample analysis, followed by identification of papaya events 16-0-1 and 18-2-4. *Food Chem.* **2016**, *213*, 536–544. [CrossRef]

57. Banach, J.L.; Stratakou, I.; Van Der Fels-Klerx, H.J.; Den Besten, H.M.W.; Zwietering, M.H. European alerting and monitoring data as inputs for the risk assessment of microbiological and chemical hazards in spices and herbs. *Food Control* **2016**, *69*, 237–249. [CrossRef]

58. Van Asselt, E.D.; Banach, J.L.; Van Der Fels-Klerx, H.J. Prioritization of chemical hazards in spices and herbs for European monitoring programs. *Food Control* **2018**, *83*, 7–17. [CrossRef]

59. Matyjaszczyk, E.; Śmiechowska, M. Edible flowers. Benefits and risks pertaining to their consumption. *Trends Food Sci. Tech.* **2019**, *91*, 670–674. [CrossRef]

60. Molyneux, R.J.; Mahoney, N.; Kim, J.H.; Campbell, B.C. Mycotoxins in edible tree nuts. *Int. J. Food Microbiol.* **2007**, *119*, 72–78. [CrossRef] [PubMed]

61. Var, I.; Kabak, B.; Gök, F. Survey of aflatoxin $B_1$ in helva, a traditional Turkish food, by TLC. *Food Control* **2007**, *18*, 59–62. [CrossRef]

62. Elviss, N.C.; Little, C.L.; Hucklesby, L.; Sagoo, S.; Surman-Lee, S.; De Pinna, E.; Threlfall, E.J. Microbiological study of fresh herbs from retail premises uncovers an international outbreak of salmonellosis. *Int. J. Food Microbiol.* **2009**, *134*, 83–88. [CrossRef] [PubMed]

63. Zwietering, M.H.; Jacxsens, L.; Membré, J.-M.; Nauta, M.; Peterz, M. Relevance of microbial finished product testing in food safety management. *Food Control* **2016**, *60*, 31–43. [CrossRef]

64. Bouzembrak, Y.; Marvin, H.J.P. Impact of drivers of change, including climatic factors, on the occurrence of chemical food safety hazards in fruits and vegetables: A Bayesian Network approach. *Food Control* **2019**, *97*, 67–76. [CrossRef]

65. Dittrich, A.J.; Ludewig, M.; Rodewald, S.; Braun, P.G.; Wiacek, C. Effect of high hydrostatic pressure on *Salmonella enterica* subsp. *enterica* in Nutrient Broth and dried parsley. *LWT-Food Sci. Technol.* **2022**, *155*, 112850. [CrossRef]

66. Majer-Baranyi, K.; Zalán, Z.; Mörtl, M.; Juracsek, J.; Szendrö, I.; Székács, A.; Adányi, N. Optical waveguide lightmode spectroscopy technique–based immunosensor development for aflatoxin B1 determination in spice paprika samples. *Food Chem.* **2016**, *211*, 972–977. [CrossRef]

67. Johannessen, G.S.; Cudjoe, K. Regulatory Issues in Europe Regarding Fresh Fruit and Vegetable Safety. In *Produce Contamination Problem: Causes and Solutions*, 2nd ed.; Matthews, K.R., Sapers, G.M., Gerba, C.P., Eds.; Elsevier: San Diego, CA, USA, 2014; pp. 365–386.

68. Yanniotis, S.; Proshlyakov, A.; Revithi, A.; Georgiadou, M.; Blahovec, J. X-ray imaging for fungal necrotic spot detection in pistachio nuts. *Proc. Food Sci.* **2011**, *1*, 379–384. [CrossRef]

69. Uyttendaele, M.; Jacxsens, L.; Van Boxstael, S. Issues surrounding the European fresh produce trade: A global perspective. In *Global Safety of Fresh Produce: A Handbook of Best Practice, Innovative Commercial Solutions and Case Studies*; Hoorfar, J., Ed.; Woodhead Publishing: Oxford, UK, 2014; pp. 33–51.

70. Foong-Cunningham, S.; Verkaar, E.L.C.; Swanson, K. Microbial decontamination of fresh produce. In *Microbial Decontamination in the Food Industry: Novel Methods and Applications*; Demirci, A., Ngadi, M.O., Eds.; Woodhead Publishing: Oxford, UK, 2012; pp. 3–29.

71. Ganesan, A.R.; Rajauria, G. Cereal-Based Animal Feed Products. In *Innovative Processing Technologies for Healthy Grains*; Pojić, M., Tiwari, U., Eds.; John Wiley & Sons: Hoboken, NJ, USA, 2021; pp. 199–226.

72. De Keuckelaere, A.; Li, D.; Deliens, B.; Stals, A.; Uyttendaele, M. Batch testing for noroviruses in frozen raspberries. *Int. J. Food Microbiol.* **2015**, *192*, 43–50. [CrossRef]

73. Arroyo-Manzanares, N.; Huertas-Pérez, J.F.; Gámiz-Gracia, L.; García-Campaña, A.M. A new approach in sample treatment combined with UHPLC-MS/MS for the determination of multiclass mycotoxins in edible nuts and seeds. *Talanta* **2013**, *115*, 61–67. [CrossRef]

74. Fraiture, M.-A.; Roosens, N.H.C.; Taverniers, I.; De Loose, M.; Deforce, D.; Herman, P. Biotech rice: Current developments and future detection challenges in food and feed chain. *Trends Food Sci. Tech.* **2016**, *52*, 66–79. [CrossRef]

75. Petrović, T. Prevalence of viruses in food and the environment. In *Viruses in Food and Water: Risks, Surveillance and Control*; Cook, N., Ed.; Woodhead Publishing: Cambridge, UK, 2013; pp. 19–46.

76. Petrović, T.; D'Agostino, M. Viral Contamination of Food. In *Antimicrobial Food Packaging*; Barros-Velazquez, J., Ed.; Elsevier: London, UK, 2016; pp. 65–79.

77. Clavel, D.; Brabet, C. Mycotoxin contamination of nuts. In *Improving the Safety and Quality of Nuts*; Harris, L.J., Ed.; Woodhead Publishing: Oxford, UK, 2013; pp. 88–118.

78. Perrin, A.; Loutreul, J.; Boudaud, N.; Bertrand, I.; Gantzer, C. Rapid, simple and efficient method for detection of viral genomes on raspberries. *J. Virol. Methods* **2015**, *224*, 95–101. [CrossRef]

79. Fraisse, A.; Coudray-Meunier, C.; Martin-Latil, S.; Hennechart-Collette, C.; Delannoy, S.; Fach, P.; Perelle, S. Digital RT-PCR method for hepatitis A virus and norovirus quantification in soft berries. *Int. J. Food Microbiol.* **2017**, *243*, 36–45. [CrossRef]

80. Bucchini, L.; Guzzon, A.; Poms, R.; Senyuva, H. Analysis and critical comparison of food allergen recalls from the European Union, USA, Canada, Hong Kong, Australia and New Zealand. *Food Addit.Contam. Part A* **2016**, *33*, 760–771. [CrossRef]

81. Soon, J.M.; Brazier, A.K.M.; Wallace, C.A. Determining common contributory factors in food safety incidents—A review of global outbreaks and recalls 2008–2018. *Trends Food Sci. Tech.* **2020**, *97*, 76–87. [CrossRef]

82. Fraiture, M.-A.; Herman, P.; Taverniers, I.; De Loose, M.; Deforce, D.; Roosens, N.H. An innovative and integrated approach based on DNA walking to identify unauthorised GMOs. *Food Chem.* **2014**, *147*, 60–69. [CrossRef]

83. Pegels, N.; González, I.; García, T.; Martín, R. Authenticity testing of wheat, barley, rye and oats in food and feed market samples by real-time PCR assays. *LWT-Food Sci. Technol.* **2015**, *60*, 867–875. [CrossRef]

84. Cajka, T.; Sandy, C.; Bachanova, V.; Drábová, L.; Kalachova, K.; Pulkrabová, J.; Hajslova, J. Streamlining sample preparation and gas chromatography–tandem mass spectrometry analysis of multiple pesticide residues in tea. *Anal. Chim. Acta* **2012**, *743*, 51–60. [CrossRef]

85. Bourdichon, F.; Betts, R.; Dufour, C.; Fanning, S.; Farber, J.; McClure, P.; Stavropoulou, D.A.; Wemmenhove, E.; Zwietering, M.H.; Winkler, A. Processing environment monitoring in low moisture food production facilities: Are we looking for the right microorganisms? *Int. J. Food Microbiol.* **2021**, *356*, 109351. [CrossRef]

86. Casado, N.; Morante-Zarcero, S.; Sierra, I. The concerning food safety issue of pyrrolizidine alkaloids: An overview. *Trends Food Sci. Tech.* **2022**, *120*, 123–139. [CrossRef]

87. Brandão, D.; Liébana, S.; Pividori, M.I. Multiplexed detection of foodborne pathogens based on magnetic particles. *New Biotechnol.* **2015**, *32*, 511–520. [CrossRef] [PubMed]

88. Rubio, L.; Sarabia, L.A.; Ortiz, M.C. Standard addition method based on four-way PARAFAC decomposition to solve the matrix interferences in the determination of carbamate pesticides in lettuce using excitation-emission fluorescence data. *Talanta* **2015**, *138*, 86–99. [CrossRef] [PubMed]

89. Bosch, A.; Pintó, R.M.; Guix, S. Foodborne viruses. *Curr. Opin. Food Sci.* **2016**, *8*, 110–119. [CrossRef] [PubMed]

90. Marín, S.; Ramos, A.J. Molds and mycotoxins in nuts. In *Food Hygiene and Toxicology in Ready-to-Eat Foods*; Kotzekidou, P., Ed.; Elsevier: San Diego, CA, USA, 2016; pp. 295–312.

91. Hakme, E.; Lozano, A.; Uclés, S.; Gómez-Ramos, M.M.; Fernández-Alba, A.R. High-throughput gas chromatography-mass spectrometry analysis of pesticide residues in spices by using the enhanced matrix removal-lipid and the sample dilution approach. *J. Chromatogr. A* **2018**, *1573*, 28–41. [CrossRef]

92. Pothimon, R.; Krusong, W.; Daetae, P.; Tantratian, S.; Gullo, M. Determination of antifungal volatile organic compounds of upland rice vinegar and their inhibition effects on *Aspergillus flavus* in dried chili pepper. *Food Biosci.* **2022**, *46*, 101543. [CrossRef]

93. Ortiz-Solà, J.; Viñas, I.; Colás-Medà, P.; Anguera, M.; Abadias, M. Occurrence of selected viral and bacterial pathogens and microbiological quality of fresh and frozen strawberries sold in Spain. *Int. J. Food Microbiol.* **2020**, *314*, 108392. [CrossRef]

94. Pan, Y.; Ren, Y.; Luning, P.A. Factors influencing Chinese farmers' proper pesticide application in agricultural products—A review. *Food Control* **2021**, *122*, 107788. [CrossRef]

95. Alcántara-Durán, J.; Moreno-González, D.; García-Reyes, J.F.; Molina-Díaz, A. Use of a modified QuEChERS method for the determination of mycotoxins residues in edible nuts by nano flow liquid chromatography high resolution mass spectrometry. *Food Chem.* **2019**, *279*, 144–149. [CrossRef]

96. Ouf, S.A.; Ali, E.M. Does the treatment of dried herbs with ozone as a fungal decontaminating agent affect the active constituents? *Environ. Pollut.* **2021**, *277*, 116715. [CrossRef]

97. Zhao, X.; Liu, D.; Zhang, L.; Zhou, Y.; Yang, M. Development and optimization of a method based on QuEChERS-dSPE Followed by UPLC-MS/MS for the simultaneous determination of 21 mycotoxins in nutmeg and related products. *Microchem. J.* **2021**, *168*, 106499. [CrossRef]

98. Thanushree, M.P.; Sailendri, D.; Yoha, K.S.; Moses, J.A.; Anandharamakrishnan, C. Mycotoxin contamination in food: An exposition on spices. *Trends Food Sci. Tech.* **2019**, *93*, 69–80. [CrossRef]

99. Vila-Donat, P.; Marín, S.; Sanchis, V.; Ramos, A.J. New mycotoxin adsorbents based on tri-octahedral bentonites for animal feed. *Anim. Feed Sci. Tech.* **2019**, *255*, 114228. [CrossRef]

100. Mörtl, M.; Klátyik, S.; Molnár, H.; Tömösközi-Farkas, R.; Adányi, N.; Székács, A. The effect of intensive chemical plant protection on the quality of spice paprika. *J. Food Compos. Anal.* **2018**, *67*, 141–148. [CrossRef]

101. Piątkowska, M.; Jedziniak, P.; Olejnik, M.; Żmudzki, J.; Posyniak, A. Absence of evidence or evidence of absence? A transfer and depletion study of Sudan I in eggs. *Food Chem.* **2018**, *239*, 598–602. [CrossRef]

102. Sebaei, A.S.; Youssif, M.I.; Ghazi, A.A.-M. Determination of seven illegal dyes in Egyptian spices by HPLC with gel permeation chromatography clean up. *J. Food Compos. Anal.* **2019**, *84*, 103304. [CrossRef]

103. Celik, D.; Kabak, B. Assessment to propose a maximum permitted level for ochratoxin A in dried figs. *J. Food Compos. Anal.* **2022**, *112*, 104705. [CrossRef]

104. Ortiz-Solà, J.; Viñas, I.; Aguiló-Aguayo, I.; Bobo, G.; Abadias, M. An innovative water-assisted UV-C disinfection system to improve the safety of strawberries frozen under cryogenic conditions. *Innov. Food Sci. Emerg.* **2021**, *73*, 10275. [CrossRef]

105. Ribeiro, S.R.; Garcia, M.V.; Copetti, M.V.; Brackmann, A.; Both, V.; Wagner, R. Effect of controlled atmosphere, vacuum packaging and different temperatures on the growth of spoilage fungi in shelled pecan nuts during storage. *Food Control* **2021**, *128*, 108173. [CrossRef]

106. Srednicka, P.; Juszczuk-Kubiak, E.; Wójcicki, M.; Akimowicz, M.; Roszko, M.Ł. Probiotics as a biological detoxification tool of food chemical contamination: A review. *Food Chem. Toxicol.* **2021**, *153*, 112306. [CrossRef]

107. Recknagel, S.; Bresch, H.; Kipphardt, H.; Koch, M.; Rosner, M.; Resch-Genger, U. Trends in selected fields of reference material production. *Anal. Bioanal. Chem.* **2022**, *414*, 4281–4289. [CrossRef]

108. Galván, A.I.; Hernández, A.; De Guía Córdoba, M.; Martín, A.; Serradilla, M.J.; López-Corrales, M.; Rodríguez, A. Control of toxigenic *Aspergillus* spp. in dried figs by volatile organic compounds (VOCs) from antagonistic yeasts. *Int. J. Food Microbiol.* **2022**, *376*, 109772. [CrossRef] [PubMed]

*Article*

# Research and Development of a New Sustainable Functional Food under the Scope of Nutrivigilance

Irina Mihaela Matran [1], Monica Tarcea [1,*], Diana Carmen Rus [2], Rares Voda [2], Daniela-Lucia Muntean [3] and Daniela Cirnatu [4]

[1] Department of Community Nutrition and Food Safety, G.E. Palade University of Medicine, Pharmacy, Science and Technology of Targu Mures, 540142 Mures, Romania; irina.matran@umfst.ro

[2] Faculty of Medicine, George Emil Palade University of Medicine, Pharmacy, Science and Technology of Targu Mures, 540142 Mures, Romania

[3] Department of Analytical Chemistry, George Emil Palade University of Medicine, Pharmacy, Science and Technology of Targu Mures, 540142 Mures, Romania

[4] Department of Medicine, Faculty of Medicine, Vasile Goldis Western University of Arad, 310025 Arad, Romania

* Correspondence: address: monica.tarcea@umfst.ro

**Abstract:** Background: The New Global Economy is represented by a series of major features, such as the use of green energy, the reduction of the carbon footprint in all industrial and civil fields, as well as finding alternative food resources. Our main objective was the research of a sustainable food product with a special nutritional purpose in the vision of nutrivigilance, developed in Romania, as an adjuvant in the repair of gastric mucosa. Methods: The materials used in the research and development of the new food are the following: inulin, lactoferrin, sericin, and sodium bicarbonate. The new adjuvant food product in the repair of the gastric mucosa was added to certain foods in order to prevent the patients from being satiated by a single food from a sensory point of view. The resulting food products were organoleptically and physico-chemically analyzed. Results: The new food is sustainable and has versatile uses. It can be hydrated with water, non-carbonated drinks, mixed with cottage cheese, or with fruit puree and oatmeal. It is stable under normal storage conditions and microbiologically safe. Conclusions: Through its versatile use, the new food product for special nutritional conditions represents a worldwide novelty. Through the development of forestry for the cultivation of white or black mulberry (*Morus alba* and *Morus nigra*), the raising of silkworms (*Bombyx mori*), the processing of fibroin to obtain natural silk and the processing of sericin resulting as a residue in the textile industry, the new food product developed actively contribute to the global economy II.

**Keywords:** food; gastric mucosa repair; sericin; lactoferrin; inulin; sodium bicarbonate; sustainable; global economy II

**Citation:** Matran, I.M.; Tarcea, M.; Rus, D.C.; Voda, R.; Muntean, D.-L.; Cirnatu, D. Research and Development of a New Sustainable Functional Food under the Scope of Nutrivigilance. *Sustainability* **2023**, *15*, 7634. https://doi.org/10.3390/su15097634

Academic Editor: Dimitris Skalkos

Received: 20 March 2023
Revised: 27 April 2023
Accepted: 28 April 2023
Published: 6 May 2023

## 1. Introduction

The New Global Economy is represented by a number of major features, such as the use of green energy, reducing the carbon footprint in all industrial and civil areas, as well as finding alternative food resources. Added to this is the rapid development of artificial intelligence in all economic sectors (agriculture, horticulture, animal husbandry, the food industry, pharmaceuticals, hospitals, the automotive and aircraft construction industries, and other industries-textiles, footwear, etc.), which has led and will lead to the loss of jobs and the decrease in people's quality of life, due to stress. Other causes that can increase the negative effects of stress are pandemics, such as the one caused by the SARS-CoV-2 (COVID-19) virus, and lifestyle changes.

Recent research [1] has shown that the COVID-19 pandemic has negatively affected the entire world economy, much more than the economic crisis of 2008. Although after its end, the world economy began to grow slowly, the geo-political conditions due to the context of the war in Ukraine started to change the global economic poles, towards China and India [2]. As presented by Ungureanu AV [3], regardless of the economic sector, an important part of the technological profile of the new global economy is driven by innovation and the entrepreneurial initiative that is at the heart of innovative business strategies. In Asian countries, such as China, Korea, India, and Vietnam, silk production is the largest in the world. This has great economic value [4]. In order to provide the raw material (fibroin) for obtaining silk, a very large number of *Bombyx mori* cocoons and a very large amount of mulberry leaves (*Morus alba*) as food for silkworms were required. Like any agricultural crop, including *Morus alba*, it requires specific agro-pedo-climatic conditions, such as soil composition, temperature, and rainfall. The white mulberry (*Morus alba*) grows well in temperate or subtropical regions of Asia, Africa, Europe, and North America, on light, loose, sufficiently moist soils. It also withstands beaten soils. Withstands transient flooding.

Due to the fact that the root is pivoting-trailing, the white mulberry (*Morus alba*) pulls its salts from the soil from certain depths, depending on the age, and has a long life (100 years [5]), in addition to the production of leaves needed as food for *Bombyx mori*, the white mulberry (*Morus alba*) culture contributes to the prevention of landslides (land). Mulberry wood (*Morus alba* and *Morus nigra*) is hard, resistant, and durable and processes and polishes well, being a particularly important material in the fields of the economy such as carpentry, handicrafts, office furniture, and musical instruments [5]. The fruits of the two varieties of mulberry (*Morus alba* and *Morus nigra*) are used in the food industry or in biotherapy.

In addition to these exceptional benefits, mulberry leaves (*Morus folium*) produce oxygen necessary for humans and animals, which leads to the possibility of establishing protection zones for localities or silvosteppe (intermediate vegetation zone between a steppe and a deciduous forest), but they are also used in the food of silkworms. The tender leaves (*Morus folium*) are picked without the petiole in the months of May–June and can be dried in the shade in a thin layer. Silkworms can be fed on tender leaves (*Morus folium*) of *Morus alba* and *Morus nigra*, and old leaves only of *Morus alba* [5].

The authors Matran IM et al. showed [6] the structure and processing method of the threads secreted by the silkworm gland in the textile industry and the importance of processing the sericin that results as a residue from this industry.

Currently, both in Romania and in other countries from the European Union (EU), as well as non-EU countries, functional foods are analyzed by manufacturers only regarding allergies. Currently, in the EU, nutrivigilance is regulated by law and implemented in France, Italy, Belgium, Slovenia, the Czech Republic, and Ireland. In these countries, risk assessments are carried out at the post-sale stage of food and food supplements. In France, the following are analyzed and monitored: food supplements (food supplements containing melatonin, food supplements containing spirulina, food supplements for pregnant women, food supplements for athletes, food supplements containing red rice yeast, food supplements containing p-synephrine), energy drinks, nutrient concentrates, plants or other substances in measured doses, foods or fortified drinks: foods supplemented with vitamins, minerals or other substances, amino acids or plant extracts, such as so-called energy drinks, vitamin D-enriched milk, certain nutrient-enriched vegetarian products, new foods and new ingredients: foods that were not consumed in Europe before 1997 or that were produced from new sources, with new substances or technologies, such as guar gum, noni juice, fruit pulp dehydrated baobab, products intended for food for specific categories of the population: preparations for infants, products for patients suffering from metabolic disorders or malnutrition, etc.

In Italy, natural products, herbal products, preparations from traditional Chinese or Ayurvedic medicine, dietary supplements, vitamins and probiotics, homeopathic medicines, medicinal preparations, or galenic masters are monitored. In Belgium, authorized food supplements are monitored.

Considering the recent research of sericin in the pharmaceutical and biomedical field, such as tissue engineering, wound healing, drug administration, and cosmetics [7], in this paper, we present the research of a sustainable food product with a special nutritional purpose in the vision of nutrivigilance, developed in Romania, as an adjuvant in the repair of the gastric mucosa. Another objective was to evaluate the versatility of usage of this new product, to prevent the appearance of sensory boredom (saturation) in adult patients, in the form of a single mono dose per day.

## 2. Materials and Methods

### 2.1. The Materials Used in the Research—Development of the New Food

2.1.1. Raw Materials and Food Ingredients

The materials used in the research and development of the new food are the following: inulin, lactoferrin, sericin, and sodium bicarbonate. The choice of raw materials was made based on the assessment of the state of knowledge using the PubMed and ResearchGate databases.

In order to be able to achieve the traceability of the finished product and the raw materials used, all related information (name of the product), batch/batch, manufacturing company, or importing company was recorded.

The information necessary to achieve the traceability of the finished product and the raw materials used (inulin, lactoferrin, sericin, and sodium bicarbonate) in the research and development of the new adjuvant food in the repair of the gastric mucosa is presented in Table 1.

**Table 1.** Traceability of the finished product (the new adjuvant food product in the repair of the gastric mucosa) and of the raw materials used.

| Name of the Raw Material | Identification Data | Manufacturer/Importer Company |
| --- | --- | --- |
| Inulin | Batch: RHBGD1BGD1 | Adams Vision SRL, Targu Mures |
| Lactoferrin 1 | Batch: 107CLXP | Frisland Campina/KUK Romania [1] |
| Sericin | Batch: S1911251 | Sollice Biotech, France |
| Sodium bicarbonate | Batch: A 02L 05 | Dr. Oetker, Romania |

[1] Lactoferrin is a newly authorized food ingredient in the European Union (EU) [8].

2.1.2. The Equipment Used

The equipment used was:

- Electronic balance with two decimal places, the brand "Digital Scale", capacity 500 g/0.01 g;
- pH-meter (0–14 pH), brand Adwa, manufacturer Adwa kft Romania;
- Electronic probe type thermometer, brand Checktemp, manufacturer Hanna Instruments Romania;
- Electronic refractometer (0–85% Brix), model HI 96801, manufacturer Hanna Romania;
- Magnetic stirrer, model Nachita, model no. 690/1, maximum capacity 2000 mL, Romania;
- Berzelius glasses with various volumes were used to carry out the experimental samples, Romania.

### 2.2. Methods

The sensory characteristics and physicochemical quality (Brix (refractometric soluble dry matter and pH)) of all the previously mentioned raw materials were analyzed. In addition, the stability of the finished product was checked under normal storage conditions (temperature between 20 to 25 °C, relative air humidity maximum 75%, and protection from direct sunlight or sources of frost).

Table 2 shows the methods and references that were applied for the sensory analysis and the physicochemical parameters (refractometric soluble dry substance (Brix) and pH) for the raw materials used in this research.

**Table 2.** The methods and references that were applied for the sensory analysis and the physicochemical parameters (refractometric soluble dry substance (Brix) and pH) for the raw materials used.

| Name of the Raw Material | Analysis Method | Analyzed Parameters | Reference |
| --- | --- | --- | --- |
| Inulin | 5% solution | Sensory Physicochemical | The quality document from the supplier |
| Lactoferrin | 2% solution | Sensory Physicochemical | The quality document from the supplier |
| Sericin | 10% solution | Sensory Physicochemical | The quality document from the supplier |

Tests such as acidity, product oxidation state, antioxidant activity, color measurement, protein content, fiber content, and carbohydrate content were not performed in this research. These will be carried out within a financing project to be submitted.

The other raw materials used (inulin, sericin, and sodium bicarbonate) were purchased from suppliers or local stores. When determining the amount of lactoferrin that must be added to the new adjuvant food in the repair of the gastric mucosa, the following aspects were taken into account: the maximum dose allowed per day, according to the Decision to place lactoferrin on the EU market as a new food ingredient [8], the goal is for the new food to be in the form of a single mono dose per day, and for the patients to be adults. According to this legislative regulation, in food products for special medical purposes, lactoferrin can be added at a maximum of 3 g/day. The preparation of this new adjuvant product in the repair of the gastric mucosa is shown in Figure 1.

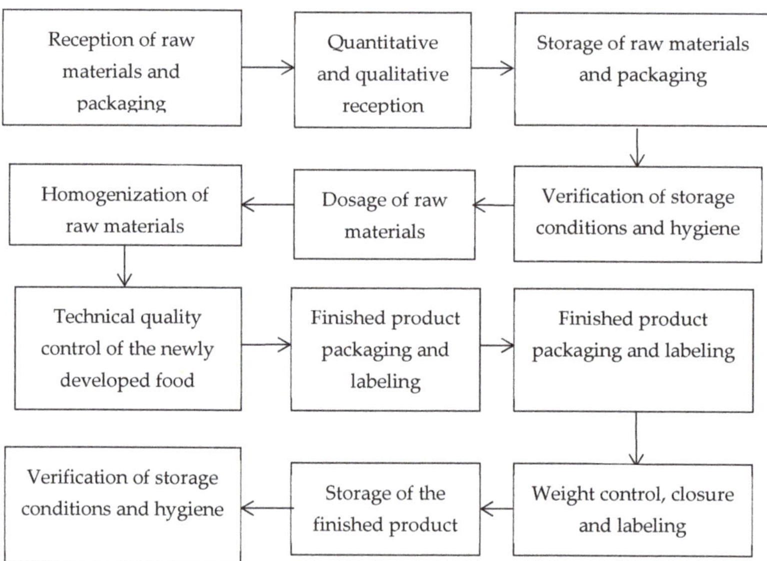


**Figure 1.** The preparation of the new adjuvant food in the repair of the gastric mucosa.

In the case of the quantitative and qualitative reception of raw materials and packaging, depending on the degree of non-conformity (minor or major), the treatment of the non-conforming product can re-submit a statement of findings or Non-conformity Sheet, including resolution of the complaint to the supplier or scrapping (destruction). The packaging is stored in different spaces compared to the raw materials. If the storage conditions (storage method, temperature, and relative air humidity) and hygiene conditions are non-compliant (both for raw materials, packaging, and the finished product), they will be remedied as soon as possible. Likewise, in the case of weight control, closing and labeling of packages with the finished product, and the adjuvant food used in the repair of the gastric mucosa.

To check the shelf life of the finished product (of the final recipe), a sample was analyzed microbiologically, at an accredited external laboratory, within the National Sanitary Veterinary and Food Safety Authority, Sanitary-Veterinary and Food Safety Directorate Mures, Romania, subordinated to the Ministry of Agriculture and Rural Development and the Government of Romania (Table 3). These analyses were carried out for the freshly prepared product, and will be repeated every 6 months, for 12 months, to verify whether, under normal storage conditions of the final product, microbiological changes have occurred or not.

When designing the adjuvant food in the repair of the gastric mucosa, from the point of view of sustainability, we considered that its manufacture, storage, and transport should not require special temperature conditions in order to prevent the use of related resources (e.g., gas, water).

**Table 3.** The analysis methods for the microbiological analysis related to the finished product applied by the accredited external laboratory within the National Sanitary Veterinary and Food Safety Authority, Sanitary-Veterinary and Food Safety Directorate Mures, Romania.

| The Analyzed Parameters | Analysis Method |
| --- | --- |
| Beta glucose-positive *Escherichia coli* | SR ISO 16649-2:2007 * (RA *) |
| *Enterobacteriaceae* | SR EN ISO 21528-2/2017 * (RA) |
| *Staphylococcus* coagulase-positive | SR EN ISO 6888-1:2021 * (RA) |
| Yeasts and molds | SR ISO 21527-2:2009 * (RA) |

* SR ISO 16649-2:2007—Romanian standard: Microbiology of food and feed. Horizontal method for the enumeration of beta-glucuronidase positive *Escherichia coli*, RA—RENAR accredited (Accreditation Association from Romania), SR EN ISO 21528-2/2017—Microbiology of the food chain—Horizontal method for the detection and enumeration of *Enterobacteriaceae*—Part 2: Colony-count technique, Microbiology of the food chain. Horizontal method for counting coagulase-positive staphylococci (*Staphylococcus aureus* and other species). Part 1: Method using Baird-Parker agar medium, Microbiology of food and animal foodstuff—Horizontal method for the enumeration of yeasts and molds—Part 2: Colony count technique in products with water activity less than or equal to 0.95 (ISO 21527-2:2008).

In addition to the sensory, physicochemical, and microbiological analyses, the finished product (new adjuvant food product in the repair of the gastric mucosa) was tested in simulated gastric fluid and physiological serum to analyze and verify its dissolution. The simulated gastric liquid formula was 7 mL concentrated HCl (36–37%), 2 mg NaCl, and the difference to 1000 mL distilled water. In both solutions, the newly developed food product completely dissolves. From the point of view of versatile uses, the following variants were made: the new product and cottage cheese, the newly developed food and non-carbonated drink, and the new product, mashed bananas, and oatmeal. The choice of versatile uses of a food for special nutritional conditions, thus designed, must be made in accordance with the European in force.

Following the tests (Brix and pH), technological tests in terms of processing and versatile use by patients, and also microbiological analyses, the final recipe was registered at the State Office for Inventions and Trademarks in Romania in order to obtain the Patent of Invention. Invention Patent Application registration number A/00589/07.11.2022.

For the validation of the laboratory model, on a reduced or increased scale, as appropriate, with the reproduction by the similarity of the real operating conditions (TRL 5) [9,10] and the following TRLs (6–9) [9,10], the control points (CP) were also analyzed and critical control points (CCPs).

## 3. Results

The other raw materials used (inulin, sericin, and sodium bicarbonate) were purchased from suppliers or local stores. When determining the amount of lactoferrin that must be added to the new adjuvant food product in the repair of the gastric mucosa, the following aspects were taken into account: the maximum dose allowed per day, according to the Decision to place lactoferrin on the EU market as a new food ingredient [8], the goal is for the new food to be in the form of a single mono dose per day, and for the patients to be adults. According to this legislative regulation, in food products for special medical purposes, lactoferrin can be added at a maximum of 3 g/day. In the case of the quantitative and qualitative reception of raw materials and packaging, depending on the degree of non-conformity (minor or major), the treatment of the non-conforming product can re-submit a statement of findings or Non-conformity Sheet, including resolution of the complaint to the supplier or scrapping (destruction).

The packaging is stored in different spaces compared to the raw materials. If the storage conditions (storage method, temperature, and relative air humidity) and hygiene conditions are non-compliant (both for raw materials, packaging, and the finished product), they will be remedied as soon as possible. Likewise, in the case of weight control, closing and labeling of packages with the finished product, the adjuvant food can be used in the repair of the gastric mucosa.

The results of sensory and physicochemical analyses (Brix and pH) of the raw materials used (inulin, lactoferrin, and sericin) can be followed in Tables 4–6. Sodium bicarbonate was not analyzed for taste nor for physicochemical parameters, because it is an additive food and has consistent quality, as well as conforming to the product specifications of each manufacturer or supplier.

**Table 4.** Results of sensory and physicochemical analyses (Brix and pH) of the inulin compound.

| Name of the Raw Material | Sensory Characteristics | Physico-Chemical Analyses [1] |
|---|---|---|
| Inulin | Fine powder, free of clumps and free of foreign particles<br>Color: White, homogeneous<br>Smell/Taste: Pleasant, specific | Brix: 4.73 (4.73, 4.72, 4.74)<br>pH [2]: 6.61 (6.61, 6.62, 6.62) |

[1] The values of the results of the physicochemical analyses represent the arithmetic mean of three consecutive determinations at the reference temperature of 20 °C. [2] The pH of inulin was analyzed on 5% solution.

**Table 5.** Results of sensory and physicochemical analyses (Brix and pH) of the lactoferrin compound.

| Name of the Raw Material | Sensory Characteristics | Physico-Chemical Analyses |
|---|---|---|
| Lactoferrin | Fine powder, free of clumps and free of foreign particles<br>Color: Light pink, homogeneous<br>Smell/Taste: Pleasant, specific | Brix: 2.43, 2.42, 2.44<br>pH [1]: 6.23, 6.24, 6.22 |

[1] The pH of lactoferrin was analyzed on a 2% solution, according to the applicable legislative regulation.

**Table 6.** The results of sensory and physicochemical analyses (Brix and pH) of the sericin.

| Name of the Raw Material | Sensory Characteristics | Physico-Chemical Analyses [1] |
|---|---|---|
| Sericin | Fine powder, free of clumps and free of foreign particles<br>Color: Light yellow, homogeneous<br>Smell/Taste: Pleasant, specific | Brix: 11.70, 11.60, 11.80<br>pH [1]: 5.45, 5.44, 5.46 |

[1] The pH of sericin was analyzed on a 10% solution, according to the product specification/quality certificate received from the supplier.

Sericin and lactoferrin have anti-inflammatory action on several interleukins (e.g., IL-1, IL-4, IL-5, IL-6, IL-8, IL-10, IL-13, IL-17, IL-31); these leading to humoral and eosinophil inflammation, mucosal damage and the production of adaptive cellular inflammation—tumor necrosis factor-alpha (TNF-$\alpha$) and interferon-gamma (IFN-$\gamma$), cytotoxic T lymphocytes (CTLA-8) and activated CD4+ T cells, and more specifically, CD4+CD45RO+ T cells, and CD19 and CD56 cells. Sodium bicarbonate has a systemic, rapid antacid action.

Table 7 shows the recipe for the new product. This recipe is protected at State Office for Inventions and Trademarks, and we are about to receive the Patent of the Invention.

**Table 7.** The recipe of the new adjuvant food product in the repair of the gastric mucosa.

| Ingredient | Quantity, g |
|---|---|
| Inulin | 10 |
| Lactoferrin | 3 |
| Sericin | 9 |
| Sodium bicarbonate | 6 |
| Total ingredients | 28 [1] |

[1] This is the amount of a single dose that can be consumed in one day.

Figure 2 shows the new adjuvant food product developed by us.

**Figure 2.** The newly developed food is adjuvant in the repair of the gastric mucosa. Version V 3.4. represents the notation during research.

Physicochemical analyses for the new food product can be found in Table 8 and Figure 3. Hydration was carried out with 100 mL of water at a temperature of 20–25 °C.

**Table 8.** Physicochemical analyses for the new food.

| The Moment of Analysis | Brix | pH |
|---|---|---|
| Immediately after moisturizing | 22.30, 22.10, 22.30 | 7.41, 7.42, 7.42 |
| After 5 h of hydration | 22.60, 22.50, 22.70 | 7.8, 7.7, 7.9 |

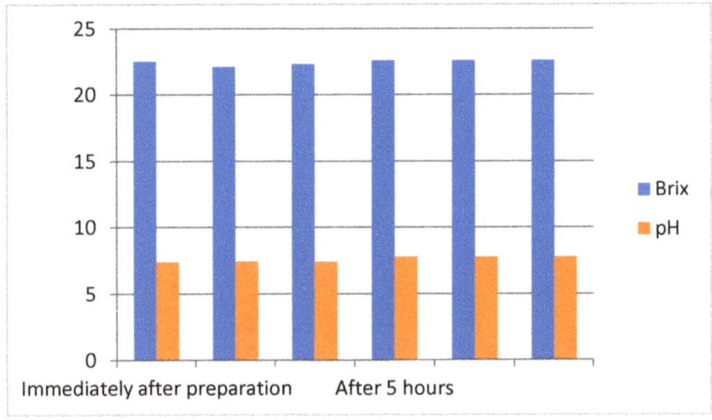


**Figure 3.** Changes in the physicochemical parameters of the newly developed food.

Under normal storage conditions (temperature between 20 and 25 °C, relative air humidity maximum 75%, and protection from direct sunlight or sources of frost), the adjuvant food was stable and does not form agglomerations or crystals.

In order to establish the validity period of the new product, microbiological analyses were carried out at the Mures Sanitary-Veterinary and Food Safety Laboratory. The results obtained are presented in Table 9.

**Table 9.** Results of the microbiological analyses carried out for the newly developed product after it has been prepared.

| Analyzed Parameters | Result, ufc/g [1] |
|---|---|
| *Escherichia coli beta-glucuronidase positive* | <10 |
| *Enterobacteriaceae* | <10 |
| *Coagulase-positive staphylococci* | <10 |
| Yeasts and molds | <10 |

[1] Analysis bulletin no. 21964 of 10 October 2022.

In addition to the sensory, physicochemical, and microbiological analyses, the finished product (the new adjuvant food in the repair of the gastric mucosa) was tested in simulated gastric fluid and physiological serum, to analyze and verify its dissolution. The simulated gastric liquid formula is: 7 mL concentrated HCl (36–37%), 2 mg NaCl and the difference to 1000 mL, distilled water. In both solutions, the newly developed food completely dissolves.

From the point of view of versatile uses, the following variants were made: the new product and cottage cheese (Figures 4 and 5), the newly developed food, non-carbonated drink (Figures 6 and 7 and Table 10), and the new food, banana puree, and oat flour (Figures 8 and 9 and Table 11).

**Figure 4.** The newly developed food is mixed with cottage cheese, raspberry jam with inulin, and no preservatives. The leaves in the picture are mint leaves (*Mentha piperita*). It was added for chromotherapy (color) and aroma diversification.

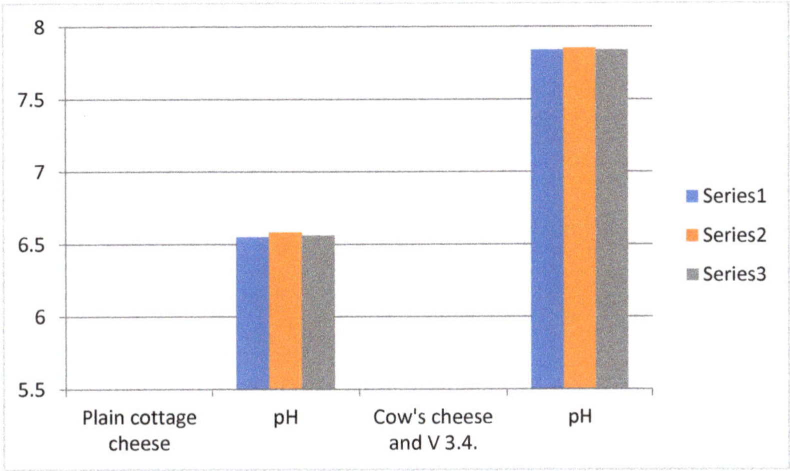



**Figure 5.** Comparative analysis of the pH of cottage cheese and cottage cheese mixture and the newly developed food. Series 1..3 represent the three values measured for the PH of Plain cottage chesse and cow's cheese and V 3.4.

**Table 10.** The results of the physicochemical analyses of the non-carbonated drink used and the new food developed, hydrated with this drink.

| Food | Brix | pH |
|---|---|---|
| Non-carbonated drink | 5.25; 5.2; 5.3 | 4.04; 4.04; 4.04 |
| The newly developed food, hydrated with the non-carbonated drink | 26.4; 26.4; 26.35 | 8.01; 8.02; 8.0 |

**Table 11.** The results of the physicochemical analyses of the newly developed food mixed with mashed bananas and oat flour.

| Food [1] | Brix | pH |
|---|---|---|
| Mashed bananas | 18.8; 19.3; 19.8 | 6.48; 6.45; 6.46; |
| The new mixed food with mashed bananas and oat flour | 39.96; 39.95; 40.5 | 8.18; 8.16; 8.17 [1] |

[1] For oat flour, we were unable to analyze refractometric soluble dry matter (Brix) and pH.

The finished product (mixture) has a lower viscosity compared to cottage cheese. The consistency of the final product is optimal for the enteral nutrition of small children and patients with dysphagia or certain dental pathologies. It is ideal for dietary/functional desserts or regular foods. Lactoferrin and sodium bicarbonate are authorized in the EU, for the food category: food for special medical purposes, cheese-based products, cakes, and pastries.

This product (mixture of cottage cheese and V.3.4) can also be used as a dessert in the hospital for patients with dysphagia. The sensory characteristics of the finished product are a creamy appearance, a golden/light yellow color, with a specific smell and taste of fresh cow's cheese. The product was kept cold until the 2nd day. In the case of the cow's cheese and the mixture with the new product, the refractometer soluble dry matter (Brix) could not be determined due to air entrapped in the products.

**Figure 6.** The appearance and color of the used non-carbonated drink and of this drink with the new adjuvant food in the repair of the gastric mucosa developed.

The color of the liquid changed due to the change in pH (basicity). It is a normal property of natural dyes. The foam of the mixture is due to lactoferrin which has a foaming capacity. It is not stable and decreases rapidly.

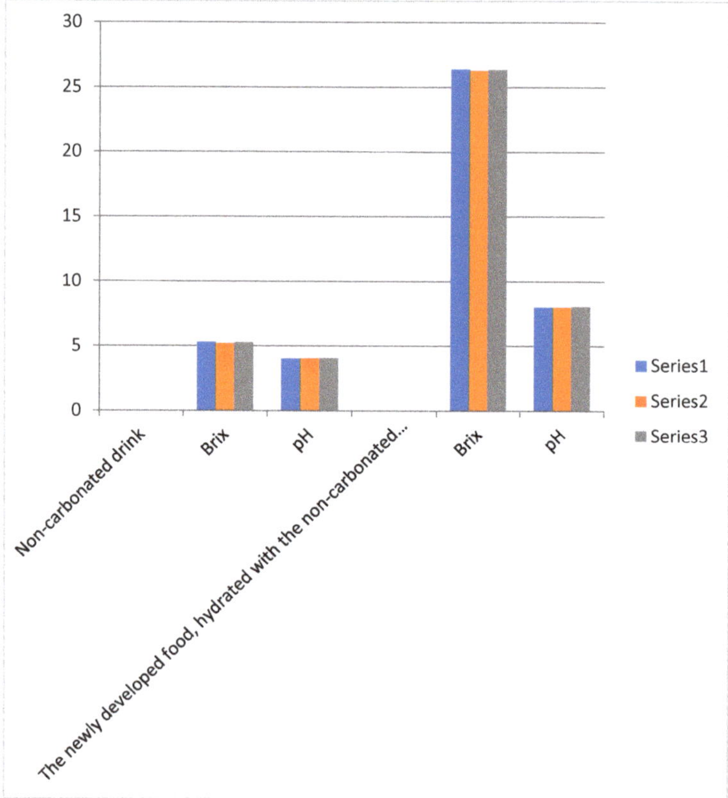


**Figure 7.** Variation of refractometric soluble dry matter and pH in the still drink and the newly developed food product hydrated with the same still drink. Series 1..3 represent the three measured values for the PH and Brix of the still drink and the newly developed food hydrated with the still drink.

The increase in refractometric dry matter (Brix) value in the newly developed food hydrated with the non-carbonated beverage is explained by the sugar content of the added beverage as well as the chemical reactions occurring between the two foods. This information is useful for health professionals (doctors, pharmacists, and dieticians) to know what to recommend to patients who have associated pathologies, including diabetes. The increase in pH is due to the composition of the newly developed food, in particular sodium bicarbonate.

The identification data related to the cottage cheese and strawberry jam with inulin used are as follows: cottage cheese, Romanian Ibanesti brand, and 32% fat. For this pod, the batch was not mentioned because the cheese was loose. For the strawberry jam with inulin, the identification data are as follows: Ver-mondo brand, produced in the EU for Lidl Discount SRL, Romania, expiration date 2 May 2025, lot 122 B3. For the "Ciao" brand non-carbonated drink: batch 204023 RS D and expiring date 4 February 2023. This drink contains fruit juices (apples and raspberries), with sugar and sweeteners: water, sugar and/or glucose-fructose syrup, fruit juices (4%) obtained from concentrated apple juice (3%) and raspberries (1%), acidity correctors: citric acid, sodium citrates, black carrot concentrate for color, vitamin C, flavor, sweeteners: Cyclamates and Saccharins.

The mixture consisting of the newly developed food, mashed bananas, and oat flour, as well as the results of the physicochemical analysis are presented in Figure 8 and Table 9.

**Figure 8.** The new food developed in a mixture of mashed bananas and oat flour.

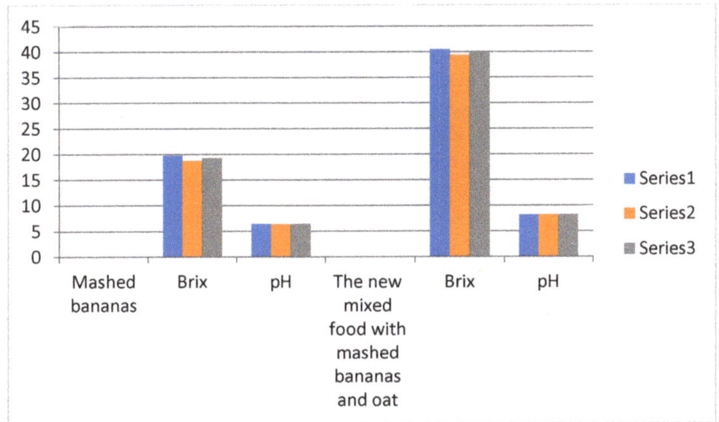


**Figure 9.** Variation of pH and Refractometric Soluble Dry Matter (Brix) for Banana Mash and the Newly Developed Food Mashed Banana and Oat flour Blend.

Dietary dessert was obtained from 100 g of mashed bananas, 28 g of the new food, and 70 g of oat flour.

Regarding the elements for achieving the traceability of the raw materials (bananas, new food, and oat flour): The bananas were purchased from the grocery store. These were bulk, without identification data for traceability, printers on the receipt from the cash register. The puree was prepared by hand from bananas. Oatmeal was obtained in the laboratory from ground oat flakes. The identification data for the oat flakes used are *Sanovita* brand. Made in Germany. Packaged and distributed by S.C. Sano Vita SRL. OF: 7 March 2023, their 124 2.

## 4. Discussion

Our research concerns a new adjuvant food for the repair of gastric mucosa used in medicine and pharmacy. Before the research and development of the new food, the state of knowledge was checked for the products sold in other countries and the existing invention patents worldwide. The results of this check can be followed in the lines below:

Alternative products marketed in other European or non-European countries are:

- "GI Repair Powder I 168g", manufacturer Vital Nutrients [11], with zinc composition and patented blend (L-glutamine powder, N-acetyl glucosamine, slippery elm bark powder, MSM (OptiMSM®), NF rutin, lactoferrin, aloe vera leaf inner fillet extract, xylitol. According to the manufacturer's recommendations, taking this formula helps support gastrointestinal health and intestinal lining cells. This supplement contains nutrients, including amino acids with botanicals and a stable probiotic. There are no mentions of verification of effectiveness in vitro or in vivo, and it contains ingredients that may cause adverse reactions and may interact with certain drugs and alter their clearance, and adversely affect drug treatment. In this sense, I exemplify aloe vera. It can cause gastrointestinal disturbances, arrhythmias, nephropathy, and edema and can interact with antidiabetic drugs, which can lead to hypoglycemia. Zinc may cause adverse effects such as nausea, vomiting, metallic taste, and sideroblastic anemia and may interact with quinolones or tetracycline and lead to decreased antibiotic absorption [12].
- MSM is the abbreviation for methylsulfonylmethane. Adverse effects observed may include bloating, constipation, indigestion, headache, fatigue, or insomnia. The most notable adverse effect recorded for this ingredient was an acute episode of bilateral iridocorneal angle closure, [13] leading to the need/recommendation to include a warning to this effect, and the product label "GI Repair Powder" does not present this information. Among the side effects of slippery elm, I mention the possibility of allergic reactions and contact dermatitis [11]. In addition, according to the label, the product "GI Repair Powder" contains many allergens (fish, shellfish, lobster, crab, and/or shrimp) and lactoferrin and is not evaluated by the FDA (Food and Drug Administration).
- Patent US20160228490 [14]—"Methods and composition for treating mucosal tissue disorders", refers to pharmaceutical compositions containing glutathione, ascorbate, and bicarbonate with or without thiocyanate and methods of using them to treat diseases and disorders in mucosal tissue. The disadvantage of this patent is that the number of patients with gastrointestinal pathologies, including those who refuse drug treatments and are adherents of alternative treatments such as food and/or food supplements, is constantly increasing, which means that the segment of patients who buy drugs is decreasing;
- A patented composition (US20190125820 [15])—"Powder for regulating intestinal flora and protecting gastric mucosa, preparation method and use thereof", has as ingredients: xylooligosaccharide, isomalto-oligosaccharide, mannitol, inulin, *Codonopsis* root, *Hypericum erinaceus* fruit extract, *Dioscorea opposita* rhizomes, *Sclerotium poria*, *Semina dolichos*, *Pericarpicerum amicium*, *Corpizum* fruit and *Hericium erinaceus* fruit. Experimental data show that the powder described in the present disclosure has the function of ameliorating gastric mucosal injury by reducing the acute ethanol-induced gastric mucosal injury effectively and has the function of regulating the gastric mucosa.

Therefore, the powder could be used to prepare a healthcare product that has the function of regulating the gastrointestinal tract and protecting the gastric mucosa. The disadvantages of this patent are represented by the multiple drug interactions it can cause, such as *Codonopsis* can slow blood clotting. Taking *Codonopsis* with medicines that slow and clot blood could increase the risk of bruising and bleeding. Another drug that may interact with this ingredient in the proprietary product is a cancer drug called Abiraterone. The interaction between *Codonopsis* and this medicine may reduce the effectiveness of Abiraterone in the treatment of cancer. In the case of people who have conditions that can be aggravated by estrogen, this patented composition containing *Codonopsis* should not be used/consumed because it contains rhizomes of *Dioscorea opposita* [16].

Phytotherapeutic products manufactured in Romania, or manufactured in other countries and marketed in our country, are:

- Reglacid—Hofigal, 60 capsules with the composition for one capsule: contains powder from 60.00 mg of sea buckthorn fruit (*Hippophae fructus*), 50.00 mg of the bird's-eye cuticle, 60.00 mg of chamomile flower (*Chamomillae flos*), 52.00 mg St. John's wort (*Hyperici herba*), 52.00 mg amaranth (*Amaranthus caudatus herba*), 1.2 mg thyme volatile oil (*Thymi vulgaris aetheroleum*), 0.8 mg lavender volatile oil (*Lavandulae aetheroleum*) and excipients (lactose, polyvinylpyrrolidone K30, magnesium carbonate, talc, magnesium stearate) up to 400.00 mg. Action: The product was intended to supplement the diet for its properties: antiseptic and anti-inflammatory in the gastrointestinal tract; to improve the symptoms of some gastrointestinal conditions and to reduce the risk of complications and evolution towards more severe forms; the product regulates gastric acidity, improves digestion, protects the gastrointestinal mucosa and stimulates its regeneration; depurative, slightly laxative and diuretic. There is no mention of verification of effectiveness in a clinical trial. Administration: 1–2 capsules three times a day or two tablets two times a day, 10–15 min before meals depending on the stage and nature of the digestive disease or on the recommendation of the doctor who evaluates the dose and the rate of administration. Storage conditions: at room temperature (15–25 °C), protected from moisture and light, in the original packaging. Manufacturer: Hofigal, Romania.
- Gastracid D100, 63 chewable tablets—Fares with the composition: clay, chamomile flowers (*Matricariae flos*), soft extract of chamomile flowers (*Matricariae flos*), soft extract of licorice root (Liquiritiae radix), tincture of propolis, calami rhizomes (*Calami rhizoma*), fennel essential oil (*Foeniculi aetheroleum*), mint essential oil (*Menthae aetheroleum*), excipients (bulking agent: cellulose, starch, anti-caking agent: talc). Action: Gastracid tablets (clay, chamomile flowers, soft extract of chamomile flowers, soft extract of licorice root, propolis tincture, rhizomes of oleander, fennel essential oil, peppermint essential oil) help to neutralize gastric acid with reduction of discomfort and unpleasant sensations from the gastroesophageal level; protects the lining of the stomach and esophagus and promotes healthy digestion. There is no mention of verification of effectiveness in a clinical trial. Administration: Take one tablet three times a day 30 min before meals or as needed. Manufacturer: Fares, Romania.
- Healthy Stomach (Ulcerofit) 7 Fares with the composition/capsule: calendula flowers (*Calendulae flos*) 40 mg, plantain leaves (*Plantaginis folium*) 40 mg, mugwort grass (*Mycelis muralis herba*) 40 mg, St. John's wort (*Hyperici herba*) 40 mg, hyssop (*Hyssopi herba*) 20 mg, licorice root (*Liquiritiae radix*) 20 mg. Action: This product acts synergistically through several mechanisms. It reduces the inflammation of the gastric mucosa through faradiol present in marigolds, active substances from plantain and robber grass (chlorogenic acid, neo chlorogenic, apigenin), glycyrrhizic acid from licorice root, amentoflavone from St. John's wort and essential oil from hyssop. They favor the healing of mucosal erosions and increase the secretion of mucus with a gastric protective role through carotenoids, mucilages, chlorogenic and neochlorogenic acid from the composition of calendula and plantain. Thieves' grass is a traditional remedy

used to heal ulcers. Added to these are the antispasmodic and stomach pain-soothing action due to the active principles of sedum grass, calendula, and licorice root. Clinical studies: Following the study carried out at the III Cluj-Napoca Medical Clinic, it became evident that Stomac Sănătos is an effective preparation in the treatment of peptic diseases (gastritis and erosive duodenitis, duodenal ulcers). Administration: Children between 6–14 years: one capsule three times a day; Adults: two capsules three times a day. The administration is done 15 min before the main meals. The duration of a cure is 4–6 weeks, according to the doctor's recommendation. Manufacturer: Fares, Romania.

Although worldwide there are invention patents for foods containing sericin, lactoferrin, inulin, and sodium bicarbonate, worldwide there is no food designed like the new food developed by us, an adjuvant in rapping the gastric mucosa with versatile use.

Existing patents are for healthy consumer (personal) foods and do not have versatile applicability. Examples of patented foods for healthy consumers are JP2000184868—Water for food and drink [17], CN107048419—Sericin drink [18], JP2000312568—Hardly digestive additive for and beverage, and auxiliary health product [19], CN107034102—Sericin wine [20], EP2025246—Multi-component dessert product [21], RU0002640872—Cream and vegetable spread with inulin [22], RU0002623739—Method of processing inulin-containing raw material with obtaining food inulin powder and method of obtaining ultra-pure inulin [23], CN1875748—Application method of lactoferrin in food [24], CN114431296—Preparation method of liquid dairy product and liquid dairy product [25].

The mixture/cream cheese with the new food developed by us has lower viscosity compared to cow's cheese. The consistency of the finished product is optimal for enteral feeding of young children and patients who have dysphagia. This food can be prepared at home, in Hotel Restaurant Canteen units, or in hospital kitchens. The finished product is ideal for dietary/functional desserts or regular foods. Lactoferrin and sodium bicarbonate are authorized in the European Union for the food categories: dairy-based food intended for young children (ready-to-eat), food for special medical purposes, cheese-based products, cakes, and pastries.

For children or to diversify the taste, any type of jam can be added. For the sensory (organoleptic) analysis, I added strawberry jam with inulin (without sugar).

The large variation in Brix related to the mixture resulting from the hydration of the new food with the non-carbonated drink is explained by the reactions that took place between the ingredients of the new food subject to the invention and the ingredients of the non-carbonated drink (water, sugar and/or glucose-fructose syrup, fruit juices (4%) obtained from concentrated apple (3%) and raspberry (1%) juices, acidity correctors: citric acid, sodium citrates, black carrot concentrate for color, vitamin C, sweeteners: cyclamate and saccharin). This drink was purchased commercially and was chosen as a possibility to hydrate the new food subject to the invention to diversify the taste.

The newly developed food is natural and easy to be administered. It does not contain preservatives, gluten, or other synthetic substances. It was presented in the form of a powder without foreign particles, with pleasant sensory characteristics specific to the ingredients, also was microbiologically safe and was stored in clean, sanitized, and disinfected spaces, away from heat sources and the direct action of the sun's rays, with a maximum temperature of 25 °C and a relative air humidity of 75%. The product was transported in clean, sanitized means of transport at a maximum temperature of 25 °C. It does not pollute the environment, and as a result, no special precautions are required for the disposal of residues. The food can be administered once a day, after meals, or when needed.

By applying the invention, the following advantages are obtained:

- Synergistic effect both on the oral cavity and on the stomach and large intestine [26];
- It has a systemic action to quickly reduce acidity, both in the oral cavity and in the stomach, thanks to sodium bicarbonate [26];
- Stimulates remineralization of tooth enamel, thanks to sodium bicarbonate which increases the pH in the oral cavity [26];

- Stimulates the absorption of minerals and including calcium, thanks to sodium bicarbonate [26];
- Prevents/treats constipation, including that caused by an adverse reaction by drugs (e.g., opiates or other drugs), thanks to sericin and lactoferrin [27];
- It is prebiotic and improves the intestinal microflora and the gut-brain connection, thanks to inulin [28];
- It is natural and easy to administer;
- It has versatile uses, being able to be used for food preparation, including by patients at home, or in hospital kitchens, for patients suffering from gastroesophageal reflux disease, gastritis, ulcers, constipation, or dysphagia;
- Does not contain preservatives, gluten, or other synthetic substances (e.g., preservatives);
- Does not require special storage conditions (e.g., refrigeration/freezing);
- The ingredients used do not show adverse reactions or interactions with over-the-counter medicines;
- It can also be consumed by people who have diabetes because it has a low glycemic index;
- Reduces the level of plasma lipids, reducing the risk of atherosclerosis due to inulin [29,30];
- Prevents cancer, thanks to lactoferrin, sericin, inulin, and sodium bicarbonate [30];
- Regulates the immune system thanks to sericin, inulin, and lactoferrin [30,31];
- Is an alternative to drugs such as proton inhibitors, which have numerous side effects and contraindications [30]. It does not precipitate in an aqueous environment due to the ratio of the components;
- It has anti-Gram + and anti-Gram negative—antibacterial, also antiviral actions thanks to sericin and lactoferrin [30];
- It has an anti-inflammatory effect, reducing pro-inflammatory cytokines, tumor necrosis factor-alpha (TNF-$\alpha$), and interferon-gamma (IFN-$\gamma$), thanks to sericin and lactoferrin [31]

Currently, there are no proprietary foods for the repair of gastric mucosa, patented and with versatile applications.

Because dental erosions are a common symptom of gastroesophageal reflux disease, it is necessary to address this extraesophageal manifestation of this disease. According to the state of the art presented previously, no product/supplement acts on the protection of the improvement of the intestinal microflora and of the gut-brain axis and does not prevent/treat constipation, which can result in the appearance of gastroesophageal reflux and gastric acidity, with the appearance other complications (e.g., ulcer) [10,12].

From the point of view of nutrivigilance, the label of the new food must contain information about possible allergens (lactoferrin), and health professionals must inform patients how this food should be consumed, including from the point of view of possible drug interactions. We recommend that their administration be done at an interval of one hour after the ingestion of the new food. As possible unwanted but harmless side effects, we mention flatulence or abdominal bloating, in sensitive patients, due to inulin. Another possible adverse reaction is an increased frequency of defecation (laxative effect) [3].

The research of the new adjuvant food in the repair of gastric mucosa represents industrial research. Future research directions are the preclinical evaluation of the efficacy of the new food in Wistar rats and randomized double-blind versus placebo clinical evaluation.

Regarding the Technology Maturity Level (TRL), for this research was 4. This grade 4 is defined as TRL 4 validation under laboratory conditions of the components and/or the assembly/system. Its description is: The main components of the technology are integrated to establish the functionality of the assembly. This approach may have a relatively low degree of fidelity compared to the real system. For example, separate components are integrated into the laboratory, and tests are carried out in a range of operating conditions. Deliverables include test results for the component assembly, highlighting proximity (or differences) to expected functionality and performance. TRL 4–6 is the bridge between scientific research and engineering/practical application. TRL 4 is the first step in determining whether the assembly of individual components is functioning properly as a system.

The lab system will most likely be a mix of existing (more general purpose) equipment and components that require handling, calibration, alignment, etc., specific to become functional [9,10].

In our case, the deliverables that validated TRL 4, in laboratory conditions of our final recipe, the analyses, the technological samples on various applications/foods, the photos taken of the foods that contain the food developed by us, and the bulletin as well, were carried out of analyses received from the Mures Sanitary-Veterinary and Food Safety Laboratory.

## 5. Conclusions

Through its versatile use, the new food product for a special nutritional state represents a sustainable worldwide novelty. Due to the development of forestry for the cultivation of white or black mulberry (*Morus alba* and *Morus nigra*), the development of their wood processing, the raising of silkworms (*Bombyx mori*), the processing of fibroin to obtain natural silk and the processing of sericin resulting as a residue in the textile industry, the new food actively contribute to the global economy II.

## 6. Patents

The work reported in this manuscript is intellectually protected by patent application number A/00589/07.11.2022.

**Author Contributions:** Conceptualization, I.M.M.; methodology, I.M.M.; M.T. and D.-L.M. software; validation, M.T., D.-L.M. and D.C.; formal analysis, I.M.M.; investigation, D.C.R.; resources, I.M.M., M.T., D.-L.M. and D.C.; data cleaning, M.T.; writing—original preparation of the project, R.V. and D.C.R.; writing, M.T.; visualization, D.-L.M. and D.C.; supervision, I.M.M.; project management, I.M.M. All authors have read and agreed to the published version of the manuscript.

**Funding:** This research received no external funding.

**Institutional Review Board Statement:** Not applicable.

**Informed Consent Statement:** Not applicable.

**Data Availability Statement:** Not applicable.

**Acknowledgments:** All authors thank the KUK Company Romania for the sample of lactoferrin offered for research.

**Conflicts of Interest:** The authors declare no conflict of interest.

## References

1. Boshkoska, M.; Jankulovski, N. Coronavirus Impact on the Global Economy. Annals of the "Constantin Brâncuşi" University of Târgu Jiu, Economy Series, Issue 4/2020, pp. 18–24. Available online: http://eprints.uklo.edu.mk/6221/1/02_Boshkoska-Published.pdf (accessed on 5 March 2023).
2. Siddiqui, K. The Ukraine-Russia War and Its Impact on the Global Economy. *World Finance Rev.* **2022**, *48*, 22–34.
3. Ungureanu, A.V. Entrepreneurship in the New Global Economy. The Role of Innovation in Economic Development. *An. Stiintifice Ale Univ. Ovidius Constanta Ser. Econ.* **2020**, *XX*, 541–548.
4. Dong, Z.; Zhao, P.; Zhang, Y.; Song, Q.; Zhang, X.; Guo, P.; Wang, D.; Xia, Q. Analysis of proteome dynamics inside the silk gland lumen of *Bombyx mori*. *Sci. Rep.* **2016**, *6*, 21158. [CrossRef] [PubMed]
5. Parvu, C. White mulberry. Black mulberry. In *The World of Plants*, 1st ed.; Asab: Bucharest, Romania, 2006; pp. 231–233.
6. Matran, I.M.; Matran, C.; Tarcea, M. Sustainable Prebiotic Dessert with Sericin Produced by *Bombyx mori* Worms. *Sustainability* **2023**, *15*, 110. [CrossRef]
7. Andreia, S.; Costa, E.C.; Reis, S.; Spencer, C.; Calhelha, R.C.; Miguel, S.P.; Ribeiro, M.P.; Barros, L.; Vaz, J.A.; Coutinho, P. Silk Sericin: A Promising Sustainable Biomaterial for Biomedical and Pharmaceutical Applications. *Polymers* **2022**, *15*, 4931.
8. European Commission. The Commission's Implementing Decision of 22 November 2012 Authorizing the Placing on the Market of Bovine Lactoferrin as a New Food Ingredient Pursuant to Regulation (EC) no. 258/97 of the European Parliament and of the Council (FrieslandCampina). OJ 2012, L(327), pp. 52–54. Available online: https://eur-lex.europa.eu/legal-content/RO/TXT/?uri=CELEX:32012D0727 (accessed on 5 March 2023).
9. European Commision. H2020. Work Programme 2016–2017, General Annexes, pp. 35/44. Available online: https://ec.europa.eu/programmes/horizon2020/en/what-workprogramme (accessed on 5 March 2023).

10.    US Department of Energy. Technology Readiness Assessment Guide, Department of Energy, U.S, pp. 9–10/34. Available online: https://www.directives.doe.gov/directives-documents/400-series/0413.3-EGuide-04 (accessed on 7 March 2023).

11.    SupplementHub-GI Repair Powder. Available online: https://supplementhub.com/products/gi-repair-powder-168g?variant=40819406110888&gclid=EAIaIQobChMI0tW5u6is-gIVyeR3Ch0mzwoNEAQYAiABEgICXvD_BwE (accessed on 5 March 2023).

12.    Morgovan, C.; Ghibu, S.; Juncan, A.M.; Rus, L.L.; Butuca, A.; Vonica, L.; Munteab, A.; Moș, L.; Gligor, F.; Olah, N.K. Nutrivigilance: A new activity in the field of dietary supplements. *Farmacia* **2019**, *67*, 537–544. [CrossRef]

13.    Cretu, M. Preparatele cu MSM, Efecte Terapeutice și Posibile Limitări [MSM Food Products, Therapeutic Effects and Possible Limitations]. Galenus. Available online: https://www.revistagalenus.ro/farmacoterapie/preparatele-cu-msm-efecte-terapeutice-si-posibile-limitari/ (accessed on 8 March 2023).

14.    Arnold, R.R.; Henke, D.C. Methods and Compositions for Treating Tissue Disorders. U.S. Patent 20160228490, 12 May 2022.

15.    Wang, L.; Ruiqi He, R.; Zhao, H.; Tang, Q. Powder for Regulating Intestinal Flora and Protecting Gastric Mucosa, Preparation Method and Use Thereof. U.S. Patent 20190125820, 2 May 2019.

16.    WebMD-Wild Yam-Uses, Side Effects and More. Available online: https://www.webmd.com/vitamins/ai/ingredientmono-970/wild-yam (accessed on 8 March 2023).

17.    Higeyuki, Y.; Yorikiyo, F.; Masakazu, N. Water for Food and Drink. Patent JP2000184868, 2000.

18.    Wen, Z. Sericin Drink. Patent CN107048419, 7 December 2011.

19.    Masahiro, S.; Hodeyuki, Y.; Masakazu, N. Hardly Digestive Additive for and Beverage, and Auxiliary Health Product. Patent JP2000312568, 26 December 2007.

20.    Wen, Z. Sericin Wine. Patent CN107034102, 11 August 2017.

21.    Eisner, M.D.; Van de Ven Martinus, J.M.; Eisner, M.D. Multi-component Dessert Product. Patent EP2025246, 16 January 2013.

22.    Tereshchuk, L.V.; Starovojtova, K.V.; Dolgolyuk, I.V. Cream and Vegetable Spread with Inulin. Patent RU0002640872, 2018.

23.    Starovojtov, V.I.; Mitrofanov, N.M. Method of Processing Inulin-Containing Raw Material with Obtaining Food Inulin Powder and Method of Obtaining Ultra-Pure Inulin. Patent RU0002623739, 2017.

24.    Yang, L. Application Method of Lactoferrin in Food. Patent CN1875748, 6 July 2011.

25.    Ba, G.; Wang, Y.; Jin, L. Preparation Method of Liquid Dairy Product. Patent CN114431296, 6 May 2022.

26.    Hațieganu, E.; Stecoza, C.E. Medicines used in diseases of the digestive system. Antiulcerative. In *Pharmaceutical Chemistry*; Zamfirescu, M., Șaramet, I., Eds.; Medical: Bucharest, Romania; Volume 2, pp. 381–390.

27.    Matran, I.M.; Boțan, E. Applicability of sericulture in the food industry and medicine. *Lucr. Științifice Ser. Agron.* **2017**, *1*, 183–186.

28.    Birkeland, E.; Gharagozlian, S.; Birkeland, K.I.; Valeur, J.; Måge, I.; Rud, I.; Aas, A.M. Prebiotic effect of inulin-type fructans on faecal microbiota and short-chain fatty acids in type 2 diabetes: A randomised controlled trial. *Eur. J. Nutr.* **2020**, *59*, 3325–3338. [CrossRef] [PubMed]

29.    Lai, S.; Mazzaferro, S.; Muscaritoli, M.; Mastroluca, D.; Testorio, M.; Perrotta, A.; Esposito, Y.; Carta, M.M.; Campagna, L.; Di Grado, M.; et al. Prebiotic Therapy with Inulin Associated with Low Protein Diet in Chronic Kidney Disease Patients: Evaluation of Nutritional, Cardiovascular and Psychocognitive Parameters. *Toxins* **2020**, *12*, 381. [CrossRef] [PubMed]

30.    Matran, I.M.; (Iuliu Hațieganu, University of Medicine and Pharmacy, Romania, Cluj-Napoca); Dumitrascu, D.L.; ("Iuliu Hațieganu", University of Medicine and Pharmacy, Romania, Cluj-Napoca). Personal Commnucation, 2015.

31.    Matran, I.M.; (George Emil Palade University of Medicine, Pharmacy, Science, and Technology, Romania, Targu Mures); Rotaru, M.; (Lucian Blaga University of Medicine and Pharmacy, Romania, Sibiu). Personal Communication, 2020.

 *sustainability*

*Article*

# Greek Semi-Hard and Hard Cheese Consumers' Perception in the New Global Era

**Dimitris Skalkos** [1,*], **Katerina Bamicha** [1], **Ioanna S. Kosma** [1] **and Elpida Samara** [2]

[1] Laboratory of Food Chemistry, Department of Chemistry, University of Ioannina, 45110 Ioannina, Greece; pch01419@uoi.gr (K.B.); i.kosma@uoi.gr (I.S.K.)

[2] Department of Regional Development and Cross Border Studies, University of Western Macedonia, 50100 Kozani, Greece; esamara@uowm.gr

* Correspondence: dskalkos@uoi.gr; Tel.: +30-(26)-51008345

**Abstract:** The COVID-19 pandemic is almost over but has already left its mark and is changing the world fast and drastically in all social, economic, and cultural aspects of humanity, including consumers' choices and motives for foods. Since cheese is a major dietary food consumed daily worldwide, motives for its purchase and consumption in the new era are an important parameter affecting current and future production and sustainable regional development. The aim of the study was to investigate the impact of COVID-19 on Greek consumers' motives for quality semi-hard and hard cheese, including the "Ladotyri" hard cheese. Consumers' motives were tested using variables of quality semi-hard and hard cheese, such as purchase and consumption, preference of choice, preference, and knowledge of the Ladotyri cheese. A self-response questionnaire survey was carried out in November and December 2022 on a sample of 860 participants, the majority being young people aged 18–25 (83.9%), through the Google platform. Basic statistical tools, combined with cross and chi-square tests, were used to analyze the collected data. The results indicate no significant changes in consumers' motives except a significant decline in consumption, reaching up to 8.4%. Consumers continue to purchase the semi-hard and hard cheese from the supermarket (90%), with preference for the most known kinds, such as kasseri and graviera, consuming it at home (90.9%), daily (31.8%), or two times per week (38.3%), primarily with bread and olives (57.6%), followed by meat (53%). Price remains the most important information for the selection of semi-hard and hard cheese (73.5%), taste (97%) among the organoleptic parameters, texture (70.9%) among the appearance parameters, origin of milk (63.9%) among the sustainable parameters, and value for money (85.8%) among the general characteristics of the cheese. The participants expressed similar motives for the "Ladotyri" Mytilinis hard cheese, appreciating the olive oil combined with the cheese (79.7%) and the possible production as a non-refrigerated cheese (65.2%), even though the majority of them would not buy it today (57.4%). Our findings indicate that the sustainability and growth of the quality semi-hard and hard cheese in the new era should stick to the good practices of production, promotion, and sales developed before the pandemic, exploring, however, new avenues and practices to increase consumption, which is currently declining.

**Keywords:** questionnaire survey; post-COVID-19 era; Greek semi-hard and hard cheese; Ladotyri cheese of Mytilene; consumer's purchase and consumption of cheese; quality cheese; food choice motives

**Citation:** Skalkos, D.; Bamicha, K.; Kosma, I.S.; Samara, E. Greek Semi-Hard and Hard Cheese Consumers' Perception in the New Global Era. *Sustainability* **2023**, *15*, 5825. https://doi.org/10.3390/su15075825

Academic Editor: Đurđica Ačkar

Received: 22 January 2023
Revised: 17 March 2023
Accepted: 25 March 2023
Published: 27 March 2023



## 1. Introduction

The world is changing rapidly following the COVID-19 pandemic and also the current war in Ukraine [1,2], with unforeseen challenges and outcomes in the market, including the selection process of goods, such as foods by consumers [3]. It remains to be seen whether or not the global economy will enter into a long-lasting recession and deglobalization in the near future [4]. Recent studies are evaluating the changes on food marketing in the new era in different countries such as Australia [5], New Zealand [6], Ethiopia [7], and Italy [8], with diverse findings in each case [9]. These changes are affecting the "new" propositions for consumers' food choice motives caused by the pandemic, which we have presented in a recent systematic review [10]. The aim of this paper is to identify the "new" consumers' perceptions for quality semi-hard and hard cheeses in the new era, thus providing practical directions to cheese producers for growth, expansion, competitive advantage, and regional development. The study is focused on young people (students) in order to gain a better future prospective and value of the results obtained, since it is the new generation that better shows the trends of the future.

*Literature Review*

Cheese is an important category not only as a food able to provide nutrients [11], but also as a commodity with unquestionable economic relevance for worldwide trade [12]. Semi-hard and hard cheese is a major dietary cheese category consumed daily worldwide [13,14]. The diversity in technology is enormous, varying the type of milk used, the production operations, the lactic cultures, the maturation times, and conditions, providing final products with a wide range of characteristics in terms of taste, flavor, texture, color, shape, or size [15]. Unlike the industrial semi-hard and hard cheeses, the traditional ones are also imprinted with a social and cultural heritage that makes them unique [16].

Greece has been one of the most cheese-producing EU countries since ancient times [17]. Numerous traditional cheeses are made throughout Greece today. Some of them are types of the same cheese variety, have somewhat different steps in technology or possibly the same, but are known with different local names [17]. Greek traditional cheeses, a total of 30 varieties, can be grouped, according to their technology of manufacture, as white brined (4), other brined (2), soft (5), semi-hard (3), hard (12), and whey (4) cheeses, as shown in Table 1 [17].

**Table 1.** Greek cheeses today.

| Category | Varieties | |
|---|---|---|
| **Brined Cheeses** | | |
| 1. White Brined Cheeses | 1. Feta | 2. Telemes |
| | 3. Kalathaki | 4. Touloumotyri |
| 2. Other brined Cheeses | 1. Batsos | 2. Sfela |
| 3. Soft cheeses | 1. Anevato | 2. Galotyri |
| | 3. Katiki DOmokou | 4. Kopanisti |
| | 5. Pichtogalo Chanion | |
| 4. Semi-hard cheeses | 1. Kasseri | 2. Formaella Arachovas |
| | 3. Krassotyri | |
| 5. Hard cheeses | 1. Graviera | 2. Graviera Agrafon |
| | 3. Graviera Kritis | 4. Graviera Naxou |
| | 5. Kefalotyri | 6. Kefalograviera |
| | 7. Ladotyri Mytilinis | 8. Manoura |
| | 9. Metsovone | 10. San Michali |
| | 11. Xinotyri | 12. Melichloro |
| 6. Whey cheeses | 1. Myzithra | 2. Anthotyros |
| | 3. Manouri | 4. Xinomyzithra Kritis |

Cheeses, being traditional foods (TFs), have geographical and traditional indicators in the EU for the promotion and protection of the names of quality foodstuffs, their origin, and authenticity (e.g., PDO: protected designation of origin, PGI: protected geographical indication, TGI: traditional specialty guaranteed) [18]. Greece has incorporated the provisions of the Regulation in the National Legislation with Ministerial Decree (3321/145849) issued by the Hellenic Ministry of Food and Agricultural Development since 2006 [19]. Among the registered Greek cheeses, 22 are PDOs and 1 is a PGI, with no TGIs registered so far [20].

Greek feta cheese is by far the major Greek cheese known and exported worldwide, a PDO product produced exclusively in Greece, at 134,025 tons out of the total 148,698 tons of PDO/PGI certified Greek cheeses in 2021 [20]. Second in the order of production and consumption are the Greek well-known semi-hard and hard cheeses (called in Greek traditionally as "yellow cheeses") such as kasseri, graviera, kefalotyri, and metsovone. On the island of Lesvos (also called Mytilini), a PDO hard cheese named Ladotyri is produced from local sheep milk or a mixture of sheep milk and caprine, up to a maximum of 30% (*w/w*) [21]. It is a type of good-quality Kefalotyri, with its main characteristic being that it is preserved in olive oil, as indicated also by its name, "ladi" meaning oil and "tyri" cheese. Instead of olive oil, when the cheese obtains a moisture content of lower than 40%, it can be covered with paraffin, but the name Ladotyri is still used [22]. The annual production of Ladotyri Mytilinis' cheese was only 342 tons in 2021 [20], even though there is the potential for increased sales as a unique (in olive oil) local cheese. Such an expansion in the market would be extremely beneficial for the islands' economy, since it would increase local breeding, livestock, and farming together with increased employment.

In view of this expected increase in cheese consumption, researchers have been systemically studying consumers' preferences for cheeses over the last two years (2020–2021), within the context of the COVID-19 pandemic and beyond, with results useful for academia and industry. Among educated young consumers in the Czech Republic, research on the consumption of organic cheese identified two segments, the "rationality involvement consumer" and the "non-rationality involvement consumer", with different characteristics for each [23]. Menozzi et al. report that perceived behavioral control and attitude are the significant predictors of intention to purchase protected designation of origin (PDO)-labelled cheese in France and Italy [24]. Del Toro-Gipson et al. found that consumers differentiated smoke aroma and flavor among smoked cheddar cheeses and preferred cherry wood-smoked cheeses over apple wood- or hickory-smoked [25]. Most hot pepper cheese consumers preferred their cheese with higher heat intensity and were also motivated by the visual characteristics of it [26]. A segmentation analysis conducted by Zhllima et al. revealed that local cheese is preferred to imported cheese, and the main selection criteria for food are the producer name/brand and knowing the seller, with educated female consumers buying cheese mainly in supermarkets [27]. Attitudes for sustainable mountain cheese show the influence of green consumers' values on the brand choice and the strong relationship between the values of green consumers and animal well-being [28]. The incorporation of ingredients with sensory properties familiar to East and Southeast Asian consumers offers the potential for the development of cheese products for consumers in these markets [29]. Ojeda et al. perceived that sensory quality is related to liking but is also modulated by product familiarity for the European cheeses [30]. A study by Endrizzi et al., showed that overall liking was significantly higher in cheeses presented as "mountain pasture product" both in whole panel and in consumer segments with different attitudes [31]. Consumers from Serbia, Croatia, and Spain valued artisan cheeses more than industrial in terms of healthiness and quality, but they believe that there is still much to be done in terms of proper packaging, labelling, branding, widening of assortment, and providing better availability [32]. Among a cohort of young, educated, internationally mobile Chinese consumers it was found that individuals' innovativeness was an important factor that influences their openness to cheese products when moving beyond familiar foods [33].

Quality semi-hard and hard cheeses, like the rest of TFs, as described above, have the potential to become the cheese of choice for the citizens in Europe and elsewhere. To contribute to this potential, the factors connected with consumers' perception for the quality semi-hard, hard, and Ladotyri Mytilinis cheeses today are evaluated here to identify the "new" consumers' motives, if changes have occurred. Ladotyri is included in the study as a representative, local, uniquely produced (in olive oil), hard cheese with very low consumption so far, but it has the potential to grow once the "new" consumer perceptions and attitudes for it are identified here, which will lead to the implementation of the targeted strategic promotion campaign. To accomplish the scope, following the literature on the parameters of consumers' preference, perceptions, attitudes for semi-hard, hard, and Ladotyri Mytilinis cheeses, the study examines the following three determinants of consumers' motives and preference on these quality Greek cheeses in the post-COVID-19 period:

- Consumers' motives for the purchase and consumption of Greek semi-hard and hard cheese. This involves data regarding place of purchase (including online), place of consumption, quantities purchased and consumed before and after COVID-19, as well as consumption preference on the combination of meals with different kinds of semi-hard and hard cheeses (graviera, kefalograviera, kasseri, kefalotyri, ladotyri).
- Consumers' preference for quality Greek semi-hard and hard cheese. This involves data regarding traditional parameters, organoleptic parameters, appearance, sustainability, and general characteristics.
- Consumers' preference and knowledge for the Ladotyri cheese (of Lesvos). This involves data regarding knowledge for the specific cheese, its unique characteristics, possible added value of the olive oil included, possible added value if it was produced as a non-refrigerated cheese preserved in olive oil, preference on the combination of meals, on place of purchase, and perception for Lesvos' quality foods.

## 2. Materials and Methods

### 2.1. Data Collection and Sample Characterization

This survey was based on a questionnaire prepared to investigate the information that influences consumers' motives and preference on Greek semi-hard and hard cheese in the new era. The questionnaire was built up in four parts and it is presented in Table S1 in the Supplementary Section. Each question was created in such a way that it could provide the best possible information for each section. The parts were built up using a similar previous study [34]. The first part included questions about the sociodemographic characteristics of the respondents, specifically gender, age, level of education, civil state, job situation, and permanent residency in different parts of Greece. The second part consisted of ten questions designed to assess the motives for the purchase and consumption of Greek semi-hard and hard cheese in the post-COVID era. The third part included five questions focused on the participants' preferred choice for quality Greek cheese. Finally, the fourth part consisted of ten questions about the knowledge and preference of the "Ladotyri" Greek semi-hard and hard cheeses. To guarantee the quality of the data obtained through the application of the questionnaire, this was pretested with 50 respondents. This phase was pivotal to ensure that the questions were clear and understandable, so that respondents could answer them easily. The research was carried out using electronic questionnaires as it was easier to distribute and collect. The distribution method chosen was by e-mail, as similarly performed in recent papers investigating consumer behaviors [35–37]. The sample of the population is very well distributed among the 5 geographic parts of Greece, with emphasis, however, on students.

A higher rate for female respondents, recorded at 76.1%, is similar to the observation by other papers as well [38–41], leading to the conclusion that women, even students, respond more willingly to food-related surveys as they are primarily involved in household organization. The research questionnaire was created through the Google Platform and the Google Forms function due to the ability of direct export of the results to an Excel sheet for further processing. The geographical context for the present study was all the Greek regions divided in five parts. Respondents received e-mails explaining the purpose of the research and the importance of their participation, while there was an attached link that led to the electronic form of the questionnaire. Responses were anonymous and no personal information was collected or correlated with any of the responses to ensure the protection of participants.

The survey took place during the period of November-December 2022, at the decline of the pandemic, and consisted of 860 participants (Table 2).

**Table 2.** Sociodemographic characterization of the sample.

| Variable | Groups | (%) |
|---|---|---|
| *Gender* | Male | 23.9 |
| | Female | 76.1 |
| *Age* | 18–25 | 83.9 |
| | 26–35 | 4.3 |
| | 36–45 | 4.0 |
| | 46–55 | 5.7 |
| | 56+ | 2.1 |
| *Level of education* | None/Primary school | 0.1 |
| | Secondary school | 10.7 |
| | High school | 0.0 |
| | University | 89.1 |
| *Civil state* | Single | 85.9 |
| | Married | 10.8 |
| | Divorced | 2.8 |
| | Widow/widower | 0.5 |
| *Job situation* | Employed | 15.8 |
| | Unemployed | 1.1 |
| | Student | 82.6 |
| | Retired | 0.6 |
| *Permanent resident in Greece* | NORTH GREECE (regions of Macedonia—Thrace) | 29.0 |
| | WEST GREECE (region of Epirus—Etoloakarnania prefecture) | 37.2 |
| | CENTRAL GREECE (including Athens) | 20.4 |
| | SOUTH GREECE (region of Peloponnese) | 5.0 |
| | ISLANDS (Ionian and Aegean) | 8.3 |

Regarding Shouthe spatial distribution, 37.2% of participants were permanent residents of west Greece, 20.4% of central Greece (including the capital of Athens), 29% residents of north Greece, 8.3% residents of the Greek islands, and 5% of south Greece, leading to a wide geographical distribution. The vast majority of the participants were aged 18–25 (83.9%) followed by 46–55, 26–35, and 36–45 years (5.7%, 4.3%, 40%, respectively). Regarding the level of education, most of the participants had higher education (university, 89.1%), while the employment status category was dominated by students (82.6%) followed by employed (15.8%) participants. Regarding the civil state of the participants, most were single at 85.9%, followed by married at 10.8% and divorced at 02.8% and only 0.5% were widows.

*2.2. Data Analysis*

The exploratory analysis of the data was achieved through basic statistical tools. The survey was prepared in Greek and divided into four parts, as detailed above:

Part 1. Sociodemographic data;

Part 2. Purchase and consumption of Greek semi-hard and hard cheese in the post-COVID-19 era;

Part 3. Preference of choice for quality Greek semi-hard and hard cheese in the post-COVID-19 era;

Part 4. Knowledge and preference of "Ladotyri" in post-COVID-19 era.

The sociodemographic characteristics were collected in the first part of the questionnaire (six questions—one dichotomous, one ordinal variable, and four nominal variables).

The second part recorded information concerning the purchase and consumption motives of participants (ten questions—two ordinal variables, three nominal variables, two dichotomous, and three multiple choices with each response considered as dichotomous variables). The third part consisted of five questions (ordinal variables) recording the preference of choice for quality Greek semi-hard and hard cheese of the participants, and finally, the fourth part (ten questions—two multiple choices with each response considered as dichotomous variables, six dichotomous, and two nominal variables) recorded information about the knowledge and preference of "Greek semi-hard and hard cheese".

Data analysis was performed using IBM SPSS Statistics for Windows (Version 25.0, IBM Corp., Armonk, NY, USA), as described by Skalkos et al. [42]. The nonparametric tests were used. A nonparametric chi-square test was performed to test the distribution of variables of each group and response based on the hypothesized equal proportions for each variable. The chi-square independence test was used to determine whether there is an association between variables. Post hoc tests for the chi-square independence test were used. The pairwise comparisons (z-tests) for independent proportions, followed by a Bonferroni correction, were applied to the data. In order to measure the strength of association (when it is present between two variables), the Phi, Cramer's V, or Kendall's tau-b test were used. The Cramer's V coefficient used in the chi-square tests, ranging from 0 to 1, can be interpreted as follows: $V \approx 0.1$ is a weak association, $V \approx 0.3$ is a moderate association, and $V \approx 0.5$ or over is a strong association. In all the tests performed, the level of significance considered was 5% ($p < 0.05$).

## 3. Results

Table 3 presents the participants' motives on purchase and consumption of Greek semi-hard and hard cheese. The results show that most of the participants before the pandemic purchased semi-hard and hard cheese very often (70.9%) and often (21.0%) from the supermarket, while they purchased from the neighboring grocery store only very often (8.8%) and often (19.9%), whereas online purchases were very low (0.3% very often and 0.2% often). These results seem to be the same in the post-COVID-19 era, as the very often purchase from the supermarket answer remained 71.0.% and the often answer 21.4%, with purchases from grocery store also remaining similar (20.3% often answer). Only the online increased slightly to 1.6% from 0.5% (0.8% very often and 0.8% often). Regarding the quantities and the money spent for semi-hard and hard cheese per month, one Kg (65.2%) and EUR 10 (55.5%) were the most popular answers. The majority of the participants consume less semi-hard and hard cheese today (58.4%) as compared with the before the COVID-19 period; daily (31.8%) and two times per week (38.3%) are the most popular frequencies of consumption.

The participants, among the Greek semi-hard and hard cheeses, exhibit high preference for the well-known kasseri (59.8%), and graviera (57.3%), less preference for kefalotyri (41.4%) and kefalograviera (34.9%), and very limited preference for ladotyri (3.0%), the quality cheese of reference in this study. They consume slightly more Greek semi-hard and hard cheese, by 52.4%, as compared with the imported varieties (i.e., mozzarella, cheddar, edam, etc.). The participants today consume semi-hard and hard cheese at home (90.9%) on different occasions such as during dinner (34.1%), during lunchtime (12.2%), occasionally (19.4%), with friends (11.8%), and only at a restaurant when they go out (24%). In terms of preference of meals with semi-hard and hard cheese, bread and olives (57.6%), meat (53.0%), chicken (45.2%), wine (42.3%), and alone (22.3%) are the most preferable accompaniment meals.

**Table 3.** Participants' motives on purchase and consumption of Greek semi-hard and hard cheese.

| From Where DID YOU PURCHASE the Greek Semi-Hard and Hard Cheese You Consumed before COVID-19? | Never | Very Seldom | Seldom | Often | Very Often | | |
|---|---|---|---|---|---|---|---|
| From supermarket | 2.0 * | 2.1 | 3.9 | 21.0 | 70.9 | | |
| From the neighborhood grocery store | 26.5 | 19.1 | 25.7 | 19.9 | 8.8 | | |
| From open market | 82.4 | 9.4 | 5.2 | 2.3 | 0.8 | | |
| Via online | 96.5 | 2.3 | 0.8 | 0.2 | 0.3 | | |
| **From where DO YOU PURCHASE the GREEK SEMI-HARD AND HARD CHEESE you consume now?** | Never | Very seldom | Seldom | Often | Very often | | |
| From supermarket | 1.9 | 1.9 | 3.8 | 21.4 | 71.0 | | |
| From the neighborhood grocery store | 31.6 | 16.8 | 22.6 | 20.3 | 8.6 | | |
| From open market | 85.5 | 7.2 | 5.1 | 1.1 | 1.1 | | |
| Via online | 93.6 | 3.4 | 1.4 | 0.8 | 0.8 | | |
| **How much GREEK SEMI-HARD AND HARD CHEESE do you buy per month today (ONLY one answer)** | 1 kg per month | 2 kg per month | 3 kg per month | 4 kg per month | 0 kg per month | | |
| | 65.2 | 21.0 | 6.0 | 2.7 | 5.2 | | |
| **How much MONEY do you spend MONTHLY for the purchase of GREEK SEMI-HARD AND HARD CHEESE** | <EUR 10 | EUR 10–20 | EUR 20–30 | <EUR 30 | | | |
| | 55.5 | 32.9 | 9.0 | 2.6 | | | |
| **How often do you consume GREEK SEMI-HARD AND HARD CHEESE** | Every day | Once a week | Two times per week | Once every two weeks | Once per month | | |
| | 31.8 | 14.4 | 38.3 | 7.9 | 7.6 | | |
| **Do you consume MORE or LESS GREEK SEMI-HARD AND HARD CHEESE TODAY as compared with the period BEFORE COVID-19** | More | Less | | | | | |
| | 41.6 | 58.4 | | | | | |
| **Do you consume MORE GREEK SEMI-HARD AND HARD CHEESE as compared with IMPORTED SEMI-HARD AND HARD CHEESE (i.e., cheddar, pecorino, edam, etc.)** | More | Less | | | | | |
| | 52.4 | 47.6 | | | | | |
| **Which KINDS OF GREEK SEMI-HARD AND HARD CHEESES do you consume TODAY** | Graviera | Kefalograviera | Ladotyri | Kaseri | Kefalotiri | Others | |
| | 57.3 | 34.9 | 3.0 | 59.8 | 41.4 | 26.1 | |
| **With what do you consume THE GREEK SEMI-HARD AND HARD CHEESE TODAY** | Meat | Fish | Wine | Chicken | Fruits | Bread and Olives | Alone |
| | 53.0 | 3.4 | 42.3 | 45.2 | 10.2 | 57.6 | 22.3 |
| **Where do you consume mostly the SEMI-HARD AND HARD CHEESE TODAY?** | At home | At the restaurant | With friends | During lunchtime | During the dinner | Occasionally | |
| | 90.9 | 24.0 | 11.8 | 12.2 | 34.1 | 19.4 | |

* Values represent %.

The results of the chi-square test in Table S2 showed significant differences between consumers' motives on purchase and consumption of Greek semi-hard and hard cheeses in terms of:

1. Purchase of cheese before COVID-19.

   From supermarket: between level of education ($x^2$ = 51.174, $p$ = 0.000).
   From the neighborhood grocery store: between residency ($x^2$ = 27.677, $p$ = 0.035).
   From open market: between residency ($x^2$ = 53.786, $p$ = 0.000).
   Via online: between age ($x^2$ = 63.711, $p$ = 0.001), level of education ($x^2$ = 325.401, $p$ = 0.000), civil state ($x^2$ = 83.932, $p$ = 0.000), job situation ($x^2$ = 40.661, $p$ = 0.001), and residency ($x^2$ = 46.313, $p$ = 0.000).

2. Purchase of cheese today.

   From supermarket: between gender ($x^2$ = 21.641, $p$ = 0.013) and level of education ($x^2$ = 53.735, $p$ = 0.001).
   From the neighborhood grocery store: between residency ($x^2$ = 33.018, $p$ = 0.007).
   From open market: between residency ($x^2$ = 41.879, $p$ = 0.000).
   Via online: between level of education ($x^2$ = 134.631, $p$ = 0.001), civil state ($x^2$ = 35.527, $p$ = 0.001), and job situation ($x^2$ = 23.211, $p$ = 0.026).

3. Quantity of cheese purchased per month.

   One kg: between gender ($x^2$ = 6.912, $p$ = 0.013) and residency ($x^2$ = 15.865, $p$ = 0.003).
   Four kg: between gender ($x^2$ = 4.987, $p$ = 0.026), level of education ($x^2$ = 37.171, $p$ = 0.000), civil state ($x^2$ = 35.691, $p$ = 0.000), and residency ($x^2$ = 10.457, $p$ = 0.033).

4. Money spent per month.
   Up to EUR 10: between gender ($x^2$ = 15.895, $p$ = 0.001), age ($x^2$ = 53.769, $p$ = 0.001), civil state ($x^2$ = 34.771, $p$ = 0.001), job situation ($x^2$ = 59.505, $p$ = 0.001), and residency ($x^2$ = 25.823, $p$ = 0.000).
   Between EUR 10 to 20: between age ($x^2$ = 16.068, $p$ = 0.003), civil state ($x^2$ = 13.480, $p$ = 0.004), and job situation ($x^2$ = 18.486, $p$ = 0.000).
   Between EUR 20 to 30: between gender ($x^2$ = 8.820, $p$ = 0.003), age ($x^2$ = 14.338, $p$ = 0.006), civil state ($x^2$ = 10.278, $p$ = 0.016), job situation ($x^2$ = 16.673, $p$ = 0.001), and residency ($x^2$ = 19.487, $p$ = 0.001).
   More than EUR 30: between age ($x^2$ = 52.805, $p$ = 0.000), level of education ($x^2$ = 38.768, $p$ = 0.000), civil state ($x^2$ = 24.230, $p$ = 0.000), job situation ($x^2$ = 23.300, $p$ = 0.000), and residency ($x^2$ = 11.935, $p$ = 0.018).

5. Kinds of cheese consumed.
   Graviera: between age ($x^2$ = 11.419, $p$ = 0.022), job situation ($x^2$ = 14.762, $p$ = 0.002), and residency ($x^2$ = 27.703, $p$ = 0.000).
   Kefalograviera: between job situation ($x^2$ = 9.770, $p$ = 0.021) and residency ($x^2$ = 16.059, $p$ = 0.003).
   Ladotyri: between gender ($x^2$ = 5.004, $p$ = 0.025).
   Kaseri: between age ($x^2$ = 37.966, $p$ = 0.000), civil state ($x^2$ = 13.117, $p$ = 0.004), job situation ($x^2$ = 22.123, $p$ = 0.000), and residency ($x^2$ = 125.493, $p$ = 0.001).
   Kefalotyri: between age ($x^2$ = 20.001, $p$ = 0.000) and civil state ($x^2$ = 11.117, $p$ = 0.008).

6. Accompaniment meals.
   Fish: between gender ($x^2$ = 5.229, $p$ = 0.022) and level of education ($x^2$ = 28.651, $p$ = 0.000).
   Chicken: between age ($x^2$ = 9.807, $p$ = 0.044) and residency ($x^2$ = 10.590, $p$ = 0.032).
   Bread and olives: between civil state ($x^2$ = 14.085, $p$ = 0.003) and residency ($x^2$ = 15.560, $p$ = 0.004).
   Alone: age ($x^2$ = 12.202, $p$ = 0.016) and civil state ($x^2$ = 9.896, $p$ = 0.019).

7. Where do you consume wine today.

At home: between gender ($x^2 = 4.450, p = 0.035$).
At the restaurant: between civil state ($x^2 = 8.000, p = 0.046$).
With friends: between level of education ($x^2 = 7.449, p = 0.024$) and civil state ($x^2 = 12.382, p = 0.006$).
During lunchtime: between age ($x^2 = 10.436, p = 0.034$), level of education ($x^2 = 7.203, p = 0.027$), and civil state ($x^2 = 20.972, p = 0.000$).
During dinner: civil state ($x^2 = 10.214, p = 0.017$).

Table 4 represents the frequencies concerning preference of choice for quality Greek semi-hard and hard cheese in the post-COVID-19 era. Participants find much and very much importance in the price (73.5%), the branding of the cheese (37.8%), the date of production (44.5%), the geographical origin (30.7%), and the existence of quality certificates (44.1%) for the selection of a quality Greek semi-hard and hard cheese. The organoleptic parameter that most seems to affect the selection of semi-hard and hard cheese by far is the taste (75.3%—very much), followed to a lesser extent by odor (38.3%—very much), aroma (36.7%—much), and hardness (34.6%—much). Among the appearance parameters with much and very much preference, the texture (70.9%) is by far the first choice by the participants, followed by the overall appearance (59.6%), the color (54.3%), and to a lesser extent the size of the package (33.3%) and the package appearance (22.7%). The sustainable characteristics seem to be of medium level of concern for the selection of semi-hard and hard cheese, with much and very much selection choice; the origin of milk by far the most important parameter (63.9%), followed by nutritional indications (51.2%), the percentage of fats (43.2%), the organic nature (30.6%), and low salt content (28.5%). Finally, from the general characteristics, only the rational value for money concerns the participants (51.0%—very much and 34.8%—much), while there is less concern for the other parameters: timeless but also modern (23.5%—much), added value for the production area (20.3%—much), uniqueness (19.4%—much), and a myth behind the cheese (8.4%—much).

The results of the chi-square test presented in Table S3 showed that there were significant differences between consumers' preference for quality Greek semi-hard and hard cheeses in terms of:

1.  Importance of information for the selection of Greek semi-hard and hard cheese.
    Price: between gender ($x^2 = 10.981, p = 0.027$).
    Branding: between age ($x^2 = 37.379, p = 0.002$), level of education ($x^2 = 15.622, p = 0.048$), and job situation ($x^2 = 23.657, p = 0.023$).
    Date of Production: between gender ($x^2 = 15.703, p = 0.003$).
    Geographical origin: between age ($x^2 = 88.629, p = 0.000$), civil state ($x^2 = 76.495, p = 0.000$), job situation ($x^2 = 51.648, p = 0.001$), and residency ($x^2 = 46.411, p = 0.000$).
    Quality certificates: between age ($x^2 = 32.008, p = 0.010$), civil state ($x^2 = 25.906, p = 0.011$), and residency ($x^2 = 28.139, p = 0.030$).

2.  Importance of organoleptic parameters.
    Taste: between gender ($x^2 = 10.687, p = 0.030$).
    Aroma: between gender ($x^2 = 21.411, p = 0.001$).
    Odor: between gender ($x^2 = 30.228, p = 0.000$) and level of education ($x^2 = 15.554, p = 0.049$).

3.  Importance of appearance parameters.
    Color: between gender ($x^2 = 16.675, p = 0.002$) and age ($x^2 = 29.091, p = 0.023$).
    Appearance: between gender ($x^2 = 16.348, p = 0.003$).
    Texture: between gender ($x^2 = 32.647, p = 0.001$).
    Package appearance: between civil state ($x^2 = 21.279, p = 0.046$).
    Size of the package: between civil state ($x^2 = 21.053, p = 0.050$).

4.  Importance of sustainable characteristics.

Milk origin: between gender ($x^2 = 10.162$, $p = 0.038$), age ($x^2 = 28.540$, $p = 0.027$), civil state ($x^2 = 21.109$, $p = 0.049$), job situation ($x^2 = 22.528$, $p = 0.032$), and residency ($x^2 = 35.650$, $p = 0.003$).

Organic: between civil state ($x^2 = 22.497$, $p = 0.032$) and residency ($x^2 = 35.436$, $p = 0.003$).

Nutritional indications: between gender ($x^2 = 10.145$, $p = 0.038$) and civil state ($x^2 = 25.654$, $p = 0.012$).

Fat quantity: between age ($x^2 = 27.125$, $p = 0.040$) and civil state ($x^2 = 24.187$, $p = 0.019$).

Low salt: between gender ($x^2 = 25.565$, $p = 0.000$) and age ($x^2 = 30.060$, $p = 0.018$).

5.  Importance of general characteristics.

Rational value for money: between level of education ($x^2 = 50.230$, $p = 0.000$) and civil state ($x^2 = 35.347$, $p = 0.000$).

Unique and special: between age ($x^2 = 31.799$, $p = 0.011$).

Added value for the production area: between age ($x^2 = 30.367$, $p = 0.016$), civil state ($x^2 = 24.264$, $p = 0.019$), and residency ($x^2 = 34.478$, $p = 0.005$).

**Table 4.** Frequencies regarding the preference of choice for quality Greek semi-hard and hard cheese.

| How Important Are for You the Following INFORMATION for the Selection of QUALITY GREEK SEMI-HARD AND HARD CHEESE | Not at All | Little | Medium Level | Much | Very Much |
|---|---|---|---|---|---|
| The price of the semi-hard and hard cheese | 2.4 * | 3.2 | 20.9 | 37.0 | 36.5 |
| The branding | 11.4 | 15.7 | 35.1 | 27.9 | 9.9 |
| The date of production | 12.9 | 17.7 | 24.9 | 26.5 | 18.0 |
| The geographical origin | 18.1 | 21.8 | 29.3 | 21.0 | 9.7 |
| The existence of quality certificates such as PDO (Protected Designation of Origin), etc. | 11.3 | 16.0 | 28.6 | 29.4 | 14.7 |
| **How important are the following ORGANOLEPTIC PARAMETERS for the selection of QUALITY GREEK SEMI-HARD AND HARD CHEESE** | **Not at all** | **Little** | **Medium Level** | **Much** | **Very much** |
| The Taste | 0.5 | 0.6 | 1.9 | 21.7 | 75.3 |
| The aroma | 3.3 | 5.8 | 25.2 | 36.7 | 29.1 |
| The hardness | 3.1 | 8.6 | 32.4 | 34.6 | 21.2 |
| The odor | 1.8 | 4.8 | 18.2 | 36.9 | 38.3 |
| **How important are the following APPEARANCE PARAMETERS for the selection of QUALITY GREEK SEMI-HARD AND HARD CHEESE** | **Not at all** | **Little** | **Medium Level** | **Much** | **Very much** |
| The color | 5.0 | 11.9 | 28.7 | 31.7 | 22.6 |
| The appearance | 4.2 | 11.2 | 25.0 | 35.3 | 24.3 |
| The texture | 2.1 | 5.8 | 21.1 | 39.4 | 31.5 |
| The package appearance | 16.5 | 24.8 | 36.0 | 15.2 | 7.5 |
| The size of the package (i.e., 200 g, 400 g, 0.5 kg, 1 kg, etc.) | 8.0 | 14.6 | 33.1 | 26.5 | 17.8 |
| **How important are the following SUSTAINABLE CHARACTERISTICS for the selection of QUALITY GREEK SEMI-HARD AND HARD CHEESE** | **Not at all** | **Little** | **Medium Level** | **Much** | **Very much** |
| Origin of the milk (cow, goat, sheep, or mixture) | 4.5 | 8.3 | 23.3 | 34.6 | 29.3 |
| Organic | 18.5 | 20.9 | 29.9 | 21.0 | 9.6 |
| Nutritional indications | 8.2 | 13.1 | 27.4 | 33.3 | 17.9 |
| Percentage of fats | 10.4 | 16.1 | 30.3 | 26.2 | 17.0 |
| Low salt | 19.4 | 21.4 | 30.6 | 17.3 | 11.2 |
| **How important are the following GENERAL CHARACTERISTICS for the selection of QUALITY GREEK SEMI-HARD AND HARD CHEESE** | **Not at all** | **Little** | **Medium Level** | **Much** | **Very much** |
| Rational value for money | 0.9 | 2.1 | 11.1 | 34.8 | 51.0 |
| Unique and special | 10.1 | 24.2 | 37.0 | 19.4 | 9.4 |
| Added value for the region where it is produced | 11.0 | 23.9 | 38.5 | 20.3 | 6.3 |
| A myth (historical narrative) | 33.8 | 28.1 | 25.8 | 8.4 | 3.9 |
| Timeless but also modern | 16.3 | 17.0 | 32.5 | 23.5 | 10.6 |

* Values represent %.

Table 5 represents the frequencies concerning the knowledge and preference of the "Ladotyri" cheese in the post-COVID-19 era. Only 42.7% of the participants know the cheese, and the majority of them have never tasted or consumed it (72.2%), while half of them know where it is produced, namely Lesvos (Mytilene) island (49.2%). The participants do not know the cheese's unique characteristics (37.4%), while the rest of them consider its unique flavor as its major characteristic (34.6%), followed by PDO label (32.6%), unique aroma (17.4%), and healthy properties (11.3%). They strongly perceive as an added value the storage of the cheese in olive oil, at 79.7%, as well as the possibility of its production as non-refrigerated cheese, preserved by the oil (65.2%), even though they are not willing to buy and consume such an innovative cheese (57.4%—no answer). In terms of preference with meals with ladotyri, the same order of choice with semi-hard and hard cheeses is recorded: bread and olives (54.9%), meat (33.60%), wine (27.1%), and alone (17.0%). Finally, participants would like to purchase ladotyri from the supermarket (61.2%), and they believe that Lesvos Island is indeed producing quality cheeses (66.1%).

The results of the chi-square test presented in Table S4 showed that there were significant differences between consumers' knowledge and preference for Ladotyri cheese in terms of:

1. Knowledge of Ladotyri cheese: between age ($x^2 = 54.305$, $p = 0.000$), civil state ($x^2 = 34.735$, $p = 0.000$), job situation ($x^2 = 36.026$, $p = 0.000$), and residency ($x^2 = 14.772$, $p = 0.005$).
2. Ever tasted Ladotyri: between gender ($x^2 = 9.475$, $p = 0.002$), age ($x^2 = 60.361$, $p = 0.000$), civil state ($x^2 = 32.148$, $p = 0.001$), job situation ($x^2 = 30.697$, $p = 0.001$), and residency ($x^2 = 12.151$, $p = 0.016$).
3. Knowledge for Ladotyri production area.
   Epirus: between age ($x^2 = 10.736$, $p = 0.030$) and residency ($x^2 = 16.936$, $p = 0.002$).
   Samos island: between level of education ($x^2 = 44.796$, $p = 0.000$), civil state ($x^2 = 10.730$, $p = 0.013$), and job situation ($x^2 = 10.932$, $p = 0.012$).
   Lesvos island: between age ($x^2 = 19.216$, $p = 0.001$), civil state ($x^2 = 14.602$, $p = 0.002$), and residency ($x^2 = 12.384$, $p = 0.015$).
   Lemnos island: between job situation ($x^2 = 10.884$, $p = 0.012$).
4. Knowledge for Ladotyri's unique characteristics.
   Bitter taste: between level of education ($x^2 = 15.796$, $p = 0.001$), job situation ($x^2 = 8.466$, $p = 0.037$), and residency ($x^2 = 9.487$, $p = 0.050$).
   Unique aroma: between job situation ($x^2 = 12.638$, $p = 0.005$).
   Unique flavor: between age ($x^2 = 17.054$, $p = 0.002$).
   Low salt: between age ($x^2 = 14.248$, $p = 0.007$).
   PDO product: between age ($x^2 = 17.691$, $p = 0.001$), civil state ($x^2 = 12.958$, $p = 0.005$), and job situation ($x^2 = 8.217$, $p = 0.042$).
   Ignorance: between age ($x^2 = 15.244$, $p = 0.004$) and job situation ($x^2 = 8.659$, $p = 0.034$).
5. Added value for Ladotyri—the fact of olive oil's addition: between age ($x^2 = 12.158$, $p = 0.016$) and job situation ($x^2 = 9.094$, $p = 0.028$).
6. Added value for Ladotyri—the fact that is a non-refrigerated cheese: between age ($x^2 = 10.673$, $p = 0.031$), civil state ($x^2 = 9.570$, $p = 0.023$), and job situation ($x^2 = 8.431$, $p = 0.038$).
7. Preference or intention of purchasing a non-refrigerated cheese: between gender ($x^2 = 8.048$, $p = 0.005$) and residency ($x^2 = 12.981$, $p = 0.011$).
8. Accompaniment meals with Ladotyri.
   Meat: between age ($x^2 = 12.595$, $p = 0.013$).
   Fish: between gender ($x^2 = 4.528$, $p = 0.033$), age ($x^2 = 19.509$, $p = 0.001$), civil state ($x^2 = 23.785$, $p = 0.001$), and job situation ($x^2 = 12.013$, $p = 0.007$).
   Wine: between age ($x^2 = 9.521$, $p = 0.049$), and job situation ($x^2 = 11.455$, $p = 0.010$).
   Bread and olives: between gender ($x^2 = 4.483$, $p = 0.034$) and residency ($x^2 = 13.104$, $p = 0.011$).

**Table 5.** Frequencies regarding the knowledge and preference of Ladotyri.

| Do You Know the LADOTYRI CHEESE? | Yes | No | | | | | | |
|---|---|---|---|---|---|---|---|---|
| | 42.7 * | 57.3 | | | | | | |
| **Have you tasted LADOTYRI or are you consuming it occasionally?** | Yes | No | | | | | | |
| | 27.8 | 72.2 | | | | | | |
| **Where do you think is LADOTYRI produced?** | Epirus region | Samos island | Macedonia region | Creta island | Peloponnese region | Lesvos island | Lemnos island | None of the above |
| | 17.9 | 2.2 | 1.3 | 14.3 | 6.2 | 49.2 | 3.9 | 4.9 |
| **Which do you think are the unique characteristics of LADOTYRI?** | Bitter taste | Unique aroma | Unique flavor | Low salt | Healthy | POD product | Don't know | |
| | 6 | 17.4 | 34.6 | 8.6 | 11.3 | 32.6 | 37.4 | |
| **Do you think it is an added value the OLIVE OIL in which LADOTYRI is inside?** | Yes | No | | | | | | |
| | 79.7 | 20.3 | | | | | | |
| **Do you think it will be added value if LADOTYRI is a NON-REFRIGERATED GREEK CHEESE preserved by the olive oil it is in?** | Yes | No | | | | | | |
| | 65.2 | 34.8 | | | | | | |
| **Would you buy/prefer a non-refrigerated CHEESE TODAY, after COVID-19 pandemic?** | Yes | No | | | | | | |
| | 42.6 | 57.4 | | | | | | |
| **What would you like to eat with the LADOTYRI CHEESE if you had the chance?** | Meat | Fish | Wine | Bread and olives | Fruits | Nuts | Alone | |
| | 33.6 | 2.5 | 27.1 | 54.9 | 6.5 | 10.1 | 17 | |
| **Where would you like to purchase LADOTYRI cheese if you had the chance TODAY** | From supermarket | From grocery store | From open market | Via on line | | | | |
| | 61.2 | 35.5 | 2.4 | 1 | | | | |
| **Do you believe that LESVOS' Island is producing quality cheeses or not, compared with the rest of Greece** | Yes | No | | | | | | |
| | 66.1 | 33.9 | | | | | | |

* Values represent %.

## 4. Discussion

In the new era after the COVID-19 pandemic and the current war in Ukraine, the food consumer is emerging with unprecedented perceptions and motives. We investigate in this study the consumer's motives for quality semi-hard and hard cheese, namely Greek cheese, mainly young Greek consumers. As a reference of quality semi-hard and hard cheese, the relatively unknown traditional Greek semi-hard and hard cheese "Ladotyri" was chosen as part of the study for comparison reasons with the rest of Greek semi-hard and hard cheeses [21,22]. The sociodemographic characteristics of the study, presented in

Table 2, exhibited a suitable distribution between the different categories, except the age of the participants, the majority being 18–25 (83.9%) and students (82.6%), for better future prospective and validity of the results obtained.

Participants' choices regarding the places of purchase for semi-hard and hard cheese before and after the COVID-19 pandemic did not change, with the supermarket being by far (more than 90% often and very often) the place of choice, followed by the grocery store (28.8%), with only a minor decrease for the open market (−0.9%) and an increase for online (+1.1%), as shown in Table 3. The results of the chi-square test, shown in Table S2, indicate that there are significant differences, with strong association for the "level of education" regarding purchase online before COVID (V = 0.5) and moderate association after COVID (V = 0.324), with weak to moderate associations varying from V = 0.100 to V = 0.208 for "gender" regarding the supermarket after COVID, "age" regarding online purchase before COVID, "civil state" and "job situation" regarding online purchase before and after COVID, and "residency" regarding purchase from grocery stores, open market, online before and after COVID. Our results indicate that the purchase selection of cheese by the consumers has not changed through the pandemic, since the cross-shopping behavior of consumers for food studied for more than a decade or so [43] provides the supermarket as the first choice, even reaching 76.4% for cheese [44], comparable with our finding of 90% after the pandemic. Another study within the pandemic in Albania also proved the first selection choice of the supermarket for traditional local cheese purchase among educated male and female participants, with lower percentages, however, around 40% [27]. Laguna et al. report a reduction in shopping frequency but no changes in shopping location during the pandemic [45].

Regarding the consumption of cheese, participants consume less Greek semi-hard and hard cheese today (−8.4%), mostly 1 kg per month, spending up to EUR 10, eating cheese one or two times per week, with a slight preference for Greek semi-hard and hard cheese (+2.4%) as compared to imported cheese. They prefer the well-known Greek semi-hard and hard cheeses, primarily kasseri, graviera, and as a second choice, kefalotyri and kefalograviera. Most participants consume the cheese at home, mainly during dinner, and only sometimes at a restaurant, with preferred accompaniment meals in the following order: bread and olives, meat, chicken, and wine. The results of the chi-square test, presented in Table S2, indicate that there are significant differences with moderate association only for "level of education" and "civil state" regarding the consumption rate of 4 kg monthly (V = 0.208/0.205), while for most of the sociodemographic variables, the significant differences showed a weak association, varying from V = 0.110 to V = 0.250 for the questions about money spent. Some of the sociodemographic variables exhibited significant differences with weak association, varying from V = 0.077 to V = 211 for the kinds of cheese consumed, with only "residence" with a moderate association (V = 0.384) for kasseri. For the questions of accompaniment meals and place of consumption, some sociodemographic variables show significant differences with weak association, varying from V = 0.092 to V = 0.136. Our findings on the frequency of cheese consumption are in agreement with the reported by Planzer et al. for Brazilian cheeses, reaching 85.4% weekly, 53.8% daily, and 31.8% once per week [44]. The recorded consumption of 1 kg per month for yellow Greek cheese (12 kg annually) appears to be a reasonable and adequate quantity for one kind of cheese only, considering the 18.44 kg average annual consumption per person worldwide in 2020 [46]. Studies on food and cheese pairing are in the framework of diets and health, such as the recent study by Iglesias et al. [47]. Finally, there is no other study comparing cheese consumption before and after the COVID-19 pandemic to evaluate the rest of our findings with reported literature on this subject matter.

Regarding the participants' preference for quality Greek semi-hard and hard cheese in terms of the provided information, price was the most important motive of choice followed by the date of production, the quality certificates, the branding, and the geographic origin (Table 4). The results of the chi-square test, shown in Table S3, indicate that there are significant differences with weak association, varying from V = 0.092 to V = 0.176 for all sociodemographic variables and selected choice parameters.

The participants in terms of organoleptic cheese selections chose by far the taste, followed by the odor, the aroma, and hardness (Table 4), with the chi-square test indicating significant differences with weak association, varying from V = 0.097 to V = 0.191 only for "gender" regarding taste, aroma, and odor and for "level of education" regarding odor, as shown in Table S3.

In terms of the appearance parameters, the order of cheese selection was texture first, followed by appearance, color, size of package, and package appearance (Table 4), with significant differences with weak association (V = 0.093 to V = 0.197) for "gender" regarding color, appearance, texture, "age" regarding color, and "civil state" regarding package appearance and size (Table S3).

In terms of the sustainable characteristics, the order of selection by the participants was origin of milk first, followed by nutritional indications, percentage of fats, organic nature, and low salt (Table 4), with significant differences with weak association (V = 0.092 to V = 0.176) for all sociodemographic variables except "level of education" and selected choice parameters, as shown in Table S3.

Finally, in terms of the general cheese characteristics, the value for money was the first choice, followed by timeless but also modern, added value for the region, uniqueness, and a myth behind the cheese, with significant differences with weak association (V = 0.096 to V = 0.173) for "age" regarding value for money and added regional value, "level of education" regarding value for money, "civil state" regarding value for money and regional added value, and "residency" regarding regional added value. Our results regarding the preference of choice for quality semi-hard and hard cheese indicate that consumers have kept the same motives in the post-COVID-19 era as their motives before the pandemic, since according to the studies before the pandemic, price [48], taste [49,50], texture [51], origin of milk [52], and value for money [53] were major food choice motives for cheeses.

Regarding the participants' knowledge and preference of the Lesvos (Mytilene) "Lado-tyri" semi-hard and hard cheese in the post-COVID-19 era, most of them do not know it (57%) and have never tasted or consumed it (72.2%), but they know that the island of origin produces quality cheeses (Table 5). They do not know about its unique characteristics (37.4%), with the flavor believed to be the major asset (34.6%). They consider as added value the immersion of the cheese in olive oil (79.7%) and the possible production as a non-refrigerated cheese (65.2%), even though they would not buy and consume a cheese which is not placed and stored in the refrigerator for themselves (57.4%). Like the rest of the semi-hard and hard cheeses, they prefer to purchase it from supermarket, eat it with bread and olives, followed by meat, and drink it with wine (Table 5). Finally, the majority of participants believe that the island of Lesvos produces quality foods. The results of the chi-square test, presented in Table S4, indicate significant differences with medium to weak association for the sociodemographic variables regarding knowledge and taste of the cheese, with V = 0.106 to V = 2.67, weak associations regarding place of production, unique characteristics, added value, non-refrigerated production and its purchase, and accompaniment meals, varying from V = 0.073 to V = 0.150, as shown in Table S4. The importance of cheese familiarity for the preferred choice motives has also been reported by others with similar results. Nacef et al. report that consumers familiar with the cheese based their hedonic judgment mainly on intrinsic cues (tasting), whereas consumers unfamiliar were more influenced by extrinsic cues [54], with similar result reported for Turkish consumer purchase decisions [55]. Furthermore, Van Loo et al. report that the level of consumer ethnocentrism affects visual attention paid to origin labeling [56]. There are no studies today reporting on consumers' preference for cheeses within olive oil or non-refrigerated cheese preserved with other processes to compare our results.

Overall, our generic findings indicate no significant changes in consumers' preference for quality Greek semi-hard and hard cheese, even in the young generation, in the post-COVID era, as compared with the period before. Consumers' selection criteria, such as motives on purchase of consumption (place of purchase, association with meals, kinds of cheeses, place of consumption) and preferences of choice (such as organoleptic characteristics, general information, appearance, sustainability, and other characteristics) remain the same today as before. Only the overall consumption has decreased today, which is reasonable considering the economic crisis worldwide, including in Greece. Similar consumer perception for a specific, relatively unknown Greek semi-hard and hard cheese "Ladotyri" is also recorded.

## 5. Conclusions

P. Kottler, the pioneer in marketing, predicts that the "new" consumer in the new post-COVID-19 era will be "anti-consumer", grouped in five distinctive categories [57], namely the *Climate activists*, the *Degrowth activists*, the *Life simplifiers*, the *Food choosers*, and the *Conservation activists*. Despite the expected dramatic changes in consumers' perception for food, our study, conducted very recently in Fall 2022, on consumers' preferences for semi-hard and hard cheese indicates that they continue to select, buy, and consume this type of cheese the same way today as compared with the period before the pandemic, with minor changes recorded. The most significant change recorded is the dramatic decrease in cheese consumption, already reaching −8.4%, which may decrease further in the long run due to the foreseen global economic crisis. Concerning cheese purchase, the supermarket is still the source of choice for 90% on a daily (31.8%) or twice weekly basis (38.3%), eating the cheese at home (90.9%) and selecting it primarily based on the price (73.5%). Other selection criteria in order of significance are taste (97%), value for money (85.8%), texture (70.9%), and origin of milk (63.9%). They primarily eat the cheese with bread and olives (57.6%). The study recorded similar consumer motives for a specific Greek type of local traditional hard cheese with a unique formation and stored in olive oil, the Ladotyri Mytilinis cheese, with participants appreciating the olive oil storage (79.7%) and the possible production as a non-refrigerated cheese (65.2%), even though they would not buy it today (57.4%). The survey study focused on youngsters aged 18–25 (83.9%) on purpose to predict the future trends in a more reliable way.

The constraints of the study include the majority of female participants, of Greek nationality only, with the use of Greek cheese only during a period just after the pandemic. The results should be used as a primary roadmap for the future growth and development of the industry of semi-hard and hard cheeses in the new global economic era. Additional research with more questionnaires is required to better clarify the parameters of consumers' motives for quality semi-hard and hard cheese in the "new normality".

**Supplementary Materials:** The following supporting information can be downloaded at: https://www.mdpi.com/article/10.3390/su15075825/s1, Table S1: Questionnaire consumers' perception for quality semi-hard and hard cheese in the Post COVID-19 era. Table S2: Associations between motives on purchase and consumption of Greek semi-hard and hard cheese and the sociodemographic variables. Table S3: Associations between preference of choice for quality Greek semi-hard and hard cheese and the sociodemographic variables. Table S4: Associations between knowledge and preference of Ladotyri and the sociodemographic variables.

**Author Contributions:** Conceptualization, supervision, methodology, D.S.; writing—original draft preparation, D.S. and I.S.K.; investigation, K.B.; review and editing, D.S. and E.S. All authors have read and agreed to the published version of the manuscript.

**Funding:** This research received partial funding by the "INNOCHEESE" ERDF-North Aegean region funded program 2014–2020 (BAP2-0060569) of the E. THIMELIS DAIRY S.A, of Lesvos.

**Institutional Review Board Statement:** Not applicable.

**Informed Consent Statement:** Not applicable.

**Data Availability Statement:** Data are contained within the article.

**Conflicts of Interest:** The authors declare no conflict of interest.

## References

1. Sidhu, G.S.; Rai, J.S.; Khaira, K.S.; Kaur, S. The Impact of COVID-19 Pandemic on Different Sectors of The Indian Economy: A Descriptive Study. *Int. J. Econ. Financ. Issues* **2020**, *10*, 113–120. [CrossRef]
2. Fu, H.; Hereward, M.; Macfeely, S.; Me, A.; Wilmoth, J. How COVID-19 is changing the world: A statistical perspective from the Committee for the Coordination of Statistical activities. *Stat. J. IAOS* **2020**, *36*, 851–860. [CrossRef]
3. Kim, R.Y. The Impact of COVID-19 on Consumers: Preparing for Digital Sales. *IEEE Eng. Manag. Rev.* **2020**, *48*, 212–218. [CrossRef]
4. Abdal, A.; Ferreira, D.M. Deglobalization, globalization, and the pandemic current impasses of the capitalist world-economy. *J. World Syst. Res.* **2021**, *27*, 202–230. [CrossRef]
5. Martino, F.; Brooks, R.; Browne, J.; Carah, N.; Zorbas, C.; Corben, K.; Saleeba, E.; Martin, J.; Peeters, A.; Backholer, K. The nature and extent of online marketing by big food and big alcohol during the COVID-19 pandemic in Australia: Content analysis study. *JMIR Public Health Surveill.* **2021**, *7*, e25202. [CrossRef]
6. Gerritsen, S.; Sing, F.; Lin, K.; Martino, F.; Backholer, K.; Culpin, A.; Mackay, S. The Timing, Nature and Extent of Social Media Marketing by Unhealthy Food and Drinks Brands During the COVID-19 Pandemic in New Zealand. *Front. Nutr.* **2021**, *8*, 645349. [CrossRef]
7. Hirvonen, K.; de Brauw, A.; Abate, G.T. Food Consumption and Food Security during the COVID-19 Pandemic in Addis Ababa. *Am. J. Agric. Econ.* **2021**, *103*, 772–789. [CrossRef]
8. Cavallo, C.; Sacchi, G.; Carfora, V. Resilience effects in food consumption behaviour at the time of COVID-19: Perspectives from Italy. *Heliyon* **2020**, *6*, e05676. [CrossRef]
9. Hidayah, I.N.; Rohmah, N.F.; Saifuddin, M. Effectiveness of Digital Platforms as Food and Beverage Marketing Media During the COVID-19 Pandemic. *Airlangga J. Innov. Manag.* **2021**, *2*, 122. [CrossRef]
10. Skalkos, D.; Kalyva, Z. Exploring the impact of COVID-19 pandemic on Food Choice Motives: A systemtic review. *Sustainability* **2023**, *15*, 1606. [CrossRef]
11. Robinson, R.K.; Wilbey, R.A. Importance of cheese as a food. In *Cheesemaking Practice*; Springer: Boston, MA, USA, 1998; pp. 9–18.
12. Johnson, M.E. A 100-Year Review: Cheese production and quality. *J. Dairy Sci.* **2017**, *100*, 9952–9965. [CrossRef]
13. Guiné, R.P.F.; Florenca, S.G. The economic and social importance of cheese. In *Cheeses Around the World*; Ferrão, A.C., dos Reis Correia, P.M., de Pinho Ferreira Guiné, R., Eds.; NOVA Science Publishers: New York, NY, USA, 2019; pp. 1–16; ISBN 978-1-53615-419-1.
14. Ferrão, A.C.; Guine, R.P.F. Cheese: Nutritional Aspects anf Health effects. In *Cheeses Around the World*; Ferrão, A.C., dos Reis Correia, P.M., de Pinho Ferreira Guiné, R., Eds.; NOVA Science Publishers: New York, NY, USA, 2019; pp. 17–44; ISBN 978-1-53615-419-1.
15. Zheng, X.; Shi, X.; Wang, B. A Review on the General Cheese Processing Technology, Flavor Biochemical Pathways and the Influence of Yeasts in Cheese. *Front. Microbiol.* **2021**, *12*, 703284. [CrossRef]
16. Nájera, A.I.; Nieto, S.; Barron, L.J.R.; Albisu, M. A review of the preservation of hard and semi-hard cheeses: Quality and safety. *Int. J. Environ. Res. Public Health* **2021**, *18*, 9789. [CrossRef]
17. Dimitreli, G.; Exarhopoulos, S.; Goulas, A.; Antoniou, K.D. Traditional Greek cheeses. In *Cheeses around the World*; Ferrão, A.C., dos Reis Correia, P.M., de Pinho Ferreira Guiné, R., Eds.; NOVA Science Publishers: NY, USA, 2019; pp. 329–378; ISBN 978-1-1-53615-419-1.
18. *2082/92*; Certificates of specific character for agricultural products and foodsttuffs. European Union: Brussels, Belgium, 1992; pp. 9–14.
19. Greek Traditional products (PDO-PGI-TSG). Hellenic Ministry of Agricultural Development and Food. 2021. Available online: http://www.minagric.gr/index.php/el/for-farmer-2/2012-02-02-07-52-07 (accessed on 27 November 2022).
20. Pappa, E.C.; Kondyli, E. Descriptive characteristics and cheesemaking technology of Greek cheeses not listed in the EU geographical indications registers. *Dairy* **2023**, *4*, 43–67. [CrossRef]
21. Vakoufaris, H. The impact of ladotyri mytilinis PDO cheese on the rural development of Lesvos Island, Greece. *Local Environ.* **2010**, *15*, 27–41. [CrossRef]
22. Tsiboukas, K.; Karanikolas, P.; Stachtiaris, S.; Kominakis, A.; Spilioti, M. A niche strategy for geographical indication products, by valorising local resources: The Greek cheese Ladotyri Mytilinis. *Int. J. Agric. Resour. Gov. Ecol.* **2022**, *18*, 160–181. [CrossRef]
23. Hrubá, R.; Sadílek, T. Behavioural differences among educated young consumers in the Czech Republic: The case of organic cheese consumption. *Int. J. Food Syst. Dyn.* **2020**, *11*, 117–126. [CrossRef]
24. Menozzi, D.; Giraud, G.; Saïdi, M.; Yeh, C.H. Choice drivers for quality-labelled food: A cross-cultural comparison on pdo cheese. *Foods* **2021**, *10*, 1176. [CrossRef]
25. Del Toro-Gipson, R.S.; Rizzo, P.V.; Hanson, D.J.; Drake, M.A. Consumer perception of smoked Cheddar cheese. *J. Dairy Sci.* **2021**, *104*, 1560–1575. [CrossRef]
26. Racette, C.M.; Drake, M.A. Consumer perception of natural hot-pepper cheeses. *J. Dairy Sci.* **2022**, *105*, 2166–2179. [CrossRef]
27. Zhllima, E.; Mehmeti, G.; Imami, D. Consumer preferences for cheese with focus on food safety—A segmentation analysis. *Sustainability* **2021**, *13*, 2524. [CrossRef]
28. Mazzocchi, C.; Orsi, L.; Sali, G. Consumers' attitudes for sustainable mountain cheese. *Sustainability* **2021**, *13*, 1743. [CrossRef]

29. Ouyang, H.; Kilcawley, K.N.; Miao, S.; Fenelon, M.; Kelly, A.; Sheehan, J.J. Exploring the potential of polysaccharides or plant proteins as structuring agents to design cheeses with sensory properties focused toward consumers in East and Southeast Asia: A review. *Crit. Rev. Food Sci. Nutr.* **2022**, *62*, 4342–4355. [CrossRef] [PubMed]

30. Ojeda, M.; Etaio, I.; Valentin, D.; Dacremont, C.; Zannoni, M.; Tupasela, T.; Lilleberg, L.; Pérez-Elortondo, F.J. Effect of consumers' origin on perceived sensory quality, liking and liking drivers: A cross-cultural study on European cheeses. *Food Qual. Prefer.* **2021**, *87*, 104047. [CrossRef]

31. Endrizzi, I.; Cliceri, D.; Menghi, L.; Aprea, E.; Gasperi, F. Does the 'mountain pasture product' claim affect local cheese acceptability? *Foods* **2021**, *10*, 682. [CrossRef] [PubMed]

32. Miloradovic, Z.; Blazic, M.; Barukcic, I.; Font i Furnols, M.; Smigic, N.; Tomasevic, I.; Miocinovic, J. Serbian, Croatian and Spanish consumers' beliefs towards artisan cheese. *Br. Food J.* **2022**, *124*, 3257–3273. [CrossRef]

33. Ouyang, H.; Li, B.; McCarthy, M.; Miao, S.; Kilcawley, K.; Fenelon, M.; Kelly, A.; Sheehan, J.J. Understanding preferences for and consumer behavior toward cheese among a cohort of young, educated, internationally mobile Chinese consumers. *J. Dairy Sci.* **2021**, *104*, 12415–12426. [CrossRef]

34. Van Rijswijk, W.; Frewer, L.J.; Menozzi, D.; Faioli, G. Consumer perceptions of traceability: A cross-national comparison of the associated benefits. *Food Qual. Prefer.* **2008**, *19*, 452–464. [CrossRef]

35. Palmieri, N.; Perito, M.A.; Macrì, M.C.; Lupi, C. Exploring consumers' willingness to eat insects in Italy. *Br. Food J.* **2019**, *121*, 2937–2950. [CrossRef]

36. Palmieri, N.; Suardi, A.; Pari, L. Italian consumers' willingness to pay for eucalyptus firewood. *Sustainability* **2020**, *12*, 2629. [CrossRef]

37. Palmieri, N.; Perito, M.A.; Lupi, C. Consumer acceptance of cultured meat: Some hints from Italy. *Br. Food J.* **2020**, *123*, 109–123. [CrossRef]

38. De Leeuw, A.; Valois, P.; Ajzen, I.; Schmidt, P. Using the theory of planned behavior to identify key beliefs underlying pro-environmental behavior in high-school students: Implications for educational interventions. *J. Environ. Psychol.* **2015**, *42*, 128–138. [CrossRef]

39. Pappalardo, G.; Lusk, J.L. The role of beliefs in purchasing process of functional foods. *Food Qual. Prefer.* **2016**, *53*, 151–158. [CrossRef]

40. Chinnici, G.; D'Amico, M.; Pecorino, B. A multivariate statistical analysis on the consumers of organic products. *Br. Food J.* **2002**, *104*, 187–199. [CrossRef]

41. Giampietri, E.; Verneau, F.; Del Giudice, T.; Carfora, V.; Finco, A. A Theory of Planned behaviour perspective for investigating the role of trust in consumer purchasing decision related to short food supply chains. *Food Qual. Prefer.* **2018**, *64*, 160–166. [CrossRef]

42. Skalkos, D.; Kosma, I.S.; Chasioti, E.; Skendi, A.; Papageorgiou, M.; Guiné, R.P.F. Consumers' Attitude and Perception toward Traditional Foods of Northwest Greece during the COVID-19 Pandemic. *Appl. Sci.* **2021**, *11*, 4080. [CrossRef]

43. Skallerud, K.; Korneliussen, T.; Olsen, S.O. An examination of consumers' cross-shopping behaviour. *J. Retail. Consum. Serv.* **2009**, *16*, 181–189. [CrossRef]

44. Planzer, S.B.; Da Cruz, A.G.; Sant'ana, A.S.; Silva, R.; Moura, M.R.L.; De Carvalho, L.M.J. Food safety knowledge of cheese consumers. *J. Food Sci.* **2009**, *74*, M28–M30. [CrossRef]

45. Laguna, L.; Fiszman, S.; Puerta, P.; Chaya, C.; Tárrega, A. The impact of COVID-19 lockdown on food priorities. Results from a preliminary study using social media and an online survey with Spanish consumers. *Food Qual. Prefer.* **2020**, *86*, 104028. [CrossRef]

46. Stadista Research Department Per Capita Consumption of Cheese Worlwide, by Country in Kilograms in 2020. Available online: https://www.statista.com/statistics/527195/consumption-of-cheese-per-capita-worldwide-country (accessed on 20 December 2022).

47. Iglesia, I.; Intemann, T.; De Miguel-Etayo, P.; Pala, V.; Hebestreit, A.; Wolters, M.; Russo, P.; Veidebaum, T.; Papoutsou, S.; Nagy, P.; et al. Dairy consumption at snack meal occasions and the overall quality of diet during childhood. Prospective and cross-sectional analyses from the idefics/i.family cohort. *Nutrients* **2020**, *12*, 642. [CrossRef]

48. Silvestri, C.; Aquilani, B.; Piccarozzi, M.; Ruggieri, A. Consumer Quality Perception in Traditional Food: Parmigiano Reggiano Cheese. *J. Int. Food Agribus. Mark.* **2020**, *32*, 141–167. [CrossRef]

49. De Barros, C.P.; Rosenthal, A.; Walter, E.H.M.; Deliza, R. Consumers' attitude and opinion towards different types of fresh cheese: An exploratory study. *Food Sci. Technol.* **2016**, *36*, 448–455. [CrossRef]

50. Schouteten, J.J.; de Steur, H.; de Pelsmaeker, S.; Lagast, S.; de Bourdeaudhuij, I.; Gellynck, X. Impact of health labels on flavor perception and emotional profiling: A consumer study on cheese. *Nutrients* **2015**, *7*, 5533. [CrossRef] [PubMed]

51. Hidalgo-Milpa, M.; Arriaga-Jordán, C.M.; Cesín-Vargas, A.; Espinoza-Ortega, A. Characterisation of consumers of traditional foods: The case of Mexican fresh cheeses. *Br. Food J.* **2016**, *118*, 915–930. [CrossRef]

52. Imami, D.; Zhllima, E.; Merkaj, E.; Chan-Halbrendt, C.; Canavari, M. Albanian consumer preferences for the use of powder milk in cheese-making: A conjoint choice experiment. *Agric. Econ. Rev.* **2016**, *17*, 20–33.

53. Boatto, V.; Rossetto, L.; Bordignon, P.; Arboretti, R.; Salmaso, L. Cheese perception in the North American market: Empirical evidence for domestic vs imported Parmesan. *Br. Food J.* **2016**, *118*, 1747–1768. [CrossRef]

54. Nacef, M.; Lelièvre-Desmas, M.; Symoneaux, R.; Jombart, L.; Flahaut, C.; Chollet, S. Consumers' expectation and liking for cheese: Can familiarity effects resulting from regional differences be highlighted within a country? *Food Qual. Prefer.* **2019**, *72*, 188–197. [CrossRef]
55. Topcu, Y.; Dağdemir, V. Turkish Consumer Purchasing Decisions Regarding PGI-labelled Erzurum Civil Cheese. *Alınteri Zirai Bilim. Derg.* **2017**, *32*, 69–80. [CrossRef]
56. Van Loo, E.J.; Grebitus, C.; Roosen, J. Explaining attention and choice for origin labeled cheese by means of consumer ethnocentrism. *Food Qual. Prefer.* **2019**, *78*, 103716. [CrossRef]
57. Kottler, P. The Consumer in the Age of Coronavirus. *J. Creat. Value* **2020**, *6*, 12–15. [CrossRef]

*Article*

# Sustainable Ring-Opening Reactions of Epoxidized Linseed Oil in Heterogeneous Catalysis

Andrei Iulian Slabu [1,2], Ionut Banu [2], Octavian Dumitru Pavel [3], Florina Teodorescu [1,]* and Raluca Stan [2,]*

[1] "C. D. Nenitzescu" Institute of Organic and Supramolecular Chemistry of the Romanian Academy, 202 B Spl. Independenței, S6, 060023 Bucharest, Romania
[2] Faculty of Chemical Engineering and Biotechnologies, University Politehnica of Bucharest, 1-7 Polizu Street, 011061 Bucharest, Romania
[3] Faculty of Chemistry, Department of Inorganic Chemistry, Organic Chemistry, Biochemistry and Catalysis, University of Bucharest, 4-12 Regina Elisabeta Av., S3, 030018 Bucharest, Romania
[*] Correspondence: florina.teodorescu@ccocdn.ro (F.T.); raluca.stan@upb.ro (R.S.)

**Abstract:** In this study, renewable products with potentially interesting properties and applications were synthesized by functionalizing linseed oil via epoxidation and epoxy ring-opening with carboxylic acids and anhydrides. LDHs (Layered Double Hydroxides), a well-known class of materials used for a wide range of reactions, are the catalysts used in this study, with the overall advantages of facile separation and reusability. In our study, different types of carboxylic acids and anhydrides were employed as reactants with the advantage of leading to sustainable products that can replace petrochemical compounds. Following the optimization of the reaction conditions, including the basicity of the catalyst, at 170 °C a quasi-total conversion of the epoxy groups was achieved for all the ring-opening reagents.

**Keywords:** sustainable; vegetable oil; heterogeneous catalysis

**Citation:** Slabu, A.I.; Banu, I.; Pavel, O.D.; Teodorescu, F.; Stan, R. Sustainable Ring-Opening Reactions of Epoxidized Linseed Oil in Heterogeneous Catalysis. *Sustainability* **2023**, *15*, 4197. https://doi.org/10.3390/su15054197

Academic Editor: Dimitris Skalkos

Received: 28 January 2023
Revised: 16 February 2023
Accepted: 23 February 2023
Published: 25 February 2023



## 1. Introduction

Vegetable oils are a plentiful renewable raw material and are being investigated as a replacement for petroleum-based compounds used in industry in order to solve the pollution problem while also respecting the principles of green chemistry [1,2]. Triglycerides, which are glycerol esters with saturated or unsaturated fatty acids, are the main component of these raw materials. Vegetable oils' fatty acid composition impacts their properties and, as a result, their use as raw materials [3]. The presence of double bonds in the alkyl chain of unsaturated fatty acids allows for the creation of a wide range of molecules via functionalization reactions [4].

A significant amount of the compounds generated through functionalization reactions of vegetable oil such as metathesis, hydroformylation, or epoxidation can be employed as intermediates in the materials industry [5]. The epoxidation reaction is particularly interesting because it can be carried out on an industrial scale [6] and leads to epoxidized vegetable oil (EVO) that can be further functionalized to obtain a wide range of products using epoxy ring-opening reactions [7–9]. The double bonds also negatively influence the physical properties of the vegetable oil, so converting the unsaturation to epoxy groups modifies the oxidative and thermal stability of the vegetable oil, making it suitable for being used in demanding conditions, such as high-temperature lubricant applications [10].

Through epoxy ring-opening reactions, EVO can be converted to monomers such as acrylic resin precursors [9] and polyols that can then be further used to obtain polyurethanes [11] or vegetable oil-based waterborne polyurethane dispersions [12]. The polymers obtained from vegetable oil can be further used for obtaining natural composites such as bio-based materials reinforced with natural fiber mats (flax, hemp) and plant oil-based acrylic monomers as matrices [13] or a porous collagen-polyurethane composite with potential biomedical applications [14].

When the ring-opening reaction is performed with monocarboxylic acid anhydrides, the obtained products can be used as biolubricants [6]. However, using dicarboxylic acid anhydrides leads to curing, with the obtaining of resins that can be used in green composites [15], having mechanical properties comparable to those of resins obtained from traditional petrochemical sources [16].

As far as we know, most of the epoxy ring-opening reactions are performed using homogenous catalysts (boron trifluoride diethyl etherate [6], tertiary amines [17], or quaternary ammonium salts [18]). The use of these homogeneous catalysts leads to some environmental issues, corrosion, difficult separation from the chemical compound's mixture, etc. As an alternative to afterwards having to deal with these drawbacks, the use of heterogeneous catalysts presents several advantages: separation from the reaction mixture by simple filtration, stability at high temperatures, reuse, etc. In the last decades, several heterogeneous catalysts were considered for epoxy ring-opening reactions, such as graphene oxide/ZnO [19], solid acid catalysts (Nafion, Amberlyte, zeolite H-Y, montmorillonite) [20], and layered double hydroxides (LDH) [21]. These special catalytic materials have a ditopic character (displayed as both acid and base active sites) that were well characterized and used in the last decades [22]. They have the general formula $[M^{2+}_{1-x}M^{3+}_x(OH)_2]^{x+}[A^{n-}_{x/n}]\cdot mH_2O$, where $M^{2+}$ and $M^{3+}$ are divalent and trivalent cations in the brucite-type layers. A is an interlayer anion with an n charge which balances the exceeding charge that occurred by the isomorphic substitution of $M^{2+}$ by $M^{3+}$, x is the fraction of the trivalent cation, and m is the crystallization of water [23]. This formula does not only refer to the presence of a single bivalent and another trivalent cation but allows the inclusion of a large number of cations, thus generating LDH of the ternary or quaternary type. On the same note, the type of incorporated anion is not limited to hydroxyl or carbonate groups that have small dimensions, but even phthalocyanines or other bulky anions can be considered. In fact, all the cations that fit in the octahedral positions and that have a radius close to that of $Mg^{2+}$ (0.72 Å) [23] can generate such layered compounds. Traditionally, these materials can be synthesized by: (i) co-precipitation (at high and low supersaturation), (ii) urea hydrolysis, (iii) hydrothermal synthesis, (iv) rehydration using structural memory effect, (v) mechano-chemical, (vi) ion-exchange, (vii) exfoliation in aqueous solution, etc. [23]. Most of the common methods imply the use of inorganic alkalis as hydrolysis agents (i.e., NaOH, KOH, $Na_2CO_3$, $K_2CO_3$, $NH_4OH$, etc.), which presents some disadvantages, namely the generation of large volumes of wastewater during washing step, the use of a large number of specific vessels, high energy consumption as well as contamination with alkaline cations, etc. Therefore, their replacement with organic bases seems to be a viable option for the synthesis of these materials despite their slightly high price. These materials were used as catalytic materials in various reactions highlighted in an extensive review [24], such as double bond isomerization [25], epoxidation [26], Michael addition [27], Knoevenagel condensation [28], reduction of nitro compounds [29] or aldol condensation [30]. To the best of our knowledge, the LDH-type materials have not been considered in the ring-opening of epoxidized vegetable oil. Therefore, the aim of this research was to synthesize materials and precursors from epoxidized linseed oil in the presence of an LDH catalyst and various ring-opening reagents (carboxylic acids and anhydrides). Additionally, highlighting the benefits presented by the use of heterogeneous catalysts for functionalizing vegetable oil was considered.

## 2. Materials and Methods

### 2.1. Materials

For the epoxidation reaction, linseed oil (LO) obtained by cold-pressing process was acquired from PTG Deutschland, Flurstedt, Germany, glacial acetic acid, sulfuric acid (95–97% vol.), and technical grade toluene were purchased from Sigma Aldrich (Saint Louis, MO, USA). Hydrogen peroxide (30% vol.) was purchased from Atochim SRL (Bucharest, Romania). For the ring-opening reactions, the necessary reagents (carboxylic acids and anhydrides) were purchased from Sigma Aldrich and used as such, except for phthalic anhydride, which was purified by dehydration.

### 2.2. Catalyst Preparation and Characterization

The Mg/Al hydrotalcite with a molar ratio of 3 was synthesized by a traditional co-precipitation method at pH 10, under low supersaturation conditions. An aqueous solution of the corresponding nitrates at a feed rate of 60 $cm^3 \cdot min^{-1}$ at room temperature was mixed at 600 rpm with another base solution of $Na_2CO_3/NaOH$ (at a molar ratio of 1/3 and a solution concentration of 1 M in $Na_2CO_3$). The obtained gel was aged for 18 h at 75 °C, cooled to room temperature, filtered, and washed with bi-distilled water until a neutral pH was reached. It was then dried for 24 h at 90 °C, with the obtaining of the MgAl LDH. In order to increase the base character of hydrotalcite, a trivalent cation was considered, e.g., La, that presents an electronegativity in a Pauling scale of 1.1, a lower value than that of Al (1.61). Therefore, a similar method was applied to synthesize a La-modified MgAl hydrotalcite at a $Al^{3+}/La^{3+}$ of 1 molar ratio, MgAlLa LDH [31]. The characterization of both materials was performed according to previous work [32], DRIFT, powder XRD, and determination of the base sites using adsorption of organic acids with different pKa values and $N_2$ adsorption-desorption isotherms.

Powder X-ray diffraction patterns were recorded with a Shimadzu XRD 7000 diffractometer that uses CuKα radiation (λ = 1.5418 Å, 40 kV, 40 mA) with a scanning speed of 0.10° $min^{-1}$ in the 5–75° 2theta range. DRIFTS spectra, obtained from accumulation of 400 scans in the domain 400–4000 $cm^{-1}$ with a scanning speed of 128 scans/min, triangle apodization, and a resolution of 4 $cm^{-1}$, were recorded with JASCO FT/IR-4700 spectrometer. The textural analysis of the samples was performed through $N^2$ physisorption at −196 °C using a Micromeritics ASAP 2020 analyzer where the samples were degassed under vacuum at 120 °C for 12 h. The base character of the catalysts was determined using a method based on the irreversible adsorption of acrylic acid (pKa = 4.2).

### 2.3. Characterization of the Products

### 2.3.1. Nuclear Magnetic Resonance (NMR) Spectrometry

$^1H$ NMR was used for studying the structures of linseed oil (LO), epoxidized linseed oil (ELO), and the obtained compounds. The samples were dissolved in 0.5 mL $CDCl_3$, and the spectra were recorded using a Bruker Advance III 600 MHz spectrometer, with a resonance frequency of 600.12 MHz for the $^1H$ nucleus, equipped with an indirect detection for nuclei probe head (BBI) and field gradients on Z axis. The chemical shifts are measured in ppm, the signals were calibrated using the $CDCl_3$ signal (7.26 ppm) and tetramethyl silane (TMS) as internal standard.

### 2.3.2. Fourier-Transform Infrared Spectroscopy (FTIR)

FTIR spectroscopy was performed with a Bruker VERTEX 70 instrument, equipped with a Harrick MVP2 diamond attenuated total reflectance (ATR) device. The FTIR spectra were recorded using 32 scans in 600–4000 $cm^{-1}$ wave number region.

### 2.4. Linseed Oil Epoxidation

The epoxidation of the vegetable oil was carried out using a well-established method [33], which uses a Prilezhaev-type reaction, with the in situ generation of the peracid [34]. LO (10 mL), acetic acid, and 50% (vol.) sulfuric acid are mixed in toluene solvent (15 mL). The corresponding volume of $H_2O_2$ is then added dropwise, under stirring at room temperature. After all the $H_2O_2$ is added, the temperature is raised to 60 °C, and the reaction mixture is kept under constant stirring for 24 h. The molar ratio of double bonds to acetic acid to $H_2O_2$ is 1:2:10, with the average unsaturation of the LO being 6 double bonds per triglyceride molecule, determined using a $^1H$ NMR method [35] (Figure 1). After the completion of the reaction, the organic and aqueous phases were separated. In order to remove traces of acetic acid and hydrogen peroxide, the organic phase was washed with distilled water and saturated sodium bicarbonate solution until a neutral pH was reached. The organic solvent was then evaporated under vacuum, obtaining a semisolid white product (95% quantitative yield). The epoxidation was considered quasi-total, based on comparing the $^1$H NMR spectra of the LO and ELO [36].

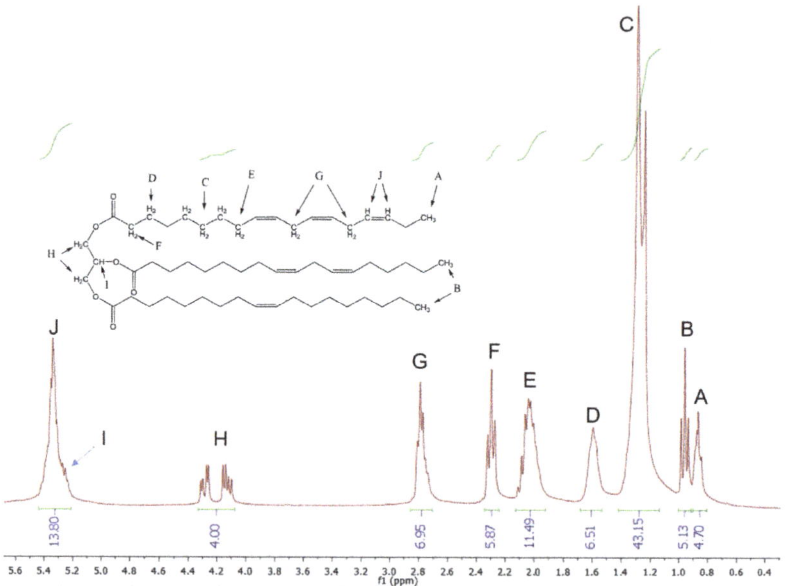

**Figure 1.** The $^1$H NMR signals used for the determination of LO unsaturation.

### 2.5. Epoxy Ring-Opening Reaction

In a typical experiment, 1 mmol (0.978 g) ELO, the ring-opening reagent, and the MgAl LDH catalyst (10% wt) were mixed in 15 mL solvent (toluene or xylene). The quantities of the ring-opening reagent were changed in accordance with the employment of various molar ratios between the epoxy groups and the ring-opening reagent. The reaction mixture was heated to temperatures from 100 to 140 °C and kept under magnetic stirring for 24 h. After the end of the reaction, the mixture was filtered for the removal of the solid catalyst, followed by the washing of the organic liquid phase with distilled water to eliminate the unreacted ring-opening reagent. Further, the solvent is evaporated under a vacuum, obtaining a liquid viscous product.

## 3. Results

### 3.1. Catalyst Characterization

The hydrotalcite sample (MgAl) showed diffraction lines corresponding to the typical structure of layered materials without the detection of other contaminating phases (Figure 2, black line). The sample modified with lanthanum displayed diffraction lines of very low intensity corresponding to the phases of $La_2O_2CO_3$ (JCPDS card 23-0320) and $La_2(CO_3)_2(OH)_2$ (JCPDS card 70-1774). The very strong electronegative character of La (1.1 in the Pauling scale) favors the formation of carbonate-type species as soon as the hydrotalcite synthesis process starts. At the same time, the large ionic radius of $La^{3+}$ (1.032 Å) prevents intercalation of the large lanthanides species in hydrotalcite galleries. The insertion of La into the octahedral positions of LDH was not performed considering the fact that the network **a**-parameter does not undergo any significant changes (3.060 Å for MgAl and 3.062 Å for MgAlLa). Instead, the decrease of the network **c**-parameter from 23.357 Å (MgAl) to 22.226 Å (MgAlLa) was not due to the presence of lanthanum, but rather to the smaller amount of aluminum that was involved in the construction of the layered structure. Mostly, the quantity of lanthanum was placed on the surface of the layered material in the carbonate oxide/hydroxide forms, but at the same time the presence of contaminants was not excluded even in the galleries, a fact supported by the $I_{006}/I_{003}$ ratios that increase from 0.43 Å to 1.23 Å. At the same time, the presence of La led to a decrease in the crystallinity from 84 Å to 38 Å.

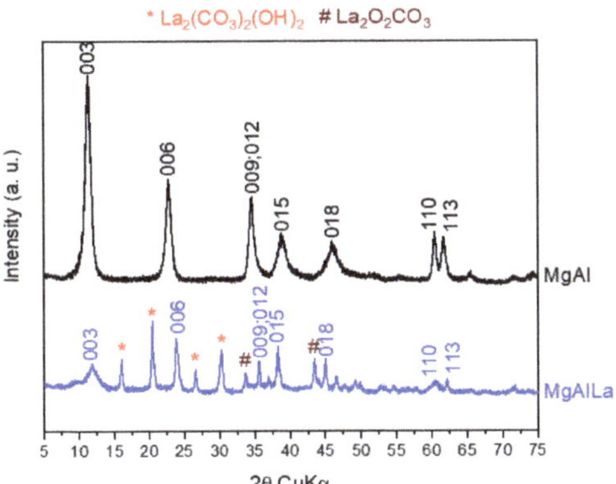


**Figure 2.** The XRD pattern of the solids (* La$_2$(CO$_3$)$_2$(OH)$_2$; # La$_2$O$_2$CO$_3$).

The DRIFT spectra (Figure S1) displayed for both samples a large band between 3700 and 3400 cm$^{-1}$ that corresponds to the OH group vibration, $\nu_{(O-H)}$, a band at 3000 cm$^{-1}$ to hydrogen bonding between water and carbonate in the interlayer space, a band between 1640–1660 cm$^{-1}$ to H$_2$O bending vibration of interlayer water. The bands at 1400 and 1200–700 cm$^{-1}$ are assigned to CO$_3^{2-}$ group vibration bands in the hydrotalcite. The cations-O bonds are attributed below 600 cm$^{-1}$. The presence of the band at 3647 cm$^{-1}$ is due to a separate phase of Mg(OH)$_2$ which is not visible in XRD. Its presence is related to the fact that a quantity of Mg that was not inserted in the layered structure (composition corresponding to the La non-inserted in the octahedral structure of LDH) is in the form of brucite as a contaminant alongside the carbonate oxide/hydroxide forms determined by XRD.

As expected, the presence of La leads to a decrease in the specific surface area from 122 to 71 m$^2$·g$^{-1}$, with the adsorption-desorption isotherms for both samples classified as type IV according to the IUPAC classification (Figure S2). However, the low electronegative character of La determined an increase in basicity from 6.73 to 8.62 mmol acrylic acid·g$^{-1}$.

*3.2. Epoxide Opening with Unsaturated Carboxylic Acids*

3.2.1. Methacrylic Acid

Acrylated and methacrylated vegetable oils obtained by the ring-opening of epoxidized vegetable oils are being used as photopolymerizable monomers, forming highly crosslinked polymer networks [9]. Furthermore, when the epoxy groups are partially opened with methacrylic acid, the remaining groups can be functionalized with another reagent, such as hydrophilic dimethacrylated poly(ethylene glycol), leading to materials such as oil-based hydrophilic monomers [37]. The goal of the present study was to test the activity of the catalysts with ditopic characters (MgAl LDH/MgAlLa LDH) for the epoxy ring-opening reaction and the total and/or partially opening of the epoxy rings with methacrylic acid (MA). Using a slight excess of MA (1:1.2 epoxy groups to MA molar ratio) led to fully opening the epoxy rings, which was confirmed by $^1$H NMR. The $^1$H NMR spectra for ELO show the following signals: 0.90 (t, terminal -**CH$_3$** from all fatty acids except linolenic acid), 1.03–0.95 (m, terminal -**CH$_3$** from linolenic acid), 1.28–1.20 (m, -**CH$_2$**- from all alkyl chains), 1.57–1.44 (m, -**CH$_2$**-CH$_2$-COO), 1.76–1.67 (m, -**CH$_2$**- between epoxy rings), 2.26 (t, -**CH$_2$**-COO acyl group), 2.91 (m, **CH** marginal protons from the epoxy ring), 3.10 (m, **CH** internal protons of the epoxy ring), 4.10–4.23 (m, -**CH$_2$**-O-CO-, glycerol protons in α positions), 5.20 (m, -**CH**-O-CO-, glycerol proton from β position), while the methacrylated product spectral data are 0.90 (t, terminal -**CH₃** from all fatty acids except linolenic acid), 1.03–0.95 (m, terminal -**CH₃** from linolenic acid), 1.28–1.20 (m, -**CH₂**- from all alkyl chains), 1.57–1.44 (m, -**CH₂**-CH₂-COO), 1.95 (s, CH₃–C=CH₂, methacrylate), 2.26 (t, -**CH₂**-COO acyl group), 4.10–4.23 (m, -**CH₂**-O-CO-, glycerol protons in α positions), 5.20 (m, -**CH**-O-CO-, glycerol proton from β position), and 6.13, 5.58 (s, **CH₂**=C-CH₃ from methacrylic group). Comparing the ¹H NMR spectra of ELO and the product shows the consumption of the epoxy groups, as evidenced by the disappearance of the corresponding signals (3.1–2.9 ppm) in the product spectra (Figure 3). Additionally, signals specific to methacrylate groups appear in the spectra of the product at chemical shifts of 5.6 and 6.2 ppm (CH₂ protons) and 1.95 ppm (CH₃ protons), which further shows that the methacrylation was successful. FTIR analysis also showed the total consumption of the epoxy rings, as evidenced by the disappearance of the epoxy group vibration (830 cm⁻¹) and the appearance of the vibration of the double of MA (1630 cm⁻¹) in the product spectrum (for details, see Figure S3).

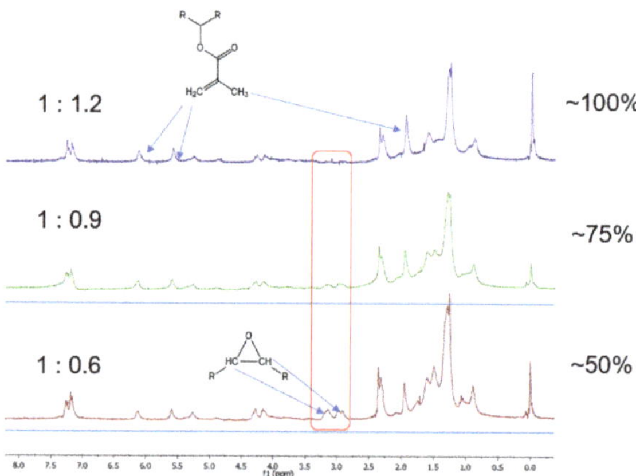


**Figure 3.** Partial ring-opening of the epoxy rings using MA.

For partial opening the epoxy rings, different molar ratios between epoxy and MA were used (Figure 3). For a ratio of 1:0.9 (epoxy groups:MA), the conversion of the epoxy groups was around 75%, while for a 1:0.6 ratio, the conversion was 50%. This trend continues at various molar ratios, as seen in Figure 4. As a general method, the conversions were calculated by comparing the peak areas of the epoxy signals in ELO and in the ring-opening products. In order to have comparable and quantifiable results, the method used is based on the integration of the signals of the glycerol protons in α positions (4.14–4.27 ppm) and β position (5.26 ppm) and using them as internal standards, due to the fact that the glycerol backbone remains unchanged during the reaction. This means that peak areas in both spectra can now be compared (for details, see Figure S4), and the conversion can be calculated using the epoxy signals with the following formula:

$$C = \frac{A}{B} \times 100$$

where $C$ is the conversion (%) of ELO, $A$ is the peak area of the epoxy signal in the product and $B$ is the peak area of the epoxy signal in ELO. The result can be confirmed by the peak area of the signals specific to the product (-CH₃ protons in the methacrylate at 1.96 ppm and =CH₂ proton of the methacrylate at 5.59 and 6.11 ppm). This general method was applied for calculating conversions for all the epoxy ring-opening reactions in the present study.

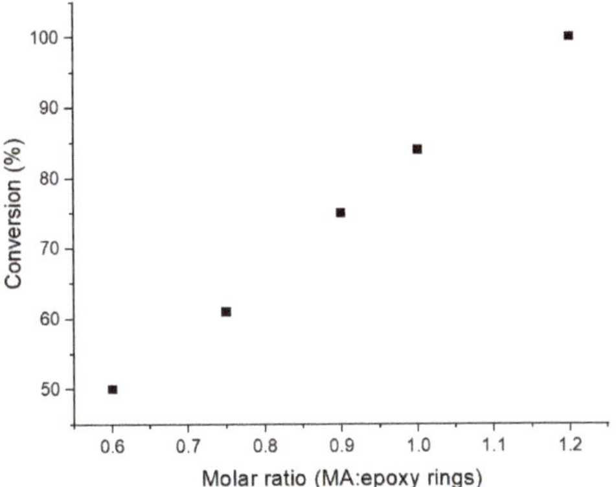

**Figure 4.** The influence of the molar ratio of the reactants on the conversion of the epoxy groups.

### 3.2.2. Undecylenic Acid

A renewable monomer that can be obtained from castor oil, undecylenic acid (UA) has numerous uses in pharmaceutical and cosmetic applications [38]. It can also be used to obtain bio-based polyurethanes, polyamides, resins [39], organogels [40], or encapsulations [41], so its potential led to it being chosen as a ring-opening reactant. The first reaction was performed in the same conditions as the ring-opening with MA, with a slight excess of UA, with a conversion of 30% for the epoxy groups. This was determined by comparing the $^1$H NMR spectra of ELO and the product and by quantifying the decrease in signal strength for the epoxy groups (2.9–3.1 ppm) in the product. Specific signals for UA can be observed, particularly for the protons adjacent to the terminal double bond, at a chemical shift of 4.9 and 5.8 ppm.

By employing a different LDH catalyst (MgAlLa LDH) with stronger basicity, as shown in the catalyst characterization results in this section, a 50% conversion of the epoxy groups was achieved. Using this catalyst (MgAlLa LDH) and increasing the temperature to 140 °C gave the best results, with the obtaining of the total conversion of the epoxy groups, as shown in the $^1$H NMR spectral data of the product (see Figure S5). Note that using the MgAl LDH at 140 °C only gave an 85% conversion, which further confirms the positive influence of the basicity of the catalyst on the conversion.

### 3.2.3. Crotonic Acid; Cinnamic Acid

A β-methyl substituted acrylic monomer, crotonid acid can be obtained from petrochemical resources via the oxidation of the crotonaldehyde or from bioresources, via acetaldehyde resulting from the fermentation of bioethanol. It has various uses in the polymer industry, such as being used in the composition of copolymers used for adhesives, paints, and coating applications [42]. Opening the epoxy groups with crotonic acid at 100 °C temperature led to conversions of 40% when using MgAl LDH and 60% when using MgAlLa LDH, further confirming the theory that the basicity of the catalyst positively influences the conversion. The obtained products have the following $^1$H NMR signals characteristic to crotonic moieties: 5.85 (m, **CH₃-CH=CH-**), 7.15 (m, **CH₃-CH=CH-**) (for details, see Figure S6).

Cinnamic acid and its esters are well-known β-phenyl substituted acrylic monomers that can be obtained from both petrochemical resources and renewable feedstock [42]. These types of compounds can give reversible [2 + 2] cycloaddition reactions initiated by UV light, leading to shape-memory polymers [43]. The ring-opening reactions with cinnamic acid gave conversions of 20% (MgAl LDH catalyst) and 70% (MgAlLa LDH catalyst) at 100 °C, and the $^1$H NMR specific signals for cinnamic acid were observed at 6.45 (d, **-CH=CH-**Ph), 7.40 (m, **-CH=CH-**Ph), 7.40–7.50 aromatic signals (for details, see Figure S7).

As in the case of undecylenic acid, total conversion of the epoxy groups can be achieved by using the MgAlLa LDH catalyst at 140 °C reaction temperature for both crotonic and cinnamic acid.

### 3.3. Epoxide Opening with Monocarboxylic Acids Anhydrides

Monocarboxylic acids and anhydrides have been previously used as ring-opening reagents for obtaining lubricant materials [6]. The reactants chosen for this study were acetic and butyric anhydride, with the aim of observing a possible influence of the length of the carbon chain on the reactivity towards epoxy ring-opening. Using the reaction conditions first employed for acids (100 °C, MgAl LDH catalyst), the conversion of the epoxy groups was total for both anhydrides. The IR spectra of the reaction product with butyric anhydride show the total conversion of epoxy groups (for details, see Figure S8), which is confirmed by the $^1$H NMR spectrum of the product, detailed below, which shows no signals characteristic of the epoxy ring: 0.90 (t, terminal -**CH$_3$** from all fatty acids except linolenic acid, terminal -**CH$_3$** from the butyric chain), 1.03–0.95 (m, terminal -**CH$_3$** from linolenic acid), 1.28–1.20 (m, -**CH$_2$**- from all alkyl chains), 1.57–1.44 (m, -**CH$_2$**-CH$_2$-COO), 2.26 (t, -**CH$_2$**-COO acyl group in the triglyceride backbone and butyric chain), 4.1–4.23 (m, -**CH$_2$**-O-CO-, glycerol protons in α positions), 5.20 (m, -**CH**-O-CO-, glycerol proton from β position). For the reaction product with acetic anhydride, the results are similar.

## 4. Discussion

Various ring-opening reagents have been used to functionalize epoxidized vegetable oil. However, unsaturated carboxylic acids are of particular interest due to the possibility of further reacting/functionalizing the double bonds. Other promising ring-opening reagents are carboxylic anhydrides, due to their high reactivity and the potential uses for the obtained products, such as lubricants in the case of vegetable oils functionalized with monocarboxylic anhydrides [6]. Unsaturated carboxylic acids and carboxylic anhydrides with interesting properties were chosen for this study (Figure 5). The ring-opening reactions using acids are further outlined (reaction conditions and conversions) in Table 1.

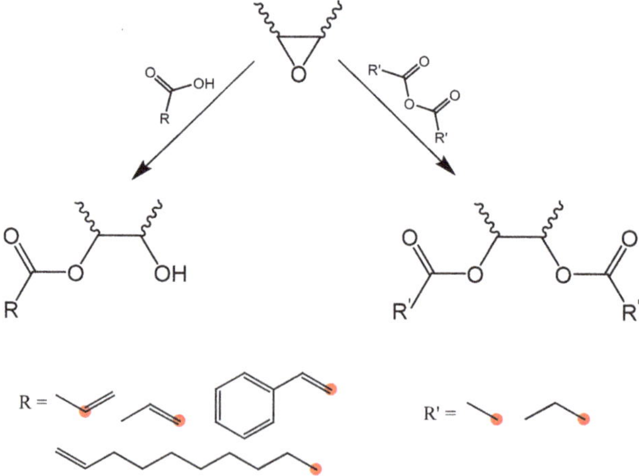

**Figure 5.** The epoxy ring-opening reaction scheme.

**Table 1.** Operating conditions and the conversions achieved for all the ring-opening reactions involving carboxylic acids.

| R | Temperature (°C) | Molar Ratio [1] | Catalyst (LDH) | Conversion (%) |
|---|---|---|---|---|
| | 100 | 1:1.2 | MgAl | 100 |
| | 100 | 1:0.9 | MgAl | 75 |
| | 100 | 1:0.6 | MgAl | 50 |
| | 100 | 1:1.2 | MgAl | 40 |
| | 100 | 1:1.2 | MgAlLa | 60 |
| | 140 | 1:1.2 | MgAlLa | 100 |
| | 100 | 1:1.2 | MgAl | 20 |
| | 100 | 1:1.2 | MgAlLa | 70 |
| | 140 | 1:1.2 | MgAlLa | 100 |
| | 100 | 1:1.2 | MgAl | 30 |
| | 100 | 1:1.2 | MgAlLa | 50 |
| | 140 | 1:1.2 | MgAlLa | 100 |
| | 140 | 1:1.2 | MgAl | 85 |

[1] epoxy groups: reactant.

The experimental results demonstrated that the increase in basicity from 6.73 mmol AA·g$^{-1}$ to 8.62 mmol AA·g$^{-1}$ for the catalyst modified with lanthanum played an important role in conversion. This significant increase was due to the fact that the electronegativity presented by La (1.1 on Pauling's scale) is much lower compared to that presented by Mg (1.31) and Al (1.61), respectively. Even if La does not isomorphously substitute Al in the octahedral sites of the layered structure, its presence in the network in the contaminant form (lanthanum hydroxycarbonate; JCPDS card 70-1774) leads to this significantly increased basicity. This base character corroborated the large ionic radius of La (1.14 Å) [44] (which can lead to larger pore size in the catalyst, facilitating the access of the reactants to the active sites), leading to a higher level of catalytic activities. In accordance with previous studies in the literature [21,45,46], a concerted reaction mechanism was proposed (Figure 6). This considers the influence of both base and acid sites, which can explain the differences in conversions between different catalysts and ring-opening reagents and further confirms that catalysts with an acid-base character such as LDH are suitable for this type of reaction.

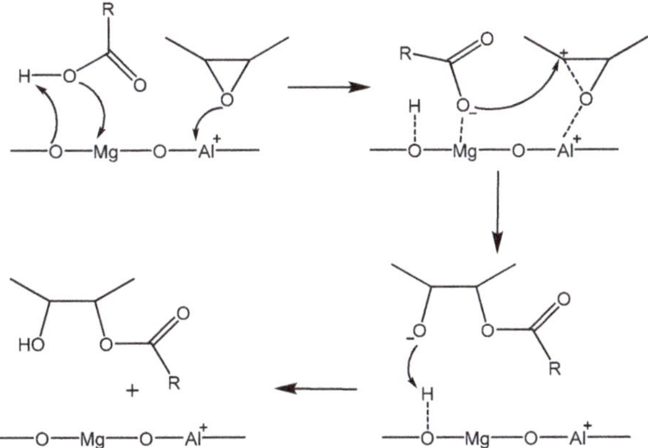

**Figure 6.** The proposed mechanism of the ring-opening reaction.

## 5. Conclusions

In the context of increasing interest in using renewable raw materials instead of petrochemical resources, obtaining intermediates and products from vegetable oils is an ongoing effort, which is also the objective of this study. Linseed oil, having a high degree of unsaturation, which leads to increased reactivity, was epoxidized and then functionalized through epoxy ring-opening with carboxylic acids and anhydrides, with the obtaining of products having potential interesting applications. This study shows that using heterogenous catalysis is a suitable method for obtaining materials and precursors from an abundant renewable resource such as vegetable oil. The influence of the basicity of the catalyst was established by comparing the activity of two materials of the same type, MgAl LDH, and MgAlLa LDH. Thus, renewable materials can be obtained using a potentially reusable heterogeneous catalyst in relatively mild conditions, in accordance with the principles of green chemistry.

**Supplementary Materials:** The following supporting information can be downloaded at https://www.mdpi.com/article/10.3390/su15054197/s1, Figure S1: The DRIFT spectra of the samples; Figure S2: The BET isotherms of the investigated samples; Figure S3: The superimposed FTIR spectra of ELO) and the reaction product of ELO with MA; Figure S4: The stacked $^1$H NMR spectra of ELO and partially methacrylated epoxidized linseed oil; Figure S5. The $^1$H NMR spectrum of ELO functionalized with undecylenic acid; Figure S6. The $^1$H NMR spectrum of ELO functionalized with crotonic acid; Figure S7. The $^1$H NMR spectrum of ELO functionalized with cinnamic acid; Figure S8. The superimposed FTIR spectra of ELO and the reaction product of ELO with butyric anhydride.

**Author Contributions:** Conceptualization, F.T., A.I.S. and R.S.; methodology, F.T., R.S. and O.D.P.; software, A.I.S. and I.B.; validation, F.T., R.S. and O.D.P.; formal analysis, A.I.S.; investigation, F.T. and A.I.S.; resources, I.B.; data curation, A.I.S.; writing—original draft preparation, A.I.S. and O.D.P.; writing—review and editing, F.T., R.S., A.I.S., O.D.P. and I.B.; visualization, F.T. and A.I.S. All authors have read and agreed to the published version of the manuscript.

**Funding:** This work has been funded by the European Social Fund from the Sectoral Operational Programme Human Capital 2014-2020, through the Financial Agreement with the title "Training of Ph.D. students and postdoctoral researchers in order to acquire applied research skills—SMART", Contract no. 13530/16.06.2022—SMIS code: 153734.

**Institutional Review Board Statement:** Not applicable.

**Informed Consent Statement:** Not applicable.

**Data Availability Statement:** Not applicable.

**Conflicts of Interest:** The authors declare no conflict of interest.

## References

1. Wang, R.; Schuman, T. Towards green, a review of recent developments in bio-renewable epoxy resins from vegetable oils. In *Green Materials from Plant Oils*; Liu, Z., Kraus, G., Eds.; The Royal Society of Chemistry, RSC Publishing: London, UK, 2015; Volume 9, pp. 202–241.
2. Xia, Y.; Larock, R.C. Vegetable Oil-Based Polymeric Materials: Synthesis, Properties and Applications. *Green Chem.* **2010**, *12*, 1893–1909. [CrossRef]
3. Zhang, C.; Garrison, T.F.; Madbouly, S.A.; Kessler, M.R. Recent Advances in Vegetable Oil-Based Polymers and Their Composites. *Prog. Polym. Sci.* **2017**, *71*, 91–143. [CrossRef]
4. Ca, V.; Lligadas, G.; Ronda, J.C.; Galia, M. Renewable Polymeric Materials from Vegetable Oils: A Perspective. *Mater. Today* **2013**, *16*, 337–343. [CrossRef]
5. Behr, A.; Westfechtel, A.; Pérez Gomes, J. Catalytic Processes for the Technical Use of Natural Fats and Oils. *Chem. Eng. Technol.* **2008**, *31*, 700–714. [CrossRef]
6. Sharma, B.K.; Liu, Z.; Adhvaryu, A.; Erhan, S.Z. One-Pot Synthesis of Chemically Modified Vegetable Oils. *J. Agric. Food Chem.* **2008**, *56*, 3049–3056. [CrossRef]
7. Karmakar, G.; Ghosh, P.; Sharma, B.K. Chemically Modifying Vegetable Oils to Prepare Green Lubricants. *Lubricants* **2017**, *5*, 44. [CrossRef]

8.  Omonov, T.S.; Curtis, J.M. Plant oil-based epoxy intermediates for polymers. In *Bio-Based Plant Oil Polymers and Composites*; Madbouly, S., Zhang, C., Kessler, M.R., Eds.; William Andrew Publishing, Elsevier Inc.: Amsterdam, The Netherlands, 2016; Volume 7, pp. 99–125.

9.  Kolot, V.; Grinberg, S. Vernonia Oil-Based Acrylate and Methacrylate Polymers and Interpenetrating Polymer Networks with Epoxy Resins. *J. Appl. Polym. Sci.* **2004**, *91*, 3835–3843. [CrossRef]

10. Adhvaryu, A.; Erhan, S.Z. Epoxidized Soybean Oil as a Potential Source of High-Temperature Lubricants. *Ind. Crops Prod.* **2002**, *15*, 247–254. [CrossRef]

11. Gaikwad, M.S.; Gite, V.V.; Mahulikar, P.P.; Hundiwale, D.G.; Yemul, O.S. Eco-Friendly Polyurethane Coatings from Cottonseed and Karanja Oil. *Prog. Org. Coat.* **2015**, *86*, 164–172. [CrossRef]

12. Lu, Y.; Larock, R.C. Soybean-Oil-Based Waterborne Polyurethane Dispersions: Effects of Polyol Functionality and Hard Segment Content on Properties. *Biomacromolecules* **2008**, *9*, 3332–3340. [CrossRef]

13. Åkesson, D.; Skrifvars, M.; Walkenström, P. Preparation of Thermoset Composites from Natural Fibres and Acrylate Modified Soybean Oil Resins. *J. Appl. Polym. Sci.* **2009**, *114*, 2502–2508. [CrossRef]

14. Oprea, S. Synthesis and Properties of the Porous Collagen/Polyurethane Composite. *J. Compos. Mater.* **2010**, *44*, 2179–2189. [CrossRef]

15. Samper, M.D.; Fombuena, V.; Boronat, T.; García-Sanoguera, D.; Balart, R. Thermal and Mechanical Characterization of Epoxy Resins (ELO and ESO) Cured with Anhydrides. *J. Am. Oil Chem. Soc.* **2012**, *89*, 1521–1528. [CrossRef]

16. Samper, M.D.; Ferri, J.M.; Carbonell-Verdu, A.; Balart, R.; Fenollar, O. Properties of Biobased Epoxy Resins from Epoxidized Linseed Oil (ELO) Crosslinked with a Mixture of Cyclic Anhydride and Maleinized Linseed Oil. *Express Polym. Lett.* **2019**, *13*, 407–418. [CrossRef]

17. Boutevin, B.; Parisi, J.P.; Robin, J.J.; Roume, C. Synthesis and Properties of Multiacrylic Resins from Epoxy Resins. *J. Appl. Polym. Sci.* **1993**, *50*, 2065–2073. [CrossRef]

18. Fogassy, G.; Pinel, C.; Gelbard, G. Solvent-Free Ring Opening Reaction of Epoxides Using Quaternary Ammonium Salts as Catalyst. *Catal. Commun.* **2009**, *10*, 557–560. [CrossRef]

19. Wanderley, K.A.; Leite, A.M.; Cardoso, G.; Medeiros, A.M.; Matos, C.L.; Dutra, R.C.; Suarez, P.A.Z. Graphene Oxide and a GO/ZnO Nanocomposite as Catalysts for Epoxy Ring-Opening of Epoxidized Soybean Fatty Acids Methyl Esters. *Braz. J. Chem. Eng.* **2019**, *36*, 1165–1173. [CrossRef]

20. Schuster, H.; Rios, L.A.; Weckes, P.P.; Hoelderich, W.F. Heterogeneous Catalysts for the Production of New Lubricants with Unique Properties. *Appl. Catal. A Gen.* **2008**, *348*, 266–270. [CrossRef]

21. Fogassy, G.; Ke, P.; Figueras, F.; Cassagnau, P.; Rouzeau, S.; Courault, V.; Gelbard, G.; Pinel, C. Catalyzed Ring Opening of Epoxides: Application to Bioplasticizers Synthesis. *Appl. Catal. A Gen.* **2011**, *393*, 1–8. [CrossRef]

22. Debecker, D.P.; Gaigneaux, E.M.; Busca, G. Exploring, Tuning, and Exploiting the Basicity of Hydrotalcites for Applications in Heterogeneous Catalysis. *Chem. A Eur. J.* **2009**, *15*, 3920–3935. [CrossRef]

23. Cavani, F.; Trifirò, F.; Vaccari, A. Hydrotalcite-Type Anionic Clays: Preparation, Properties and Applications. *Catal. Today* **1991**, *11*, 173–301. [CrossRef]

24. Sels, B.F.; De Vos, D.E.; Jacobs, P.A. Hydrotalcite-like Anionic Clays in Catalytic Organic Reactions. *Catal. Rev. Sci. Eng.* **2001**, *43*, 443–488. [CrossRef]

25. Handa, H.; Baba, T.; Yamada, H.; Takahashi, T.; Ono, Y. Double-Bond Isomerization of Olefinic Amines over Potassium Amide Loaded on Alumina. *Catal. Lett.* **1997**, *44*, 119–121. [CrossRef]

26. Ionescu, R.; Pavel, O.D.; Bîrjega, R.; Zăvoianu, R.; Angelescu, E. Epoxidation of Cyclohexene with H2O2 and Acetonitrile Catalyzed by Mg-Al Hydrotalcite and Cobalt Modified Hydrotalcites. *Catal. Lett.* **2010**, *134*, 309–317. [CrossRef]

27. Choudary, B.M.; Lakshmi Kantam, M.; Venkat Reddy, C.R.; Koteswara Rao, K.; Figueras, F. The First Example of Michael Addition Catalysed by Modified Mg–Al Hydrotalcite. *J. Mol. Catal. A Chem.* **1999**, *146*, 279–284. [CrossRef]

28. Angelescu, E.; Pavel, O.D.; Bîrjega, R.; Zăvoianu, R.; Costentin, G.; Che, M. Solid Base Catalysts Obtained from Hydrotalcite Precursors, for Knoevenagel Synthesis of Cinamic Acid and Coumarin Derivatives. *Appl. Catal. A Gen.* **2006**, *308*, 13–18. [CrossRef]

29. Kumbhar, P.S.; Sanchez-Valente, J.; Millet, J.M.M.; Figueras, F. Mg-Fe Hydrotalcite as a Catalyst for the Reduction of Aromatic Nitro Compounds with Hydrazine Hydrate. *J. Catal.* **2000**, *191*, 467–473. [CrossRef]

30. Roelofs, J.C.A.A.; van Dillen, A.J.; de Jong, K.P. Condensation of Citral and Ketones Using Activated Hydrotalcite Catalysts. *Catal. Lett.* **2001**, *74*, 91–94. [CrossRef]

31. Corma, A.; Fornés, V.; Martín-Aranda, R.M.; Rey, F. Determination of Base Properties of Hydrotalcites: Condensation of Benzaldehyde with Ethyl Acetoacetate. *J. Catal.* **1992**, *134*, 58–65. [CrossRef]

32. Teodorescu, F.; Deaconu, M.; Bartha, E.; Zăvoianu, R.; Pavel, O.D. Addition of Alcohols to Acrylic Compounds Catalyzed by Mg-Al LDH. *Catal. Lett.* **2014**, *144*, 117–122. [CrossRef]

33. Bălănucă, B.; Stan, R.; Hanganu, A.; Iovu, H. Novel Linseed Oil-Based Monomers: Synthesis and Characterization. *UPB Sci. Bull. Ser. B Chem. Mater. Sci.* **2014**, *76*, 129–140.

34. Rangarajan, B.; Havey, A.; Grulke, E.A.; Culnan, P.D. Kinetic Parameters of a Two-Phase Model for in Situ Epoxidation of Soybean Oil. *J. Am. Oil Chem. Soc.* **1995**, *72*, 1161–1169. [CrossRef]

35. Chira, N.A.; Todasca, M.C.; Nicolescu, A.; Rosu, A.; Nicolae, M.; Rosca, S.I. Evaluation of the Computational Methods for Determining Vegetable Oils Composition Using 1H-NMR Spectroscopy. *Rev. Chim.* **2011**, *62*, 42–46.

36. Slabu, A.; Banu, N.D.; Balanuca, B.; Teodorescu, F.; Stan, R. Plant-Based Resins Obtained from Epoxidized Linseed Oil Using a MgAl Hydrotalcite Catalyst. *UPB Sci. Bull. Ser. B Chem. Mater. Sci.* **2022**, *84*, 45–54.

37. Balanuca, B.; Stan, R.; Hanganu, A.; Lungu, A.; Iovu, H. Design of New Camelina Oil-Based Hydrophilic Monomers for Novel Polymeric Materials. *J. Am. Oil Chem. Soc.* **2015**, *92*, 881–891. [CrossRef]

38. Van der Steen, M.; Stevens, C.V. Undecylenic Acid: A Valuable and Physiologically Active Renewable Building Block from Castor Oil. *ChemSusChem* **2009**, *2*, 692–713. [CrossRef] [PubMed]

39. Bigot, S.; Daghrir, M.; Mhanna, A.; Boni, G.; Pourchet, S.; Lecamp, L.; Plasseraud, L. Undecylenic Acid: A Tunable Bio-Based Synthon for Materials Applications. *Eur. Polym. J.* **2016**, *74*, 26–37. [CrossRef]

40. Doll, K.M.; Walter, E.L.; Murray, R.E.; Hwang, H.S. Organogel Polymers from 10-Undecenoic Acid and Poly(Vinyl Acetate). *J. Polym. Environ.* **2018**, *26*, 3670–3676. [CrossRef]

41. Huerta-Ángeles, G.; Brandejsová, M.; Kopecká, K.; Ondreáš, F.; Medek, T.; Žídek, O.; Kulhánek, J.; Vagnerová, H.; Velebný, V. Synthesis and Physicochemical Characterization of Undecylenic Acid Grafted to Hyaluronan for Encapsulation of Antioxidants and Chemical Crosslinking. *Polymers* **2020**, *12*, 35. [CrossRef]

42. Fouilloux, H.; Thomas, C.M. Production and Polymerization of Biobased Acrylates and Analogs. Macromol. *Rapid Commun.* **2021**, *42*, 2000530. [CrossRef]

43. Lendlein, A.; Jiang, H.; Junger, O.; Langer, R. Light-Induced Shape-Memory Polymers. *Nature* **2005**, *434*, 879–881. [CrossRef] [PubMed]

44. Angelescu, E.; Pavel, O.D.; Che, M.; Bîrjega, R.; Constentin, G. Cyanoethylation of Ethanol on Mg-Al Hydrotalcites Promoted by Y3+ and La3+. *Catal. Commun.* **2004**, *5*, 647–651. [CrossRef]

45. Yamaguchi, K.; Ebitani, K.; Yoshida, T.; Yoshida, H.; Kaneda, K. Mg-Al Mixed Oxides as Highly Active Acid-Base Catalysts for Cycloaddition of Carbon Dioxide to Epoxides. *J. Am. Chem. Soc.* **1999**, *121*, 4526–4527. [CrossRef]

46. Karami, Z.; Jouyandeh, M.; Ali, J.A.; Ganjali, M.R.; Aghazadeh, M.; Paran, S.M.R.; Naderi, G.; Puglia, D.; Saeb, M.R. Epoxy/Layered Double Hydroxide (LDH) Nanocomposites: Synthesis, Characterization, and Excellent Cure Feature of Nitrate Anion Intercalated Zn-Al LDH. *Prog. Org. Coat.* **2019**, *136*, 105218. [CrossRef]

*Article*

# Investigation of the Level of Knowledge in Different Countries about Edible Insects: Cluster Segmentation

Raquel P. F. Guiné [1,*], Sofia G. Florença [1], Cristina A. Costa [1], Paula M. R. Correia [1], Manuela Ferreira [2], Ana P. Cardoso [3], Sofia Campos [3], Ofélia Anjos [4], Cristina Chuck-Hernández [5], Marijana Matek Sarić [6], Ilija Djekic [7], Maria Papageorgiou [8], José M. F. Baro [9], Malgorzata Korzeniowska [10], Maša Černelič-Bizjak [11], Elena Bartkiene [12], Monica Tarcea [13], Nada M. Boustani [14], Dace Klava [15] and Emel Damarli [16]

[1] CERNAS-IPV Research Centre, Polytechnic Institute of Viseu, 3504-510 Viseu, Portugal
[2] Health Sciences Research Unit: Nursing (UICISA: E), Polytechnic Institute of Viseu, 3504-510 Viseu, Portugal
[3] CIDEI-IPV Research Centre, Polytechnic Institute of Viseu, 3504-510 Viseu, Portugal
[4] School of Agriculture, Polytechnic Institute of Castelo Branco, 6001-909 Castelo Branco, Portugal
[5] Tecnologico de Monterrey, The Institute for Obesity Research, Monterrey 64849, Mexico
[6] Department of Health Studies, University of Zadar, 23000 Zadar, Croatia
[7] Department of Food Safety and Quality Management, Faculty of Agriculture, University of Belgrade, 11000 Belgrade, Serbia
[8] Department of Food Science and Technology, International Hellenic University, 57001 Thessaloniki, Greece
[9] BALAT Research Group, Faculty of Veterinary Medicine, University of León, 24071 León, Spain
[10] Faculty of Food Science, Wroclaw University of Environmental and Life Sciences, 51-630 Wrocław, Poland
[11] Department of Nutritional Counseling—Dietetics, Faculty of Health Science, University of Primorska, 6320 Izola, Slovenia
[12] Department of Food Safety and Quality, Lithuanian University of Health Sciences, LT-47181 Kaunas, Lithuania
[13] Department of Community Nutrition and Food Safety, George Emil Palade University of Medicine, Pharmacy, Science, and Technology of Targu Mures, 540139 Targu Mures, Romania
[14] Faculty of Business and Administration, Saint Joseph University, Beirut 1104 2020, Lebanon
[15] Faculty of Food Technology, Latvia University of Life Sciences and Technologies, LV-3001 Jelgava, Latvia
[16] Altıparmak Food Coop. Research & Development Center, Çekmeköy, 34782 İstanbul, Turkey
[*] Correspondence: raquelguine@esav.ipv.pt

**Citation:** Guiné, R.P.F.; Florença, S.G.; Costa, C.A.; Correia, P.M.R.; Ferreira, M.; Cardoso, A.P.; Campos, S.; Anjos, O.; Chuck-Hernández, C.; Sarić, M.M.; et al. Investigation of the Level of Knowledge in Different Countries about Edible Insects: Cluster Segmentation. *Sustainability* **2023**, *15*, 450. https://doi.org/10.3390/su15010450

Academic Editor: Michael A. Long

Received: 11 October 2022
Revised: 6 December 2022
Accepted: 21 December 2022
Published: 27 December 2022



**Abstract:** This study aimed to investigate the level of knowledge about edible insects (EIs) in a sample of people from thirteen countries (Croatia, Greece, Latvia, Lebanon, Lithuania, Mexico, Poland, Portugal, Romania, Serbia, Slovenia, Spain, and Turkey). Data collection was based on a questionnaire survey applied through online tools between July and November 2021. For data analysis, techniques such as factor analysis, cluster analysis, and chi-square tests were used, with a significance level of 5%. A total of 27 items were used to measure knowledge on a five-point Likert scale. Applying factor analysis with principal components and Varimax rotation, a solution that explains about 55% of variance was obtained. This accounts for four factors that retained 22 of the 27 initial items: F1 = Sustainability (8 items), F2 = Nutrition (8 items), F3 = Production Factors (2 items), and F4 = Health Concerns (4 items). Internal consistency was evaluated through Cronbach's alpha. The cluster analysis consisted of the application of hierarchical methods followed by k-means and produced three clusters (1—'fearful', 2—'farming,' and 3—'ecological' individuals). The characterisation of the clusters revealed that age did not influence cluster membership, while sex, education, country, living environment, professional area, and income all influenced the composition of the clusters. While participants from Mexico and Spain were fewer in the 'fearful' cluster, in those from Greece, Latvia, Lebanon, and Turkey, the situation was opposed. Participants from rural areas were mostly in cluster 2, which also included a higher percentage of participants with lower income. Participants from professional areas linked with biology, food, and nutrition were mostly in cluster 3. In this way, we concluded that the level of knowledge about EIs is highly variable according to the individual characteristics, namely that the social and cultural influences of the different countries lead to distinct levels of knowledge and interpretation of information, thus producing divergent approaches to the consumption of insects—some more reluctant and measuring possible risks. In contrast, others consider EIs a good and sustainable protein-food alternative.

**Keywords:** knowledge; edible insects; factor analysis; cluster analysis; sustainable food; nutritional value

## 1. Introduction

The practice of eating insects (entomophagy) has been attracting attention on a global scale for a number of reasons, among which the following stand out: environmental sustainability, nutrition properties, health benefits, and social/economic advantages [1]. Being able to achieve food security is presently one of the most challenging aspects of dealing with the world population, and being able to do so with the lowest possible environmental impact is pivotal in today's context of climate changes and societies' pressure on the ecosystems [1,2].

A wide variety of edible insect (EI) species with high nutritional value is available for human consumption. Insect consumption has been a traditional practice throughout the history of the human race [3–5]. However, their consumption is variable according to the region of the globe, with some areas where eating insects is recognised as an old practice, and EIs are a much-appreciated type of food. In contrast, for other regions, entomophagy is not seen as natural. The number of people estimated to consume insects regularly in their diets, as a traditional practice, has been estimated at over two billion worldwide [6]. Presently, a wide range of insects is consumed in Africa, Asia, Latin America, and Australia. On the other hand, in Western societies, including Europe and the United States, eating insects is not common, and there is still a high resistance to adopting such exotic dietary practices [7,8]. Neophobia is present in the minds of Western consumers still today, even despite their recognition that EIs have environmental advantages over other types of animal protein [9,10]. A recent review [11] discussing consumer perceptions of EIs, revealed that it is widely accepted that EIs are not part of the diets in Western countries, and therefore it is difficult to include them as regular foods. A way to improve consumer acceptance is by renowned chefs using them in their culinary practices or by using insects as food ingredients rather than consuming whole insects [12–14].

The nutritional quality of EIs is not inferior to that of other meats (beef, pork, chicken, turkey, and others), and sometimes the macro and micro components' balance is even more advantageous from a nutritional point of view. Insects are rich in protein and fat and are a very energy-dense food. Therefore, they could assume a leading role in a solution to mitigate hunger worldwide. Not only are EIs rich in protein, but those proteins are also high-quality proteins with a good balance of amino acids, especially essential amino acids. Additionally, most EIs contain a low saturated/total fatty acids ratio (less than 40% of the total fatty acids are saturated). On the other hand, EIs are rich in micronutrients, such as minerals (particularly zinc and iron) and vitamins (among which are vitamin E and vitamin B12) [1,15–17]. However, there are also some possible hazards and problems related to consuming insects. Some EIs can be a source of anti-nutrients, such as oxalates, hydrogen cyanides, phytic acid, and tannins, which, even though occurring naturally in foods, can compromise the digestion, absorption, and utilisation of certain nutrients [18–20]. Additionally, there is still a gap in guaranteeing food safety and antinutritional factors associated with edible insects, as discussed by Murefu et al. [21]. In the case of tannins, although tannins can act as agents limiting the absorption of some nutrients in some EIs [22,23], they can also act as antioxidants in some food matrices like wine [24,25] or cheese [26,27], and they can even be added into chitosan and cellulose-based films to provide antioxidant and antimicrobial properties [28]. Thus, tannins may have both a positive and a negative contribution to the impact of their consumption.

The consumption of edible insects is a culturally accepted practice in some parts of the world while not so readily accepted in others. Studies undertaken in some Western countries report some reluctance to adopt EIs into their diets while also stating that people tend to feel motivated to consumption by sustainable aspects. A study conducted in

Australia [9] showed barriers to consumption, but people there are more ready to accept foods containing insects or those in which the insects are "disguised," like insect-based flour or chocolate-covered ants. A study conducted among German adults [12] showed that they were generally willing to try insects and concluded that an attractive packaging design did not increase the willingness to try them. Another study conducted in Germany [29] showed that there were some important barriers to consumption that must be taken into consideration when implementing the adoption of EIs.

The consumption of EIs could be strongly influenced by both cultural influences and knowledge about their effects and their impact. However, a search in the scientific literature proved that this topic is highly understudied and requires attention from the scientific community, unlike the motivations for the consumption of EIs, which have been intensively studied [9,30–32]. Hence, the aim of this study, as part of the international project EISuFood, was to characterise the level of knowledge about EIs in a sample of people from thirteen countries and to understand how this knowledge could lead to the formation of groups of people based on their knowledge regarding various aspects related to EIs, going from nutrition and health effects to sustainability issues. Additionally, it also sought to understand how sociodemographic, geographic, or professional characteristics influenced the composition of the clusters.

## 2. Materials and Methods

### 2.1. Instrument

This research was based on a questionnaire survey using an instrument that was pre-validated for a sample of Portuguese participants [33]. The items of the questionnaire were grouped into two sets, one with the items aimed at measuring knowledge and the other with items measuring perceptions. For this study, 27 items were used to measure knowledge, as indicated in Table 1. The participants had to express their agreement on a five-point Likert scale as follows: 1 = strongly disagree, 2 = disagree, 3 = no opinion, 4 = agree, and 5 = strongly agree [34].

**Table 1.** Items used for measuring knowledge about edible insects.

| N° | Item Description |
| --- | --- |
| 1 | Entomophagy is a dietary practice that consists in the consumption of insects by humans. |
| 2 | There are thousands of species of insects that are consumed by humans in the world. |
| 3 | Insects are a more sustainable alternative when compared with other sources of animal protein. |
| 4 | Insect production for human consumption emits much less greenhouse gases than beef production. |
| 5 | Insects efficiently convert organic matter into protein. |
| 6 | The production of insect protein uses considerably less feed than beef protein. |
| 7 | Insects are a possibility for responding to the growing world demand for protein. |
| 8 | The production of chicken protein requires much less water than insect protein. * |
| 9 | The ecological footprint (impact) of insects is smaller when compared with other animal proteins. |
| 10 | The production of insect protein requires much more area than pork protein. * |
| 11 | Insects are collected as a means of pest control for some cultivated crops. |
| 12 | The loss of biodiversity is lower with insect production compared with other animal food production. |
| 13 | The energy input needed for production of insect protein is lower than for the production of other proteins from animal origin. |
| 14 | Insects have poor nutritional value. * |
| 15 | Insects are a good source of energy. |
| 16 | Insects have high protein content. |
| 17 | Insect proteins are of poorer quality compared with other animal species. * |
| 18 | Insects provide essential amino acids necessary for humans. |
| 19 | Insects contain group B vitamins. |
| 20 | Insects contain dietary fibre. |
| 21 | Insects contain minerals of nutritional interest, such as calcium, iron, and magnesium. |
| 22 | Insects contain fat, including unsaturated fatty acids. |
| 23 | Insects contain anti-nutrients, such as oxalates and phytic acid. |
| 24 | There are appropriate regulations to guarantee the food safety of edible insects. |
| 25 | Insects contain bioactive compounds beneficial to human health. |
| 26 | Insects are potential sources of allergens. |
| 27 | Aflatoxins, which are carcinogens, can be present in insects. |

* False statement.

### 2.2. Data Collection

This descriptive cross-sectional study was carried out on a non-probabilistic sample of 6899 participants from the following 13 countries: Croatia, Greece, Latvia, Lebanon,

Lithuania, Mexico, Poland, Portugal, Romania, Serbia, Slovenia, Spain, and Turkey. The questionnaire was first prepared and validated in Portugal [33] and then translated into English to send to all the partners in the project. In all participating countries, the questionnaire was translated into the corresponding native language following a standard back-translation procedure. In each country, the data were collected based on the translated questionnaire, using the native languages of all participants.

All ethical principles were strictly followed when designing the questionnaire and collecting data, especially those of the Declaration of Helsinki. The Ethics Committee of the Polytechnic Institute of Viseu approved this questionnaire survey on 25 May 2020 with reference number 45/SUB/2021.

Due to the restrictions caused by the COVID-19 pandemic, data were collected between July and November 2021 using the electronic platform Google Forms. Recruitment was done using email and social media and followed a snowball methodology in each of the participating countries. This methodology has proven more effective than multisite data collection [35]. Only adult citizens (18 years old or over) who expressed their informed consent were allowed to participate in the survey.

*2.3. Sample Characterization*

The sample of 6899 participants was distributed among the participating countries, as indicated in Table 2. The participants were recruited according to variable sociodemographic characteristics in an attempt to have individuals of different groups, such as gender, age, level of education, or living environment. Although the survey had the limitation of not having equal representation across all sociodemographic classes, representation was nevertheless ensured by including a high number of participants.

**Table 2.** Distribution of the participants by country.

| Country | N | % |
|---|---|---|
| Croatia | 686 | 9.9 |
| Greece | 636 | 9.2 |
| Latvia | 300 | 4.3 |
| Lebanon | 357 | 5.2 |
| Lithuania | 510 | 7.4 |
| Mexico | 1139 | 16.5 |
| Poland | 520 | 7.5 |
| Portugal | 527 | 7.6 |
| Romania | 492 | 7.1 |
| Serbia | 344 | 5.0 |
| Slovenia | 517 | 7.5 |
| Spain | 575 | 8.3 |
| Turkey | 296 | 4.3 |
| **Total** | **6899** | **100** |

The participants' ages varied from 18 to 88 years, with an average age of 35 ± 14 years. Most of the participants were female (63.0%). Regarding the living environment, 68.6% of the participants lived in urban areas, and a smaller percentage lived in suburban (15.9%) or rural areas (15.5%). With respect to the highest education level attained, 36.5% had completed secondary or elementary school, 32.4% had completed a university degree, and 31.1% had completed postgraduate studies (master's or doctoral degree).

*2.4. Statistical Analyses*

The software used for the statistical analysis was SPSS Version 28 from IBM, Inc. (Armonk, NY, USA). Basic descriptive statistics were used, and more complex analyses were also performed, namely Factor Analysis (FA) and Cluster Analysis (CA). As the first step, exploratory factor analysis was applied using the method of Principal Component

Analysis (PCA), aimed at determining if there was a grouping structure between the items. After this, the factors identified in the first step were submitted to cluster analysis.

Initially, the data were analysed to verify if they were appropriate for the techniques of FA using PCA. The correlation matrix between the variables was analysed to identify possible correlations. The Kaiser-Meyer-Olkin (KMO) measure of the adequacy of the sample was calculated, and Bartlett's test was performed to evaluate correlations between variables [36]. The reference values of the KMO are as follows: excellent for $0.9 \leq KMO \leq 1.0$, Good for $0.8 \leq KMO < 0.9$, Acceptable for $0.7 \leq KMO < 0.8$, Tolerable for $0.6 \leq KMO < 0.7$, Bad for $0.5 \leq KMO < 0.6$, and unacceptable for $KMO < 0.5$ [37].

To apply FA, the false items (9, 11, 15, and 18) were reversed for compatibility with the other items measuring knowledge. In this way, higher values of the score always correspond to higher knowledge.

Upon verification of the adequacy of the data, FA was applied with extraction using PCA and Varimax rotation with Kaiser Normalization. The number of components was determined based on the Eigenvalues greater than 1. The communalities were calculated to indicate the percentage of variance explained by the factors extracted [36]. Factor loadings with an absolute value lower than 0.4 were excluded [38,39]. The internal consistency of each factor was evaluated using Cronbach's alpha ($\alpha$) [36,40]. Regarding the reference values for alpha, although dependent on the authors, in general, values over 0.7 are desirable, with values over 0.8 considered very good. Nevertheless, some authors also state that values over 0.5 could be acceptable [41–43].

The CA started with the application of hierarchical methods based on the factors obtained by FA (4 variables) to establish the most adequate number of clusters. The seven methods tested were: Within Groups Linkage (WGL), Between Groups Linkage (BGL), Nearest Neighbour (NN), Furthest Neighbour (FN), Centroid (CE), Median Clustering (MC), and Ward (WA), all considering the Squared Euclidean distance for interval measurement. The coefficients obtained in the agglomeration schedule for the different methods indicated that the optimal number of groups that should be formed was three. Then, the seven methods were run again with the fixed number of clusters, and the obtained solutions were compared by means of contingency tables (crosstabs) in order to evaluate stability (Table 3). Some of the solutions showed a possible similarity of over 98% (BGL, NN, CE, MC), which is very high and indicative of potential stability. Therefore, these four clustering solutions were used as initial solutions to proceed with the analysis using the partitive method of k-means, which is particularly recommended and frequently used in CA [44].

**Table 3.** Similarity between the solutions obtained through hierarchical clustering methods.

| Methods [1] | WGL | BGL | NN | FN | CE | MC | WA |
|---|---|---|---|---|---|---|---|
| WGL | 100% | | | | | | |
| BGL | 43% | 100% | | | | | |
| NN | 43% | 99% | 100% | | | | |
| FN | 41% | 52% | 51% | 100% | | | |
| CE | 43% | 98% | 99% | 50% | 100% | | |
| MC | 44% | 99% | 99% | 52% | 98% | 100% | |
| WA | 69% | 45% | 37% | 41% | 45% | 45% | 100% |

[1] WGL = Within Groups Linkage, BGL = Between Groups Linkage, NN = Nearest Neighbour, FN = Furthest Neighbour, CE = Centroid, MC = Median Clustering, WA = Ward.

The chi-square test was used to assess differences between clusters according to the sociodemographic factors. A level of significance of 5% was used, and the Cramer's V coefficients were also calculated as a measure of the association between the categorical variables tested. The Cramer's V coefficient varies from 0 to 1; for $V \approx 0.1$, the association was considered weak; for $V \approx 0.3$, the association was moderate; and for $V \approx 0.5$ or over, the association was strong [45].

## 3. Results

### 3.1. Factor Analysis

The correlation matrix showed that there were some correlations between the variables, with 30 values higher than 0.5. The highest value in the correlation matrix was found to be 0.647, corresponding to the correlation between items 4 and 6. Based on this evidence, some relevant correlations between the variables were indicative that we could apply FA. Additionally, Bartlett's test was significant ($p < 0.0005$), confirming the rejection of the null hypothesis "H0: The correlation matrix is equal to the identity matrix". The value of KMO (0.944) can be classified as excellent, based on the classification of Kaiser and Rice [37], confirming once more the suitability for the application of PCA and FA techniques. The anti-image matrix showed that there was no value of MSA (Measure of Sampling Adequacy) below 0.5, meaning that all the variables were properly included in the analysis (the values of MSA ranged between a minimum of 0.660 for item 8 to a maximum of 0.973 for item 25).

The solution obtained by FA with PCA and Varimax rotation retained five components (eigenvalues. 8.551, 2.618, 1.750, 1.301, and 1.009). The percentages of total variance explained (VE) by the factors were: F1—19.45%, F2—13.70%, F3—9.25%, F4—6.67%, and F5—6.32%, resulting in a total variance explained of 55.39% (Table 4). Items with higher communalities were 10 (0.684, 68.4% VE), 21 (0.682, 68.2% VE), and 6 (0.674, 67.4% VE). The item with the lowest variance explained by the solution was 24 (22.1% VE). Item 24 was not included in any of the factors due to loading values lower than 0.4.

**Table 4.** Solution obtained through factor analysis.

| Factor | %VE [1] | Items | Loadings | Factor Name | Cronbach's Alpha ($\alpha$) |
|---|---|---|---|---|---|
| F1 | 19.45 | 3. Insects more sustainable than other animal proteins | 0.650 | | 0.899 |
| | | 4. Insects emit fewer greenhouse gases than cows | 0.748 | | 0.905 [2] |
| | | 5. Insects efficiently convert organic matter into protein | 0.685 | | |
| | | 6. Insects use considerably less feed than cows | 0.781 | | |
| | | 7. Insects can meet the growing demand for protein | 0.718 | Sustainability (SUS) | |
| | | 9. The footprint of insects is smaller than other animals | 0.755 | | |
| | | 11. Insect collection is a pest control mechanism | 0.528 | | |
| | | 12. Insects originate lower loss of biodiversity | 0.670 | | |
| | | 13. Insects require less energy than other animals | 0.750 | | |
| F2 | 13.70 | 18. Insects provide essential amino acids | 0.639 | | 0.832 |
| | | 19. Insects contain group B vitamins | 0.740 | | 0.844 [3] |
| | | 20. Insects contain dietary fibre | 0.694 | | |
| | | 21. Insects contain minerals of nutritional interest | 0.740 | Nutrition (NUT) | |
| | | 22. Insects contain fat, including unsaturated fatty acids | 0.711 | | |
| | | 23. Insects contain anti-nutrients | 0.499 | | |
| | | 25. Insects contain bioactive compounds | 0.444 | | |
| F3 | 9.25 | 1. Entomophagy consists in the consumption of insects | 0.545 | | 0.712 |
| | | 2. There are thousands of species of edible insects | 0.580 | | |
| | | 14. Insects have poor nutritional value (reversed) | 0.558 | Insects as Protein Food (IPF) | |
| | | 15. Insects are a good source of energy | 0.516 | | |
| | | 16. Insects have high protein content | 0.518 | | |
| | | 17. Insect proteins are of poorer quality (reversed) | 0.450 | | |
| F4 | 6.67 | 26. Insects can contain allergens | 0.790 | Health Risks (HR) | 0.577 |
| | | 27. Insects can contain aflatoxins | 0.740 | | |
| F5 | 6.32 | 8. Chickens require less water than insects (reversed) | 0.794 | Production Factors (PF) | 0.617 |
| | | 10. Insects require more area than pigs (reversed) | 0.823 | | |

[1] VE = Variance explained. [2] Alpha if item 11 is removed. [3] Alpha if item 23 is removed.

The FA solution converged in six iterations. Table 4 shows that the first group of items seems related to sustainability aspects of EIs and was named Sustainability (SUS). Items in factor F2 are associated with the nutritional aspects of edible insects and were named Nutrition (NUT). The items in factor F3 are related to the consumption of insects as a source of protein, so it was named Insects as Protein Food (IPF). Factor F4 contained only two items and was named Health Risks (HR) because the items relate to the presence of compounds harmful to human health. Finally, the last factor, F5, also contained two items, both related to insect production specifications and was, for that reason, named Production Factors (PF).

In general, the item loadings for all factors were high (for F1, varying from 0.528 to 0.781; for F2, varying from 0.444 to 0.740; for F4, varying from 0.740 to 0.790; for F5, varying from 0.794 to 0.823), with factor F3 being just a little lower (varying from 0.450 to 0.580). High loadings are indicative of the high contribution of the items to the definition of the factors. Items with the highest loadings are item 10 (loading of 0.823 into factor F5) and item 26 (loading of 0.790 into factor F4), meaning that these items are most strongly associated with the respective factors.

To validate the solution, Cronbach's alpha ($\alpha$) values were determined to measure the internal consistency within each factor [36]. The value of Cronbach's alpha for factor F1 (SUS) was 0.899 and 0.832 for factor F2 (NUT), both of which are considered very good [41–43]. However, factors F1 and F2 could have a higher internal consistency if one item were removed from each of those factors (items 11 and 23, respectively), as shown in Table 4.

The value of alpha for F3 was 0.712, which is good, being over the threshold of 0.7. The values of alpha for F4 and F5 were 0.577 and 0.617, respectively, and although lower than for the other factors, they can still be considered acceptable [41–43]. For factor F3, the value of alpha would not increase with the removal of any item, and factors F4 and F5, are also fixed for having only two items.

Considering these results, we conclude that the scale would be stronger if three items were removed [38]—11, 23, and 24, as discussed earlier—and for that reason, the final factor solution was run considering only 24 items instead of the 27 originally tested. For this group of items, the KMO was 0.942, and the significance of Bartlett's test was significant ($p < 0.0005$). This final solution (Table 5) explains 55.07% of the variance and comprises four factors. Items 1 and 2 ("Entomophagy consists in the consumption of insects" and "There are thousands of species of edible insects", respectively) were not included in any of the factors due to loading values lower than 0.4.

The first factor, named Sustainability (SUS), has the exact same eight items as in the previous solution ($\alpha = 0.905$) and accounts for items related to the sustainability of EIs as alternative protein foods. Factor F2 included the six items from the previous solution but added two new items, 15 and 16, both also related to nutritional aspects of edible insects, the first relating to energetic value and the second to protein content. So, this factor name was kept as Nutrition (NUT) because it accounted for a total of eight items related to dietary components of EIs, and its internal consistency increased as compared with the previous solution ($\alpha = 0.872$).

Factor F3 remained equal to F5 from the previous solution, Production Factors (PF) ($\alpha = 0.617$) and consists of two items that compare production factors of EIs with other sources of animal protein, specifically chickens and pigs.

The last factor, F4, added two items to the previous factor, F4. This factor accounted for items related to allergens or aflatoxins and was named Health Concerns (HC). The internal reliability of this factor is lower than 0.5, which can be explained by the negative values of the loading for items 14 and 17.

In all factors, the values of alpha would not improve with the removal of any item, so this is considered the final solution, including four factors and 22 items.

**Table 5.** Final solution obtained through factor analysis, considering 22 items.

| Factor | %VE [1] | Items (Loadings) | Factor Name | Cronbach's Alpha ($\alpha$) |
|---|---|---|---|---|
| F1 | 22.51 | Item 3 (0.696), item 4 (0.778), item 5 (0.715), item 6 (0.793), item 7 (0.758), item 9 (0.750), item 12 (0.641), item 13 (0.732) | Sustainability (SUS) | 0.905 |
| F2 | 18.27 | Item 15 (0.594), item 16 (0.560), item 18 (0.714), item 19 (0.746), item 20 (0.711), item 21 (0.788), item 22 (0.653), item 25 (0.579) | Nutrition (NUT) | 0.872 |
| F3 | 7.15 | Item 8 (0.802), item 10 (0.801) | Production Factors (PF) | 0.617 |
| F4 | 7.14 | Item 14 ($-0.401$), item 17 ($-0.492$), item 26 (0.761), item 27 (0.780) | Health Concerns (HC) | <0.5 |

[1] VE = Variance explained.

### 3.2. Cluster Analysis

Table 6 shows the results obtained for the application of k-means clustering to the four initial solutions obtained with hierarchical methods, which offered a higher probability of stability (BGL, CE, MC, and NN). From the tested initial solutions, two of them converged to the same final solution (CE and NN), as can be seen, both from the number of cases classified in each cluster and also the coordinates of the cluster centres. Additionally, because the values of ANOVA $p$-value are significant ($p < 0.0005$) and the values of the test statistic (F) are high, they indicate similarity between the cases within the groups and differentiation between groups. Furthermore, while factors F1 (SUS) and F3 (PF) contributed more to the definition of the clusters (with values of F in the same magnitude), factors F2 (NUT) and F4 (HC) contributed less (also with values of F in the same magnitude). In this way, the final solution is accepted as that originating from the initial CE and NN solutions, and the clustering analysis was more deeply influenced by the sustainability issues and lower production requirements of EIs as compared with other animal species than the nutritive aspects or health issues associated with the consumption of EIs.

**Table 6.** Results for the k-means clustering.

| Initial Solution [1] | Factors | ANOVA | | Cluster 1 | | Cluster 2 | | Cluster 3 | |
|---|---|---|---|---|---|---|---|---|---|
| | | F | $p$-Value | PC [2] | FCC [3] | PC [2] | FCC [3] | PC [2] | FCC [3] |
| BLG | F1 (SUS) | 1467 | $p < 0.0005$ | 50% | $-0.433$ | 25% | $-0.057$ | 25% | 0.901 |
| | F2 (NUT) | 2961 | $p < 0.0005$ | | $-0.526$ | | 1.129 | | $-1.068$ |
| | F3 (PF) | 1712 | $p < 0.0005$ | | $-0.334$ | | $-0.338$ | | 0.990 |
| | F4 (HC) | 34 | $p < 0.0005$ | | 0.092 | | $-0.147$ | | $-0.031$ |
| CE | F1 (SUS) | 3287 | $p < 0.0005$ | 61% | $-0.061$ | 10% | $-1.815$ | 29% | 0.741 |
| | F2 (NUT) | 169 | $p < 0.0005$ | | $-0.080$ | | $-0.402$ | | 0.304 |
| | F3 (PF) | 3776 | $p < 0.0005$ | | $-0.576$ | | 0.687 | | 0.969 |
| | F4 (HC) | 158 | $p < 0.0005$ | | 0.157 | | $-0.446$ | | $-0.177$ |
| MC | F1 (SUS) | 2098 | $p < 0.0005$ | 53% | 0.126 | 35% | 0.378 | 12% | $-1.613$ |
| | F2 (NUT) | 152 | $p < 0.0005$ | | $-0.186$ | | 0.260 | | 0.056 |
| | F3 (PF) | 841 | $p < 0.0005$ | | 0.274 | | $-0.595$ | | 0.519 |
| | F4 (HC) | 1340 | $p < 0.0005$ | | 0.502 | | $-0.565$ | | $-0.538$ |
| NN | F1 (SUS) | 3286 | $p < 0.0005$ | 61% | $-0.063$ | 10% | $-1.845$ | 29% | 0.737 |
| | F2 (NUT) | 171 | $p < 0.0005$ | | $-0.090$ | | $-0.382$ | | 0.312 |
| | F3 (PF) | 3747 | $p < 0.0005$ | | $-0.573$ | | 0.701 | | 0.966 |
| | F4 (HC) | 163 | $p < 0.0005$ | | 0.159 | | $-0.459$ | | $-0.181$ |

[1] BLG = Between Groups Linkage, NN = Nearest Neighbour, CE = Centroid, MC = Median Clustering. [2] PC = percentage of cases in the cluster. [3] FCC = final cluster centres.

Concerning the final clusters, cluster 1 includes 61% of the individuals, and those have a positive input for factor F4 (HC), meaning that they are informed about health concerns related to the consumption of EIs. Factors F1 (SUS) and F2 (NUT) have a marginally negative input for cluster 1, so these individuals are not informed about the sustainability issues or the nutritive aspects of EIs. On the other hand, they reveal a very low level of knowledge about the production characteristics of EIs (F3—PC). Individuals in cluster 2 represent a minority of only 10% and possess high knowledge about the production factors associated with EIs (positive centre for F3—PF) but low knowledge about all other aspects associated with factors F1 (SUS), F2 (NUT), and F4 (HC). Finally, individuals in cluster 3 represent nearly one-third of the participants, and these are well-informed about all aspects except for the health issues, as evidenced by the positive centres for factors F1 (SUS), F2 (NUT), and F3 (PF). Based on these results, the clusters can be defined accordingly:

- Cluster 1 ('fearful' individuals)—individuals with low knowledge about EIs, but who are aware of the possible harmful effects resulting from their consumption;
- Cluster 2 ('farming' individuals)—individuals with very low knowledge about EIs, but who are informed about their production;
- Cluster 3 ('ecological' individuals)—individuals with very high knowledge about EIs, particularly concerning sustainability aspects and the production of EIs, but who are not informed about their possible health effects.

*3.3. Characterisation of the Clusters*

After defining the clusters, it is important to characterise the individuals in each of the groups. For this, the sociodemographic, geographic, and professional variables were used as segmentation characteristics.

Table 7 presents the clusters' membership according to sociodemographic variables like sex, age, and education level. The results indicate that, with respect to sex, cluster 2 (the 'farming' individuals) had proportionally more male participants, while cluster 1 (the 'fearful') had comparatively more female participants than the other two clusters (these differences being significant, $p < 0.0005$). Concerning age, no significant differences were found ($p > 0.05$) among clusters, with a similar distribution among the three age classes: most individuals in the young adults' class, followed by the adults' class and a smaller number of individuals in the senior adults' class, following the trend of the age distribution of the study sample. As for education level, significant differences were found ($p < 0.0005$), so members of cluster 2 (the 'farming') tended to have lower levels of education than those in cluster 1 (the 'fearful') and cluster 3 (the 'ecological'). This last cluster had the highest education level, which indicates that more educated people are better informed about sustainability issues related to EI.

Table 8 shows the clusters in terms of the geographical variables, country and living environment. The results highlight a significant difference among the clusters according to country ($p < 0.0005$), with a moderate association according to the value of Cramer's coefficient ($V = 0.221$). In some countries, like Greece, Latvia, Lebanon, or Turkey, a clearly higher percentage of individuals fall into cluster 1 (the 'fearful'). There are other countries, such as Croatia or Serbia, for which more individuals are categorised in cluster 2 (the 'farming'). Finally, cluster 3 (the 'ecological') is predominant in countries like Lithuania, Mexico, Poland, Slovenia, and Spain. With respect to the living environment, people in rural areas are classified more in cluster 2 (the 'farming'), while people in urban areas are more in cluster 3 (the 'ecological'). People in suburban areas are divided equally among the three clusters.

**Table 7.** Association between cluster membership and sociodemographic variables.

| Variables | | Cluster 1 Fearful | Cluster 2 Farming | Cluster 3 Ecological | Total |
|---|---|---|---|---|---|
| Sex | Female | 65.9% | 57.5% | 58.9% | 63.0% |
| ($p < 0.0005$; | Male | 33.5% | 42.0% | 40.2% | 36.3% |
| $V = 0.054$) | Other | 0.6% | 0.5% | 0.9% | 0.7% |
| Age group | Young adults (18–30 y) | 48.4% | 46.1% | 48.8% | 48.3% |
| ($p = 0.327$; | Adults (31–50 y) | 35.8% | 36.3% | 36.7% | 36.1% |
| $V = 0.018$) | Senior adults (51 y or over) | 15.8% | 17.6% | 14.5% | 15.6% |
| Education level | Postgraduate education (master's or PhD) | 30.3% | 21.8% | 35.5% | 31.0% |
| ($p < 0.0005$; | University degree | 32.6% | 32.0% | 32.3% | 32.5% |
| $V = 0.066$) | No University degree | 37.1% | 46.2% | 32.2% | 36.5% |

**Table 8.** Association between cluster membership and geographical variables.

| Variables | | Cluster 1 Fearful | Cluster 2 Farming | Cluster 3 Ecological | Total |
|---|---|---|---|---|---|
| | Croatia | 10.9% | 16.8% | 5.3% | 9.8% |
| | Greece | 10.9% | 9.5% | 5.6% | 9.2% |
| | Latvia | 5.7% | 2.3% | 2.3% | 4.4% |
| | Lebanon | 6.3% | 2.6% | 3.7% | 5.2% |
| | Lithuania | 7.5% | 3.0% | 8.6% | 7.4% |
| Country | Mexico | 12.8% | 20.2% | 23.1% | 16.5% |
| ($p < 0.0005$; | Poland | 6.5% | 2.7% | 11.4% | 7.6% |
| $V = 0.221$) | Portugal | 7.5% | 7.7% | 7.7% | 7.6% |
| | Romania | 8.2% | 8.2% | 4.5% | 7.1% |
| | Serbia | 5.8% | 9.5% | 1.9% | 5.0% |
| | Slovenia | 6.7% | 6.0% | 9.7% | 7.5% |
| | Spain | 5.4% | 8.3% | 14.4% | 8.3% |
| | Turkey | 5.7% | 3.2% | 1.7% | 4.3% |
| Living environment | Rural | 16.0% | 18.3% | 13.2% | 15.4% |
| ($p = 0.005$; | Urban | 67.6% | 66.2% | 71.3% | 68.6% |
| $V = 0.033$) | Suburban | 16.4% | 15.5% | 15.5% | 16.0% |

In Table 9, the cluster membership is reported according to professional variables (area of work and income). The results show that individuals with professional areas of food/nutrition and biology are more prone to be in cluster 3 (the 'ecological'), but for the other professions, the distribution among the different clusters is more even. The results further indicate that participants in the agricultural sector are slightly more likely to be in clusters 2 and 3 ('farming' and 'ecological'). The participants with professional activity linked to the environment are mostly in the 'ecological' cluster, although some are also present in cluster 1 ('fearful'). People engaged in professions related to tourism tend to fit into cluster 2 ('farming'), and this is also the case for individuals with professions related to the health sector. Nevertheless, for these two types of professionals, cluster 1 ('fearful') is also representative.

**Table 9.** Association between cluster membership and professional variables.

| Variables | | Cluster 1 Fearful | Cluster 2 Farming | Cluster 3 Ecological | Total |
|---|---|---|---|---|---|
| | Food/Nutrition | 30.0% | 24.0% | 38.2% | 31.9% |
| | Agriculture | 7.8% | 8.5% | 8.6% | 8.1% |
| Professional area | Environment | 5.2% | 3.2% | 5.4% | 5.1% |
| ($p < 0.0005$; | Biology | 4.9% | 2.2% | 7.8% | 5.5% |
| $V = 0.107$) | Health | 12.4% | 16.1% | 11.4% | 12.5% |
| | Tourism | 3.1% | 3.8% | 2.1% | 2.9% |
| | Others | 36.6% | 42.2% | 26.5% | 34.0% |
| | Much below average | 6.0% | 8.7% | 5.5% | 6.1% |
| Family income | Below average | 16.6% | 19.3% | 15.5% | 16.5% |
| ($p < 0.0005$; | Average | 40.4% | 40.0% | 38.1% | 39.7% |
| $V = 0.057$) | Above average | 32.5% | 26.0% | 33.4% | 32.2% |
| | Much above average | 4.5% | 6.0% | 7.5% | 5.5% |

When it comes to income (Table 9), people with an average income are more or less equally distributed among the three clusters. However, differences become clearer for low or very low incomes, for which the individuals tend to be more in cluster 2 (the 'farming') and on the other hand, those with high or very high incomes tend to be more in cluster 3 (the 'ecological').

## 4. Discussion

### 4.1. Analysis of the Scale

The validation of the scale showed that some items included in the questionnaire were not strong enough to be part of the scale.

Item 24, "There are appropriate regulations to guarantee the food safety of edible insects", relates to a very complex issue because regulations can be highly variable according to the geographic regions, countries, and even different political environments. In many countries, the rearing of insects for human food has been restrained by regulatory measures. For example, in Europe, regulations are very strict, and the topic of edible insects is still novel, as evidenced in the EU Novel Foods Regulation (EC) No 258/97 [46], which applies to foods and food ingredients that have not been used for human consumption to a significant degree within the European Community before 15 May 1997. EIs are considered novel foods under Regulation (EU) 2015/2283 [47] and can only be commercialised after a safety assessment and authorisation. The most recent advancement in this field in Europe was the recognition of dried yellow mealworm (*Tenebrio molitor* larva) as a safe novel food by EFSA [48]. Schiel et al. [49] discuss the possible application of analytical methods to analyse the composition of EIs for the purpose of food control in Germany. In Finland, despite the need to comply with the official existing EU regulations, EI production has been a reality since 2014 [50]. In the English-speaking markets (United States, United Kingdom, Canada, Australia, and New Zealand), EIs have been approved by their food safety agencies [15,51]. In areas where insects are considered traditional foods and have been consumed over generations (Africa, Southeast Asia, and Latin America), there is still a lack of regulatory measures regarding the production and consumption of EIs [15].

Item 11, "Insects are collected as a means of pest control for some cultivated crops", refers to wild insects that populate some agricultural crops, and this practice is specific to farmers and crop producers [52]. Therefore, it is highly probable that it might be difficult for the general public to be informed about such crop protection measures. Forest biodiversity is important not only in connection with the conservation of trees but also for the continued presence of insect populations [52]. However, in areas where edible forest insects grow into vast populations that can compromise cultivated crops, they are collected as a means of pest control and are included in a planned and nutritionally more valuable diet throughout the year [53,54].

Item 23, "Insects contain anti-nutrients, such as oxalates and phytic acid", refers to particular components that can be present in EIs, and which can be considered anti-nutrients [20]. Oxalate and phytic acid are biologically active compounds which can directly chelate nutrients such as minerals and proteins, thus making them unavailable for absorption. This immobilises the nutrients in undigested food complexes or, even if digestion and absorption occur, the anti-nutrients can represent barriers to the efficient utilisation of the nutrients [55]. In this way, they will prevent the human body from obtaining the necessary amounts of these nutrients. In some cases, anti-nutrients bind to proteins, especially digestive enzymes [56]. These properties of such substances are known by some people, like doctors or nutritionists, but most likely are unknown to the majority of the general public.

Items 1 and 2 ("Entomophagy consists in the consumption of insects" and "There are thousands of species of edible insects", respectively) were also problematic and therefore were excluded. Both items refer to true statements [4,57], and they deal with the knowledge that is most likely to be present for participants originating from countries where traditional insects consumption is a common practice. However, most countries included in this research do not fall into this category and this may explain the lower relevance of these items for the final factorial solution.

The factorial structure obtained included four factors. The first factor, named Sustainability (SUS), accounts for items related to the sustainability of EIs as alternative protein foods. EIs have been pointed out as considerably more sustainable compared to other sources of animal protein. In this way, the partial replacement of meat by EIs can alleviate pressure on the environment, while contributing to the feeding of a growing world population [57]. Ordoñez-Araque and Egas-Montenegro [58] present a literature review that demonstrates the viability of EIs as an alternative that can relieve nutritional deficiencies while contributing to slowing down the rate of deterioration of the environment.

Factor F2 was named Nutrition (NUT) and included items related to dietary components of EIs. Insects are categorised as one of the pillars of future human nutrition [59]. One of the reasons that contribute to this claim is that in many places where the availability of nutritious foods is scarce, EIs are usually present abundantly, and their nutritional value must be considered [15]. Proteins constitute the highest fraction of the composition of EIs, ranging from 50 to 70% on a dry basis. Lipids represent the second largest fraction of the nutritional composition of EIs, right after proteins [60]. Additionally, EIs contain dietary fibre, minerals, vitamins, and also some bioactive compounds with beneficial health properties [60–64].

Factor F3 (Production Factors—PF) consisted of two items that compared production factors of EIs with other sources of animal protein, specifically chickens and pigs. The production of insects has a lower environmental impact when compared to other sources of animal protein, namely, beef, pork, or chicken meats. Some of the advantages include lower emissions of greenhouse gases, the need for considerably less area/land for their rearing, a more efficient use of energy, and much lower needs for feed and water [57,65].

The last factor, F4 (Health Concerns—HC), accounts for items related to allergens or aflatoxins, which can be harmful to human health, and allied to items associated with a poor nutritional value or poor protein content, which can result in deficient nutrition. This factor includes items that have been reversed because they refer to statements that were false. However, the responses of the participants are against this reversion, which implies that, in general, the perceptions of the participants are towards agreement with the false statements, thus revealing a lack of knowledge when it comes to the quality of EIs as a nutritious food and containing high-quality proteins. It has been demonstrated that EIs present high-quality proteins in interesting amounts and these proteins contain all the essential amino acids in the recommended ratios [61,66,67].

*4.2. Characterisation of the Participants' Clusters and Discussion of Sociocultural Influences*

The cluster membership showed that more male participants were categorised into cluster 2 (the 'farming' individuals) than females. Farmers and people with knowledge about agriculture and husbandry are more frequently men. This is, in fact, a sector where there is still a high gender inequality [68,69], although some countries have started to empower women in this domain, such as Europe [68,70] or Africa [71–73]. Additionally, members of cluster 2 (the 'farming') tended to have lower levels of education than those in cluster 1 (the 'fearful') and cluster 3 (the 'ecological'). This last cluster had the highest education level, which indicates that more-educated people are better informed about sustainability issues related to EIs. The work by Guiné et al. [74] studied the level of information about the sustainability of EIs in Portugal and found that the most relevant discriminating sociodemographic variable was education, with people having a university degree being considerably more informed than those with lower education levels. Additionally, the study by Palmieri et al. [67] reinforces this aspect.

The professional area of the participants was also found to be related to cluster membership. In the work by Florença et al. [75], it was found that people in the areas of nutrition, agriculture, and environment tended to have more correct perceptions about EIs than those with professions linked to food, biology, or the health sector. In our work, the participants in the agricultural sector were more prone to be included in clusters 2 and 3 ('farming' and 'ecological'), which can be explained by their close relationship with agricultural practices, the land, and natural systems from which they derive their livelihood. Participants with work related to the environment tended to be categorised into the 'ecological' cluster, which is expected given their higher consciousness about the ecological and sustainability aspects.

People with professions linked to tourism or health tended to fit into cluster 2 ('farming') and cluster 1 ('fearful'). Possible explanations can be linked to the fear of consuming novel foods and the possible adverse effects that these can have, namely the safety issues associated with the consumption of EIs. Murefu et al. [21], reviewing the safety of EIs, alerted readers to the limitations of the actual food systems around the world in controlling hazards derived from the production and processing of insects, although highlighting that Europe was at the forefront when it came to the safety of EIs.

The results further showed that the level of income also affected the distribution of the participants by clusters. Differences were major for low or very low incomes, corresponding to participants categorised as 'farmers', while participants with high or very high incomes were categorised more into the 'ecological'. People with higher incomes usually also have a higher level of education, and those are associated with a higher ecological conscience [76,77]. On the other hand, people from rural environments, such as farmers, can have a lower level of income.

With regards to the cluster distribution by country, differences were found, resulting from the different sociocultural influences. Social and cultural influences greatly shape people's attitudes and level of knowledge. In the case of EIs, aspects related to ecological or health concerns were greatly present in Western societies, even in those countries where entomophagy was not a traditional practice. Cultural and social influences were drivers that influenced consumers towards the willingness to have EIs. Bisconsin-Júnior et al. [4] discussed the social aspects related to edible insects in regions where entomophagy was not familiar. Among the factors pointed out associated with positive and negative associations with EI, the authors referred to risk perception, level of acceptance or disgust, sustainability, culture, and organoleptic characteristics. Hartmann et al. [78] addressed the psychological factors underlying the consumption of EIs in countries with very diverse cultural influences towards insects, namely an Asian country (China) and a European country (Germany). Not surprisingly, they reported that the Chinese revealed a higher willingness to consume insects compared to the Germans. Ribeiro et al. [79] referred to differences in acceptance of insects as food and feed between consumers in a Southern Europe country (Portugal) and a Northern Europe country (Norway). In a work by Florença et al. [75] studying the motivations for consuming insects in a sample of the Portuguese population, it was shown that the preservation of the environment and natural resources constituted the strongest motivations to consume EIs for people who were not usual consumers of this type of food.

Schardong et al. [80] investigated consumers' perceptions of EIs in Brazil, a Latin American country, with some regions where the consumption of insects is possibly traditional. Their survey included participants from different regions of Brazil: North, Northeast, Midwest, Southeast, and South. Their results showed that men were more willing to consume insects than women. Flour was the preferred form of consumption, but the whole insect was preferred for those participants with familiarity with the insects. Gasca-Álvares et al. [81] conducted a review of EIs as food among indigenous communities of Colombia and reported that 69 edible insects were ingested by 13 ethnic groups originally from the Amazon and Caribbean regions. With regards to African countries, the study by Ebenebe et al. [82] highlighted that the tradition of entomophagy is somehow compromised by Western dietary patterns, which have been imposed over traditional insect eating. Still, they were able to do an inventory of 17 insect species consumed in Nigeria. In Ivory Coast, a study by Ehounou et al. [83] revealed that more than half the people in Abidjan consumed insects (60%), and they identified eight insect species consumed by the participants in the survey. Additionally, the trade of insects represented an important source of income for families. In Ghana, the survey conducted by Anankware et al. [84] aimed at identifying edible insects that were still underutilised and that should be more intensively used as human food and animal feed. They identified nine edible insects that were consumed differently depending on the region of the country. In South Africa, a questionnaire survey by Hlongwane et al. [85] investigated the level of indigenous knowledge about edible insects, namely what insects were consumed and how they were collected and prepared for food among the rural people. This work revealed that, like in other African countries, the influence of Western diets is leading to a decline in entomophagy. A review by Matandirotya et al. [86] made an overview of the consumption of edible insects in African countries. Some of the most relevant conclusions of this study point out the existence of a high number of edible insect species on the continent. These were easily accessible to the communities, and the populations had an incentive to use traditional knowledge to take advantage of this sustainable food source. However, they alert us to the need to establish food safety guidelines as a way to safely consume insects and their derived food products.

A work by Ruby et al. [87] investigated the willingness to consume EIs in two different countries with different cultural backgrounds, the United States of America and India. Their results showed that in both countries, the majority of participants were willing to consider eating at least some form of insect food (72% for Uthe SA and 74% for India). In China, Liu et al. [31] studied the factors that conditioned consumption of EIs and found that buying intentions were mostly dictated by phobia and disgust, but also knowledge level, age, household size and income, as well as the geographical region had a remarked influence.

In Hungary, the willingness to consume insect-based food was found to be low; however, it was higher for men and for those with higher school levels (university degrees) [88]. According to Detilleux et al. [89], Belgian youngsters showed a willingness to consume edible insects as processed foods, and their negative perception of entomophagy was changed towards a more positive one after actually tasting food products made with insects (falafel). The work by House [90] revealed that in the Netherlands, the development of a Dutch edible insect network was ongoing, focused on the production, supply, and consumption of a variety of insect-based foods. The paper also discussed the question of frequent consumption as opposed to just trying EI-based foods sporadically. In Switzerland, a study by Penedo et al. [91] showed that the acceptability of consumers towards EIs was related to various sociodemographic and behavioural factors. Although the participants were potentially willing to consume EIs, there were some practical barriers that impeded their adoption, such as disgust. In the Czech Republic, Kulma et al. [92] reported that peoples' preferences were towards consuming EIs as ingredients in foods, and they were generally favourable to the use of EIs to feed cattle to serve as human food. Orkusz et al. [93] showed that the willingness to adopt edible insects as a meat substitute was still low among the Poles, and the main constraints were related to psychological barriers, such as neophobia and disgust. However, the authors also reported that the consumption of insect-based foods was considerably higher than that of unprocessed whole insects. Among the positive drivers to incentive consumption of EIs stood the environmental benefits. In another study in Poland, Zielinska et al. [94] revealed that people over 40 years old were more ready to possibly accept edible insects in the future. When it comes to foods containing EIs, a great majority of respondents said they would consider accepting products that were made from insect protein. A study by Gałęcki et al. [95] showed that in Poland, insect farming could become a novel branch of agriculture, and it could create new opportunities for Polish farmers.

## 5. Conclusions

The present research allowed the statistical analysis of the results obtained for a set of items aimed at measuring knowledge about EIs, producing a solution with four factors, which included 22 of the 27 initial items. The solution explains 55% of the variance, and the four factors were identified as relating to sustainability (8 items), nutrition (8 items), production factors (2 items), and health concerns (4 items). For the first two factors, the internal consistency was very high, as given by the values of Cronbach's alpha, but for the health concerns factor, the internal consistency was low. Posterior cluster analysis revealed three clusters (fearful, farming, and ecological individuals). The cluster characterisation indicated that age did not influence cluster membership, while sex, education, country, living environment, professional area, and income all influenced the composition of the clusters.

In conclusion, this work confirmed the statistical validation of the present scale used to measure knowledge about EIs. Furthermore, its application to a wide set of countries, different in nature, allows its future usage on a global scale, making it a valuable instrument for application for a wide set of circumstances in the future, with participants in different countries with different cultural backgrounds and different population segments. The measurement of knowledge about EIs is a valuable way to define strategies for the implementation of policies designed to possibly improve EIs' attractiveness to people as a way to better contribute to more sustainable food systems while also benefiting from adequate nutrition and health improvement. This is of particular relevance since EIs are considered an instrument to contribute to food security while ensuring food safety.

Although providing a great deal of new information and wide coverage in terms of the geographical distribution of the study, the present research has some limitations that are worth highlighting. One of them is related to unequal group distribution, particularly by country (more participants from Mexico), sex (more men), and living environment (more people residing in urban areas). Another limitation is related to the countries included in the study, which, by being selected based on an invitation from the project manager and past collaboration, resulted in a higher representation of European countries as compared with other regions of the globe. Finally, it is worth mentioning that the design of the instrument itself has some limitations, which resulted from the fact that the same instrument should be suitable for participants with such a diverse cultural background regarding EIs, namely some in countries where eating insects is part of the local culture since time immemorial and others where it is still seen as a strange and somehow daring practice.

**Author Contributions:** Conceptualisation, R.P.F.G., C.A.C., P.M.R.C., M.F., A.P.C., S.C. and O.A.; methodology, R.P.F.G.; software, R.P.F.G.; validation, R.P.F.G.; formal analysis, R.P.F.G.; investigation, R.P.F.G., C.A.C., P.M.R.C., S.G.F., M.F., A.P.C., S.C., O.A., C.C.-H., M.M.S., M.P., J.M.F.B., M.K., M.Č.-B., E.B., M.T., N.M.B., I.D., D.K. and E.D.; resources, R.P.F.G., C.A.C., P.M.R.C., M.F., A.P.C., S.C. and O.A.; data curation, R.P.F.G.; writing—original draft preparation, R.P.F.G. and S.G.F.; writing—review and editing, R.P.F.G., C.A.C., P.M.R.C., S.G.F., M.F., A.P.C., S.C. and O.A.; supervision, R.P.F.G.; project administration, R.P.F.G.; funding acquisition, R.P.F.G., C.A.C., P.M.R.C., M.F., A.P.C. and S.C. All authors have read and agreed to the published version of the manuscript.

**Funding:** This work was funded by the CERNAS Research Centre (Polytechnic Institute of Viseu, Portugal) in the ambit of the project EISuFood (Ref. CERNAS-IPV/2020/003). We also received funding from the FCT—Foundation for Science and Technology (Portugal) through projects Ref. UIDB/00681/2020, UIDB/05507/2020 and UIDB/007421/2020. The APC was funded by FCT through project Ref. UIDB/00681/2020, project Ref. UIDB/05507/2020, project Ref. UIDB/007421/2020.

**Institutional Review Board Statement:** This research was implemented, taking care to ensure all ethical standards and following the guidelines of the Declaration of Helsinki. The development of the study by questionnaire survey was approved on 25 May 2020 by the ethics committee of the Polytechnic Institute of Viseu (Reference No. 45/SUB/2021).

**Informed Consent Statement:** Informed consent was obtained from all subjects involved in the study.

**Data Availability Statement:** Data are available from the corresponding author upon request.

**Acknowledgments:** This work was supported by the FCT—Foundation for Science and Technology, I.P. Furthermore, we would like to thank the CERNAS, CIDEI and UCISA:E Research Centres, and the Polytechnic Institute of Viseu for their support. This research was developed in the ambit of the project "EISuFood—edible Insects as Sustainable Food", with reference CERNAS-IPV/2020/003.

**Conflicts of Interest:** The authors declare no conflict of interest.

## References

1. Lange, K.W.; Nakamura, Y. Edible Insects as Future Food: Chances and Challenges. *J. Future Foods* **2021**, *1*, 38–46. [CrossRef]
2. Davidowitz, G. Habitat-Centric versus Species-Centric Approaches to Edible Insects for Food and Feed. *Curr. Opin. Insect Sci.* **2021**, *48*, 37–43. [CrossRef] [PubMed]
3. Backwell, L.R.; d'Errico, F. Evidence of Termite Foraging by Swartkrans Early Hominids. *Proc. Natl. Acad. Sci. USA* **2001**, *98*, 1358–1363. [CrossRef] [PubMed]
4. Bisconsin-Júnior, A.; Rodrigues, H.; Behrens, J.H.; da Silva, M.A.A.P.; Mariutti, L.R.B. "Food Made with Edible Insects": Exploring the Social Representation of Entomophagy Where It Is Unfamiliar. *Appetite* **2022**, *173*, 106001. [CrossRef] [PubMed]
5. La Barbera, F.; Verneau, F.; Videbæk, P.N.; Amato, M.; Grunert, K.G. A Self-Report Measure of Attitudes toward the Eating of Insects: Construction and Validation of the Entomophagy Attitude Questionnaire. *Food Qual. Prefer.* **2020**, *79*, 103757. [CrossRef]
6. Van Huis, A.; Itterbeeck, J.V.; Klunder, H.; Mertens, E.; Halloran, A.; Muir, G.; Vantomme, P. *Edible Insects: Future Prospects for Food and Feed Security*; Food and Agriculture Organization of the United Nations: Rome, Italy, 2013; ISBN 978-92-5-107596-8.
7. Sogari, G.; Ricciolo, F.; Moruzzo, R.; Menozzi, D.; Sosa, D.A.T.; Li, J.; Liu, A.; Mancini, S. Engaging in entomophagy: The role of food neophobia and disgust between insect and non-insect eaters. *Food Qual. And Pref.* **2023**, *104*, 104764. [CrossRef]
8. Çınar, Ç.; Karinen, A.K.; Tybur, J.M. The Multidimensional Nature of Food Neophobia. *Appetite* **2021**, *162*, 105177. [CrossRef]
9. Hopkins, I.; Farahnaky, A.; Gill, H.; Newman, L.P.; Danaher, J. Australians' Experience, Barriers and Willingness towards Consuming Edible Insects as an Emerging Protein Source. *Appetite* **2022**, *169*, 105832. [CrossRef]
10. La Barbera, F.; Verneau, F.; Amato, M.; Grunert, K. Understanding Westerners' Disgust for the Eating of Insects: The Role of Food Neophobia and Implicit Associations. *Food Qual. Prefer.* **2018**, *64*, 120–125. [CrossRef]
11. Dagevos, H. A Literature Review of Consumer Research on Edible Insects: Recent Evidence and New Vistas from 2019 Studies. *J. Insects Food Feed* **2021**, *7*, 249–259. [CrossRef]
12. Bruckdorfer, R.E.; Büttner, O.B. When Creepy Crawlies Are Cute as Bugs: Investigating the Effects of (Cute) Packaging Design in the Context of Edible Insects. *Food Qual. Prefer.* **2022**, *100*, 104597. [CrossRef]
13. Mishyna, M.; Chen, J.; Benjamin, O. Sensory Attributes of Edible Insects and Insect-Based Foods—Future Outlooks for Enhancing Consumer Appeal. *Trends Food Sci. Technol.* **2020**, *95*, 141–148. [CrossRef]
14. Borges, M.M.; da Costa, D.V.; Trombete, F.M.; Câmara, A.K.F.I. Edible Insects as a Sustainable Alternative to Food Products: An Insight into Quality Aspects of Reformulated Bakery and Meat Products. *Curr. Opin. Food Sci.* **2022**, *46*, 100864. [CrossRef]
15. Baiano, A. Edible Insects: An Overview on Nutritional Characteristics, Safety, Farming, Production Technologies, Regulatory Framework, and Socio-Economic and Ethical Implications. *Trends Food Sci. Technol.* **2020**, *100*, 35–50. [CrossRef]
16. Jantzen da Silva Lucas, A.; Menegon de Oliveira, L.; da Rocha, M.; Prentice, C. Edible Insects: An Alternative of Nutritional, Functional and Bioactive Compounds. *Food Chem.* **2020**, *311*, 126022. [CrossRef]
17. Schmidt, A.; Call, L.-M.; Macheiner, L.; Mayer, H.K. Determination of Vitamin B12 in Four Edible Insect Species by Immunoaffinity and Ultra-High Performance Liquid Chromatography. *Food Chem.* **2019**, *281*, 124–129. [CrossRef] [PubMed]
18. Raheem, D.; Raposo, A.; Oluwole, O.B.; Nieuwland, M.; Saraiva, A.; Carrascosa, C. Entomophagy: Nutritional, Ecological, Safety and Legislation Aspects. *Food Res. Int.* **2019**, *126*, 108672. [CrossRef]
19. Imathiu, S. Benefits and Food Safety Concerns Associated with Consumption of Edible Insects. *NFS J.* **2020**, *18*, 1–11. [CrossRef]
20. Grdeń, A.S.; Sołowiej, B.G. Macronutrients, Amino and Fatty Acid Composition, Elements, and Toxins in High-Protein Powders of Crickets, Arthrospira, Single Cell Protein, Potato, and Rice as Potential Ingredients in Fermented Food Products. *Appl. Sci.* **2022**, *12*, 12831. [CrossRef]
21. Murefu, T.R.; Macheka, L.; Musundire, R.; Manditsera, F.A. Safety of Wild Harvested and Reared Edible Insects: A Review. *Food Control* **2019**, *101*, 209–224. [CrossRef]
22. Ojha, S.; Bekhit, A.E.-D.; Grune, T.; Schlüter, O.K. Bioavailability of Nutrients from Edible Insects. *Curr. Opin. Food Sci.* **2021**, *41*, 240–248. [CrossRef]
23. Kunatsa, Y.; Chidewe, C.; Zvidzai, C.J. Phytochemical and Anti-Nutrient Composite from Selected Marginalized Zimbabwean Edible Insects and Vegetables. *J. Agric. Food Res.* **2020**, *2*, 100027. [CrossRef]

24. Paissoni, M.A.; Bitelli, G.; Vilanova, M.; Montanini, C.; Río Segade, S.; Rolle, L.; Giacosa, S. Relative Impact of Oenological Tannins in Model Solutions and Red Wine According to Phenolic, Antioxidant, and Sensory Traits. *Food Res. Int.* **2022**, *157*, 111203. [CrossRef] [PubMed]
25. Strati, I.F.; Tataridis, P.; Shehadeh, A.; Chatzilazarou, A.; Bartzis, V.; Batrinou, A.; Sinanoglou, V.J. Impact of Tannin Addition on the Antioxidant Activity and Sensory Character of Malagousia White Wine. *Curr. Res. Food Sci.* **2021**, *4*, 937–945. [CrossRef]
26. Santillo, A.; Ciliberti, M.G.; Ciampi, F.; Luciano, G.; Natalello, A.; Menci, R.; Caccamo, M.; Sevi, A.; Albenzio, M. Feeding Tannins to Dairy Cows in Different Seasons Improves the Oxidative Status of Blood Plasma and the Antioxidant Capacity of Cheese. *J. Dairy Sci.* **2022**, *105*, 8609–8620. [CrossRef]
27. González, C.M.; Llorca, E.; Quiles, A.; Hernando, I.; Moraga, G. An in Vitro Digestion Study of Tannins and Antioxidant Activity Affected by Drying "Rojo Brillante" Persimmon. *LWT* **2022**, *155*, 112961. [CrossRef]
28. Cano, A.; Contreras, C.; Chiralt, A.; González-Martínez, C. Using Tannins as Active Compounds to Develop Antioxidant and Antimicrobial Chitosan and Cellulose Based Films. *Carbohydr. Polym. Technol. Appl.* **2021**, *2*, 100156. [CrossRef]
29. Orsi, L.; Voege, L.L.; Stranieri, S. Eating Edible Insects as Sustainable Food? Exploring the Determinants of Consumer Acceptance in Germany. *Food Res. Int.* **2019**, *125*, 108573. [CrossRef]
30. Lucchese-Cheung, T.; Aguiar, L.K.D.; Da Silva, R.F.F.; Pereira, M.W. Determinants of the Intention to Consume Edible Insects in Brazil. *J. Food Prod. Mark.* **2020**, *26*, 297–316. [CrossRef]
31. Liu, A.-J.; Li, J.; Gómez, M.I. Factors Influencing Consumption of Edible Insects for Chinese Consumers. *Insects* **2019**, *11*, 10. [CrossRef]
32. Florença, S.G.; Guiné, R.P.F.; Gonçalves, F.J.A.; Barroca, M.J.; Ferreira, M.; Costa, C.A.; Correia, P.M.R.; Cardoso, A.P.; Campos, S.; Anjos, O.; et al. The Motivations for Consumption of Edible Insects: A Systematic Review. *Foods* **2022**, *11*, 3643. [CrossRef] [PubMed]
33. Guiné, R.P.F.; Florença, S.G.; Costa, C.A.; Correia, P.M.R.; Ferreira, M.; Duarte, J.; Cardoso, A.P.; Campos, S.; Anjos, O. Development of a Questionnaire to Assess Knowledge and Perceptions about Edible Insects. *Insects* **2022**, *13*, 47. [CrossRef]
34. Likert, R. A Technique for the Measurement of Attitudes. *Arch. Psychol.* **1932**, *22*, 5–55.
35. Leighton, K.; Kardong-Edgren, S.; Schneidereith, T.; Foisy-Doll, C. Using Social Media and Snowball Sampling as an Alternative Recruitment Strategy for Research. *Clin. Simul. Nurs.* **2021**, *55*, 37–42. [CrossRef]
36. Broen, M.P.G.; Moonen, A.J.H.; Kuijf, M.L.; Dujardin, K.; Marsh, L.; Richard, I.H.; Starkstein, S.E.; Martinez–Martin, P.; Leentjens, A.F.G. Factor Analysis of the Hamilton Depression Rating Scale in Parkinson's Disease. *Park. Relat. Disord.* **2015**, *21*, 142–146. [CrossRef]
37. Kaiser, H.F.; Rice, J. Little Jiffy, Mark Iv. *Educ. Psychol. Meas.* **1974**, *34*, 111–117. [CrossRef]
38. Stevens, J.P. *Applied Multivariate Statistics for the Social Sciences*, 5th ed.; Routledge: New York, NY, USA, 2009; ISBN 978-0-8058-5903-4.
39. Rohm, A.J.; Swaminathan, V. A Typology of Online Shoppers Based on Shopping Motivations. *J. Bus. Res.* **2004**, *57*, 748–757. [CrossRef]
40. Tanaka, K.; Akechi, T.; Okuyama, T.; Nishiwaki, Y.; Uchitomi, Y. Development and Validation of the Cancer Dyspnoea Scale: A Multidimensional, Brief, Self-Rating Scale. *Br. J. Cancer* **2000**, *82*, 800–805. [CrossRef]
41. Hair, J.F.H.; Black, W.C.; Babin, B.J.; Anderson, R.E. *Multivariate Data Analysis*, 7th ed.; Prentice Hall: Hoboken, NJ, USA, 2009; ISBN 978-0-13-813263-7.
42. Maroco, J.; Garcia-Marques, T. Qual a fiabilidade do alfa de Cronbach? Questões antigas e soluções modernas? *Lab. Psicol.* **2006**, *4*, 65–90. [CrossRef]
43. Davis, F.B. *Educational Measurements Their Interpretation*; Wadsworth Pub. Co.: Belmont, CA, USA, 1964.
44. Dolnicar, S. A Review of Data-Driven Market Segmentation in Tourism. *J. Travel Tour. Mark.* **2002**, *12*, 1–22. [CrossRef]
45. Witten, R.; Witte, J. *Statistics*, 9th ed.; Wiley: Hoboken, NJ, USA, 2009.
46. EC REGULATION (EC) No 258/97 of the European Parliament and of the Couoncil of 27 January 1997 Concerning Novel Foods and Novel Food Ingredients. *Off. J. Eur. Comm.* **1997**, *L 43*, 1–9.
47. EU Regulation (EU) 2015/2283 of the European Parliament and of the Council on Novel Foods, Amending Regulation (EU) No. 1169/2011 of the European Parliament and of the Council and Repealing Regulation (EC) No. 258/97 of the European Parliament and of the Council and Commission Regulation (EC) No. 1852/2001. *Off. J. Eur. Union* **2015**, *L 327*, 1–22.
48. EFSA. Safety of Dried Yellow Mealworm (Tenebrio Molitor Larva) as a Novel Food Pursuant to Regulation (EU) 2015/2283 | EFSA. Available online: https://www.efsa.europa.eu/en/efsajournal/pub/6343 (accessed on 22 June 2022).
49. Schiel, L.; Wind, C.; Krueger, S.; Braun, P.G.; Koethe, M. Applicability of Analytical Methods for Determining the Composition of Edible Insects in German Food Control. *J. Food Compos. Anal.* **2022**, 104676. [CrossRef]
50. Arppe, T.; Niva, M.; Jallinoja, P. The Emergence of the Finnish Edible Insect Arena: The Dynamics of an 'Active Obstacle'. *Geoforum* **2020**, *108*, 227–236. [CrossRef]
51. Liceaga, A.M. Processing Insects for Use in the Food and Feed Industry. *Curr. Opin. Insect Sci.* **2021**, *48*, 32–36. [CrossRef] [PubMed]
52. Gahukar, R.T. Edible Insects Collected from Forests for Family Livelihood and Wellness of Rural Communities: A Review. *Glob. Food Secur.* **2020**, *25*, 100348. [CrossRef]
53. Gahukar, R.T. Entomophagy Can Support Rural Livelihood in India. *Curr. Sci.* **2012**, *103*, 10.

54. Gahukar, R.T. Entomophagy for Nutritional Security in India: Potential and Promotion. *Curr. Sci.* **2018**, *115*, 1078–1084. [CrossRef]
55. Etuk, E.B.; Okeudo, N.J.; Esonu, B.O.; Udedibie, A.B.I. Antinutritional Factors in Sorghum: Chemistry, Mode of Action and Effects on Livestock and Poultry. *Online J. Anim. Feed Res. (OJAFR)* **2012**, *2*, 113–119.
56. Gazuwa, S.Y.; Denkok, Y. Organic Metabolites, Alcohol Content, and Microbial Contaminants in Sorghum-Based Native Beers Consumed in Barkin Ladi Local Government Area, Nigeria. *J. Sci. Res. Rep.* **2017**, *17*, 1–8. [CrossRef]
57. Guiné, R.P.F.; Correia, P.; Coelho, C.; Costa, C.A. The Role of Edible Insects to Mitigate Challenges for Sustainability. *Open Agric.* **2021**, *6*, 24–36. [CrossRef]
58. Ordoñez-Araque, R.; Egas-Montenegro, E. Edible Insects: A Food Alternative for the Sustainable Development of the Planet. *Int. J. Gastron. Food Sci.* **2021**, *23*, 100304. [CrossRef]
59. Oonincx, D.G.; Dierenfeld, E.S. An Investigation Into the Chemical Composition of Alternative Invertebrate Prey. *Zoo Biol.* **2012**, *31*, 40–54. [CrossRef] [PubMed]
60. Zhou, Y.; Wang, D.; Zhou, S.; Duan, S.; Duan, H.; Guo, J.; Yan, W. Nutritional Composition, Health Benefits, and Application Value of Edible Insects: A Review. *Foods* **2022**, *11*, 3961. [CrossRef]
61. Longvah, T.; Mangthya, K.; Ramulu, P. Nutrient Composition and Protein Quality Evaluation of Eri Silkworm (Samia Ricinii) Prepupae and Pupae. *Food Chem.* **2011**, *128*, 400–403. [CrossRef]
62. Kulma, M.; Kouřimská, L.; Homolková, D.; Božik, M.; Plachý, V.; Vrabec, V. Effect of Developmental Stage on the Nutritional Value of Edible Insects. A Case Study with Blaberus Craniifer and Zophobas Morio. *J. Food Compos. Anal.* **2020**, *92*, 103570. [CrossRef]
63. Anaduaka, E.G.; Uchendu, N.O.; Osuji, D.O.; Ene, L.N.; Amoke, O.P. Nutritional Compositions of Two Edible Insects: Oryctes Rhinoceros Larva and Zonocerus Variegatus. *Heliyon* **2021**, *7*, e06531. [CrossRef]
64. Kouhsari, F.; Saberi, F.; Kowalczewski, P.Ł.; Lorenzo, J.M.; Kieliszek, M. Effect of the Various Fats on the Structural Characteristics of the Hard Dough Biscuit. *LWT* **2022**, *159*, 113227. [CrossRef]
65. van Huis, A.; Oonincx, D.G.A.B. The Environmental Sustainability of Insects as Food and Feed. A Review. *Agron. Sustain. Dev.* **2017**, *37*, 43. [CrossRef]
66. Belluco, S.; Losasso, C.; Maggioletti, M.; Alonzi, C.C.; Paoletti, M.G.; Ricci, A. Edible Insects in a Food Safety and Nutritional Perspective: A Critical Review. *Compr. Rev. Food Sci. Food Saf.* **2013**, *12*, 296–313. [CrossRef]
67. Palmieri, N.; Perito, M.A.; Macrì, M.; Lupi, C. Exploring Consumers' Willingness to Eat Insects in Italy. *Br. Food J.* **2019**, *121*, 2937–2950. [CrossRef]
68. Balezentis, T.; Morkunas, M.; Volkov, A.; Ribasauskiene, E.; Streimikiene, D. Are Women Neglected in the EU Agriculture? Evidence from Lithuanian Young Farmers. *Land Use Policy* **2021**, *101*, 105129. [CrossRef] [PubMed]
69. Ragsdale, K.; Read-Wahidi, M.R.; Wei, T.; Martey, E.; Goldsmith, P. Using the WEAI+ to Explore Gender Equity and Agricultural Empowerment: Baseline Evidence among Men and Women Smallholder Farmers in Ghana's Northern Region. *J. Rural Stud.* **2018**, *64*, 123–134. [CrossRef]
70. Unay-Gailhard, İ.; Bojnec, Š. Gender and the Environmental Concerns of Young Farmers: Do Young Women Farmers Make a Difference on Family Farms? *J. Rural Stud.* **2021**, *88*, 71–82. [CrossRef]
71. Ingutia, R.; Sumelius, J. Determinants of Food Security Status with Reference to Women Farmers in Rural Kenya. *Sci. Afr.* **2022**, *15*, e01114. [CrossRef]
72. Owusu, V.; Yiridomoh, G.Y. Assessing the Determinants of Women Farmers' Targeted Adaptation Measures in Response to Climate Extremes in Rural Ghana. *Weather Clim. Extrem.* **2021**, *33*, 100353. [CrossRef]
73. Sell, M.; Minot, N. What Factors Explain Women's Empowerment? Decision-Making among Small-Scale Farmers in Uganda. *Women's Stud. Int. Forum* **2018**, *71*, 46–55. [CrossRef]
74. Guiné, R.P.F.; Florença, S.G.; Anjos, O.; Correia, P.M.R.; Ferreira, B.M.; Costa, C.A. An Insight into the Level of Information about Sustainability of Edible Insects in a Traditionally Non-Insect-Eating Country: Exploratory Study. *Sustainability* **2021**, *13*, 12014. [CrossRef]
75. Florença, S.G.; Correia, P.M.R.; Costa, C.A.; Guiné, R.P.F. Edible Insects: Preliminary Study about Perceptions, Attitudes, and Knowledge on a Sample of Portuguese Citizens. *Foods* **2021**, *10*, 709. [CrossRef]
76. Zakari, A.; Toplak, J. Investigation into the Social Behavioural Effects on a Country's Ecological Footprint: Evidence from Central Europe. *Technol. Forecast. Soc. Chang.* **2021**, *170*, 120891. [CrossRef]
77. Saboya de Aragão, B.; Alfinito, S. The Relationship between Human Values and Conscious Ecological Behavior among Consumers: Evidence from Brazil. *Clean. Responsible Consum.* **2021**, *3*, 100024. [CrossRef]
78. Hartmann, C.; Shi, J.; Giusto, A.; Siegrist, M. The Psychology of Eating Insects: A Cross-Cultural Comparison between Germany and China. *Food Qual. Prefer.* **2015**, *44*, 148–156. [CrossRef]
79. Ribeiro, J.C.; Gonçalves, A.T.S.; Moura, A.P.; Varela, P.; Cunha, L.M. Insects as Food and Feed in Portugal and Norway—Cross-Cultural Comparison of Determinants of Acceptance. *Food Qual. Prefer.* **2022**, *102*, 104650. [CrossRef]
80. Schardong, I.S.; Freiberg, J.A.; Santana, N.A.; Richards, N.S.P. dos S. Brazilian Consumers' Perception of Edible Insects. *Cienc. Rural* **2019**, *49*, e20180960. [CrossRef]
81. Gasca-Álvarez, H.J.; Costa-Neto, E.M. Insects as a Food Source for Indigenous Communities in Colombia: A Review and Research Perspectives. *J. Insects Food Feed* **2022**, *8*, 593–603. [CrossRef]

82. Ebenebe, C.I.; Amobi, M.I.; Udegbala, C.U.; Ufele, A.N.; Nweze, B.O. Survey of Edible Insect Consumption in South-Eastern Nigeria. *J. Insects Food Feed* **2017**, *3*, 241–252. [CrossRef]

83. Ehounou, G.P.; Ouali-N'Goran, S.W.M.; Niassy, S. Assessment of Entomophagy in Abidjan (Cote D'ivoire, West Africa). *Afr. J. Food Sci.* **2018**, *12*, 6–14.

84. Anankware, J.P.; Osekre, E.A.; Obeng-Ofori, D.; Khamala, C.M. Identification and Classification of Common Edible Insects in Ghana. *Int. J. Entomol. Res.* **2016**, *1*, 33–39.

85. Hlongwane, Z.T.; Slotow, R.; Munyai, T.C. Indigenous Knowledge about Consumption of Edible Insects in South Africa. *Insects* **2020**, *12*, 22. [CrossRef]

86. Matandirotya, N.R.; Filho, W.L.; Mahed, G.; Maseko, B.; Murandu, C.V. Edible Insects Consumption in Africa towards Environmental Health and Sustainable Food Systems: A Bibliometric Study. *Int. J. Environ. Res. Public Health* **2022**, *19*, 14823. [CrossRef]

87. Ruby, M.B.; Rozin, P.; Chan, C. Determinants of Willingness to Eat Insects in the USA and India. *J. Insects Food Feed* **2015**, *1*, 215–225. [CrossRef]

88. Szendrő, K.; Tóth, K.; Nagy, M.Z. Opinions on Insect Consumption in Hungary. *Foods* **2020**, *9*, 1829. [CrossRef] [PubMed]

89. Detilleux, L.; Wittock, G.; Dogot, T.; Francis, F.; Caparros Megido, R. Edible Insects, What about the Perceptions of Belgian Youngsters? *Br. Food J.* **2021**, *123*, 1985–2002. [CrossRef]

90. House, J. Insects as Food in the Netherlands: Production Networks and the Geographies of Edibility. *Geoforum* **2018**, *94*, 82–93. [CrossRef]

91. Penedo, A.O.; Bucher Della Torre, S.; Götze, F.; Brunner, T.A.; Brück, W.M. The Consumption of Insects in Switzerland: University-Based Perspectives of Entomophagy. *Foods* **2022**, *11*, 2771. [CrossRef]

92. Kulma, M.; Tůmová, V.; Fialová, A.; Kouřimská, L. Insect Consumption in the Czech Republic: What the Eye Does Not See, the Heart Does Not Grieve Over. *J. Insects Food Feed* **2020**, *6*, 525–535. [CrossRef]

93. Orkusz, A.; Wolańska, W.; Harasym, J.; Piwowar, A.; Kapelko, M. Consumers' Attitudes Facing Entomophagy: Polish Case Perspectives. *Int. J. Environ. Res. Public Health* **2020**, *17*, 2427. [CrossRef]

94. Zielińska, E.; Zieliński, D.; Karaś, M.; Jakubczyk, A. Exploration of Consumer Acceptance of Insects as Food in Poland. *J. Insects Food Feed* **2020**, *6*, 383–392. [CrossRef]

95. Gałecki, R.; Zielonka, Ł.; Zasępa, M.; Gołębiowska, J.; Bakuła, T. Potential Utilization of Edible Insects as an Alternative Source of Protein in Animal Diets in Poland. *Front. Sustain. Food Syst.* **2021**, *5*, 675796. [CrossRef]

*Article*

# Is Contract Farming with Modern Distributors Partnership for Higher Returns? Analysis of Rice Farm Households in Taiwan

Ming-Feng Hsieh [1] and Yir-Hueih Luh [2,*]

1  Department of Economics, Tunghai University, 1727, Sec. 4, Taiwan Blvd., Xitun Dist., Taichung 407224, Taiwan
2  Department of Agricultural Economics, National Taiwan University, 1, Sec. 4, Roosevelt Rd., Taipei 10617, Taiwan
*  Correspondence: yirhueihluh@ntu.edu.tw

**Abstract:** This study provides empirical evidence of the economic effect of contract farming for the agriculture sector dominated by smallholder farms. In light of the association between contract farming and modern food distribution channels, we categorize the adoption decisions of contract farming and modern marketing channels into four mutually exclusive choices and investigate their economic effects through the simulated maximum likelihood estimation of the multinomial treatment effects model. The results provide empirical evidence supporting higher returns from the dual partnerships as choosing modern distributors generates more revenues for the those participating in contract farming than for those with no contract farming, and contract farming is more likely to help generate more revenues for those who have taken modern distributors as their major marketing channel compared with those relying on traditional channels. Moreover, we examine whether any distributional pattern of marginal economic effects, of either contract farming or modern marketing channel, is present among farmers at various scales by using the conditional and unconditional quantile regression models. Our findings suggest that the marginal treatment effects are generally in an increasing trend as the quantile increases, implying that the economic effects of contract farming or partnership with modern distributors are more pronounced for higher returns among rice farmers in Taiwan. This finding has great policy implications for developing sustainable agriculture and food supply when facing greater uncertainties due to global warming in the future, especially in an agriculture sector with most smallholder farmers.

**Keywords:** rice farming; contract farming; modern distributors; multinomial treatment effects; conditional quantile regression; unconditional quantile regression

**Citation:** Hsieh, M.-F.; Luh, Y.-H. Is Contract Farming with Modern Distributors Partnership for Higher Returns? Analysis of Rice Farm Households in Taiwan. *Sustainability* 2022, 14, 15188. https://doi.org/10.3390/su142215188

Academic Editor: Dimitris Skalkos

Received: 17 October 2022
Accepted: 10 November 2022
Published: 16 November 2022

**Publisher's Note:** MDPI stays neutral with regard to jurisdictional claims in published maps and institutional affiliations.

## 1. Introduction

Contract farming is an agreement, commitment, and partnership between farmers and other stakeholders in the food supply chain [1–4]. The effects of contract farming on farmers' productivity, production efficiency, and income has been well documented in the literature both for the developed and developing countries. For example, several important issues concerning contract farming were examined for some developed countries, such as Australia and Taiwan [5,6]. The majority of more recent studies aimed at studying the influences of contract farming for farmers in various developing countries. The research focusing on the economic consequences of contract farming has been deemed a trend and also with compelling reasons [1,7,8]. Some reviews of the economic outcomes of contract farming, for example, [7,8], indicated two issues to be addressed while presenting empirical evidence of benefits associated with contract farming. On the one hand, it was indicated that contract farming results in additional benefits for farmers in developing countries since their farm scale is generally small [8]. This additional benefit is likely to be due to the increase in comparative advantage of small-scale farmers in accessing "higher-end markets" through contract farming [9]. On the other hand, the economic impacts of contract farming on the farm income, especially for the smallholder farmers, have been mixed, depending on differing production scales and commodity focus [7]. Such disparities between conceptual and empirical findings motivate the present study for further exploration.

This article investigates the effect of contract farming on the sales revenue of the rice producers in Taiwan, to address the two issues mentioned above. The motivation for this study is twofold. First, the issue of lower accessibility of farmers small in production scale, compared with their large counterparts, is an issue common in both the developed and developing worlds since small farmed area constitutes a major constraint for the agriculture sector regardless of a country's economic development. Second, while some view contract farming as some kind of institutional innovation that can reduce smallholders' transaction costs and market risks [10], some hold the view that contract farming assists the modernization of smallholder farmers [8]. The economic viability of contracting with modern retailers, such as supermarkets, hypermarkets, retailers (restaurants, fast food chains), and food processors, has been increasing in the process of modernization in the agriculture sector [6,11]. Provision of quality rice meeting food safety standards or high-quality FFVs, in terms of their appearance, shape, size, etc., represents local supermarkets' response to consumer needs and strategy to compete with traditional markets [12]. In light of the association between contract farming and modern food distribution channels pinpointed in previous studies, for example [13], we also intend to examine the following hypotheses: whether contract farming paired with modern food retailers can increase farmers' sales revenue and whether such a dual partnership has different economic impacts on farmers varying by their sales revenue.

This study makes three primary contributions to the existing body of knowledge. Our first contribution concerns providing empirical evidence of the economic effect of contract farming for the agriculture sector characterized by smallholder farms. Although there has been a rising trend in the practice of contract farming in developed countries including the US, Western Europe, and Japan [14], empirical evidence in support of the economic influences of contract farming, especially that with modern food distributors, for small-scale farmers in developed countries, is as of yet largely unavailable. Taking Japan as an example, the average farm size per commercial farm in Japan increased from 1.9 ha in 2000 to 2.5 ha in 2020 [15]. Japan's farmland area per farm is generally larger than many developed countries in which farmers face even more severe farmland constraints. As for Taiwan, the majority of the farmland area per farm household in Taiwan is less than 1 ha, and about 70% of the farm households operate on land less than 0.8 ha (Figure 1). The dominance of farm households who operate on farmland that is less than 0.8 hectares is prevalent in both the 2010 and 2015 Agricultural Census. Therefore, our study contributes to the extant body of knowledge by providing empirical evidence concerning the effects of contract farming in the agriculture sector dominated by smallholder farms. Second, most previous studies addressing the association between contract farming and modern food distribution channels focused on the economic effects on fresh fruits and vegetables. To the best of our knowledge, attention to rice producers has been quite limited. In order to bridge this knowledge gap, the present study aims at understanding the effects of contracting farming with the modern food distributors for famers with a commodity focus on rice.

More importantly, we address the issue of gaps between conceptual arguments and empirical evidences. While contract farming has long been conceptually thought of as a potential to boost smallholder farmer's revenues due to the risk sharing with the contractors and for market expansion, there were, however, mixed empirical results in the existing literature. Enlightened by a potential distributional difference in economic effects among farmers, we demonstrate a profound research on the treatment effects of marketing channel choice and contract farming engagement by using more rigorous empirical methodologies including multinomial treatments model and quantile regressions. It is important to investigate how the effects of contract farming and partnership with modern distributors vary by farm scale, especially when the adoption of the two is disproportional for rice farming in Taiwan. It is an important feature to encourage more sustainable agriculture as the risk factor plays a key role in farmer's farming decisions facing uncertainties in the future, especially under extreme weather conditions induced by global warming. Hence, we not only contribute to the literature by constructing a solid empirical study for a consistent set of estimates on the treatment effects but also assess an influential policy concern in Taiwan's agriculture for sustainable farming and adequate food supply.

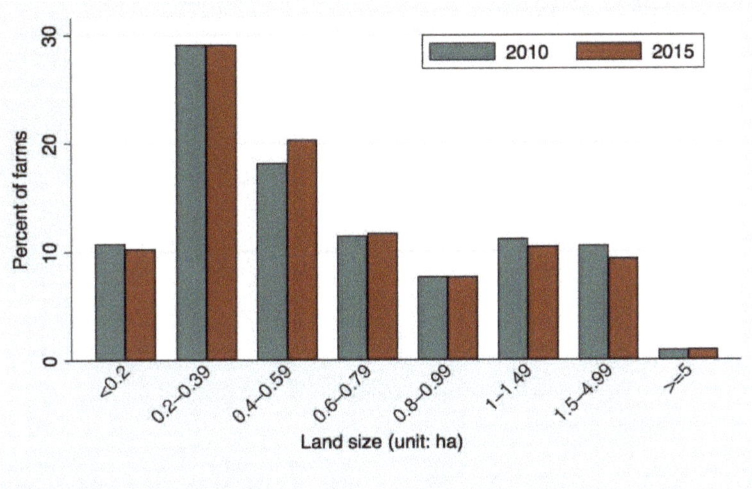


**Figure 1.** Farm size per household from the 2010 and 2015 Agricultural Census.

There are three reasons for our research focus on rice farmers. Firstly, rice has been the major staple food in Asia where the development of supermarkets and hypermarkets has been seen as the emerging trend in the era of economic growth [16]. It was indicated that in some countries, such as Thailand, almost half of the rice consumed by residents in more urbanized cities was purchased from modern marketing channels, including supermarkets, hypermarkets, and convenience stores, etc. [12]. According to the ranking of the Consumer Reach Points (CRPs) of Taiwan's retail channels for the second quarter of 2020, by Kantar Taiwan (Worldpanel Division), Quanlian supermarket ranked first with 187 million consumer touches with an increase in annual growth rate of 4%, followed by 7–ELEVEN (a branded convenience store) which has nearly 68 million CRPs with a flat annual growth rate, followed by Carrefour Group with nearly 55 million CRPs (6% annual growth rate) [17]. Kantar Taiwan also indicated that the ranking of the sales in food category is similar to that of the overall retail sales. These figures illustrate the important role of organized retailers (supermarkets, hypermarkets, convenience stores) in Taiwanese food purchases. Secondly, it was found in a study in Vietnam [18], that the low perishability of rice and consumers' lower concern over rice safety lead to different marketing channel distributions for rice and the fresh fruits and vegetables with the former relying more on wholesalers. However, contrary to what was observed in Vietnam, there are approximately 73.6% of rice farm households relying on modern distributors, including supermarkets, hypermarkets, convenience stores, processors, etc., as the major marketing channels in Taiwan. Thirdly, participation rate of contract farming is still low [6] since the government launched a production–marketing program in 2005, whereas rice is one of the top three contract farming commodities in Taiwan with a share of 5.59% out of 26,563 rice farmers participating in contract farming with food distributors (Table 1).

**Table 1.** Frequency and percentage of contract farming.

| Commodity Focus | No. of Obs. | No Contract Farming | Contract Farming |
|---|---|---|---|
| Rice | 26,563 | 94.41% | 5.59% |
| Sundry | 3661 | 86.42% | 13.58% |
| Special crops | 8509 | 96.51% | 3.49% |
| Vegetables | 34,896 | 96.95% | 3.05% |
| Fruits | 61,674 | 98.93% | 1.07% |
| Mushrooms | 782 | 98.98% | 1.02% |
| Sugarcane | 116 | 90.52% | 9.48% |
| Flowers | 2975 | 95.19% | 4.81% |
| Other crops | 1730 | 97.80% | 2.20% |
| Livestock | 4690 | 91.11% | 8.89% |
| Poultry | 3272 | 59.20% | 40.80% |
| Other raising | 123 | 100.00% | 0.00% |

Source: The primary farm household survey (PFHS).

This paper is organized as the following way. We delineate the farm household data and the econometric model used in this study in the next section, followed by the section presenting and discussing the results. The conclusion section summarizes the major findings in this article, in which we also propose the direction for possible extension in the future.

## 2. Materials and Methods

### 2.1. The Farm Household Data

Our rice producers' data is drawn from the 2013 Primary Farm Household Survey (PFHS). The primary farm households are randomly selected from the Agricultural Census based on two criteria: (1) making an annual income of more than NTD 200,000; and (2) at least one member working on the farm is under 65 years old (NTD, short for New Taiwanese Dollar, was exchanged at a rate of 0.0337 USD on average during 2013 when the PFHS was conducted). In the PFHS, farm households are categorized by their commodity focus which takes the highest share in total farm revenue from the farm produce that are not processed. Farm households whose major sales revenue is from rice are selected. There are 26,563 rice farm households in our final data set after rescaling by the sample weights.

The variable definition and descriptive statistics of the rice farm households are listed in Table 2. The average level of annual farm sales revenue is around NTD 740,000, which is around USD 25,000. Primary farm households participating in contract farming are approximately 4% [6]. Rice households participating in contract farming are about 6%. Following a broader definition of modern food distributors [19], we define modern food distributors as supermarkets, hypermarkets, convenience stores, brand stores, restaurants, and processors, which take around 74% of the rice farm households.

Approximately 88% of the principal operators of rice farm households are male. The average age of the rice farm operators is 61 years old, about two years older than the average of farm operators in the PFHS data. Principal operators' educational levels are most with elementary school and below (55%), while the rest are with junior high (23%), senior high (17%), and college degree and above (5%). On average, principal operators in the rice households have approximately 33 years of farming experience. About 42% of the farm operators do not have work experience before farming. Those having previous work experience are: 42% worked in the secondary and tertiary industries or were government employees, 11% self-employed, and 5% agriculture-related work.

The household's own and hired labor are around 2.62 and 0.02, which indicate that the majority of the rice farm households rely on their own household members. The average farmland area of rice farms is about 1.2 hectare. While the geographical distribution of the primary farm households is central (45%), south (35%), north (15%), and east (5%), there are more rice farm households located in central Taiwan (53%) than in the south (35%), the east (7%), and the north (6%).

**Table 2.** Variable definition and descriptive statistics.

| Variable | Definition | Mean | Std. Dev. |
|---|---|---|---|
| Revenue | Sales revenue from fresh produce (in thousand NTDs) | 520.88 | 633.61 |
| Contract Farming | Contract farming, yes = 1, no = 0 | 0.06 | 0.23 |
| Modern retailer | Supermarkets, hypermarkets, retailers, processors, etc., yes = 1, no = 0 | 0.74 | 0.44 |
| Male | Male operator, yes = 1, no = 0 | 0.88 | 0.33 |
| Age | Operator's age, 22–92 | 61.50 | 10.61 |
| Elementary | Elementary school degree and below, yes = 1, no = 0 | 0.55 | 0.50 |
| Junior high | Junior High school, yes = 1, no = 0 | 0.23 | 0.42 |
| Senior high | Senior High school, yes = 1, no = 0 | 0.17 | 0.37 |
| University | College degree and above, yes = 1, no = 0 | 0.05 | 0.23 |
| Experience | Years of farming experience | 33.36 | 14.95 |
| Agriculture | Previous work: agriculture, yes = 1, no = 0 | 0.05 | 0.22 |
| 2nd and 3rd industry | Previous work: secondary and tertiary industries, yes = 1, no = 0 | 0.42 | 0.49 |
| Self-employed | Self-employed, yes = 1, no = 0 | 0.11 | 0.32 |
| No previous work | No previous work experience, yes = 1, no = 0 | 0.42 | 0.49 |
| Farm workdays | Farm operators' on-farm workdays | 141.46 | 77.90 |
| Own labor | Number of household members working on the farm | 2.62 | 1.00 |
| Hired labor | Number of hired works | 0.02 | 0.36 |
| Farmland | Farmland size in area (0.01 ha) | 119.29 | 90.64 |
| North | Farm household located in northern Taiwan, yes = 1, no = 0 | 0.06 | 0.23 |
| Central | Farm household located in central Taiwan, yes = 1, no = 0 | 0.53 | 0.50 |
| South | Farm household located in southern Taiwan, yes = 1, no = 0 | 0.35 | 0.48 |
| East | Farm household located in eastern Taiwan, yes = 1, no = 0 | 0.07 | 0.26 |

Source: The primary farm household survey (PFHS).

We present the by-group descriptive statistics in Table 3. In general, farm households that participate in contract farming outperform those that do not participate in contract farming in terms of sales revenue from farm produce by about 90.8% on average. Principal farm operators that participate in contract farming are younger, with fewer years of farm experience and have a higher educational level (junior high or above). Furthermore, the farm households adopting contract farming have more farm workdays, more hired workers, larger farmland areas, and are disproportionately located in East Taiwan, which is famed for its natural and recreational scenery and especially with much less pollution due to its low degree of industrialization and urbanization.

**Table 3.** By-group descriptive statistics.

| Variable | No Contract Farming | | Contract Farming | |
|---|---|---|---|---|
| | Mean | Std. Dev. | Mean | Std. Dev. |
| Revenue | 495.6904 | 593.7813 | 945.9314 | 1017.6580 |
| Modern retailer | 0.7431 | 0.4369 | 0.6157 | 0.4866 |
| Male | 0.8734 | 0.3325 | 0.9307 | 0.2541 |
| Age | 61.9295 | 10.4958 | 54.1999 | 9.7927 |
| Elementary | 0.5734 | 0.4946 | 0.1931 | 0.3949 |
| Junior high | 0.2188 | 0.4134 | 0.3318 | 0.4710 |
| Senior high | 0.1569 | 0.3637 | 0.3661 | 0.4819 |
| University | 0.0509 | 0.2198 | 0.1090 | 0.3118 |
| Experience | 33.7738 | 14.8989 | 26.2934 | 14.1117 |
| Agriculture | 0.0518 | 0.2215 | 0.0377 | 0.1905 |
| 2nd and 3rd industry | 0.4148 | 0.4927 | 0.4711 | 0.4993 |
| Self-employed | 0.1168 | 0.3211 | 0.0552 | 0.2284 |
| No previous work | 0.4167 | 0.4930 | 0.4361 | 0.4961 |
| Farm workdays | 136.1281 | 75.6095 | 231.4603 | 58.5379 |

**Table 3.** *Cont.*

| Variable | No Contract Farming | | Contract Farming | |
|---|---|---|---|---|
| | **Mean** | **Std. Dev.** | **Mean** | **Std. Dev.** |
| Own labor | 2.6561 | 0.9971 | 2.0128 | 0.8444 |
| Hired labor | 0.0128 | 0.2349 | 0.1837 | 1.1775 |
| Farmland | 115.9513 | 80.2865 | 175.7073 | 186.3568 |
| North | 0.0611 | 0.2394 | 0.0020 | 0.0449 |
| Central | 0.5490 | 0.4976 | 0.1225 | 0.3279 |
| South | 0.3559 | 0.4788 | 0.1824 | 0.3863 |
| East | 0.0341 | 0.1815 | 0.6931 | 0.4613 |
| No. of observations | 25,077 | | 1486 | |

Source: The primary farm household survey (PFHS).

### 2.2. Empirical Specification

We categorize the adoption decisions of contract farming and modern marketing channels into four mutually exclusive choices: (1) choice = 1 (does not participate in either contract farming or modern marketing channels); (2) choice = 2 (participate in contract farming but not modern channels); (3) choice = 3 (participate in modern channels but not contract farming); (4) choice = 4 (participate in both contract farming and modern channels).

The "mtreatreg" module in STATA is used to test for our working hypothesis: whether contract farming paired with modern food retailers can contribute to higher revenue for farm households with a commodity focus on rice. The estimation of the multinomial treatment effects (MTE) model using the simulated maximum likelihood algorithm has been widely applied in the field of agricultural economics. The advantages of application of the MTE model are mainly due to the design of the model taking the endogenous treatment effects of choice variables (contract farming and/or partnership with modern distributors) into consideration when performing the regression of outcome variable (sales revenue or profit). It would lead to relatively more consistent or unbiased estimates compared to regressions without controlling for the treatment effects. For example, the MTE model was used to investigate the economic outcomes of Indian farmers' choice of marketing channels [20]. To measure the effects of adopting contract farming and modern food distribution channels, we control for farm household characteristics including major operators' gender, age, educational level, years of farming experience, farmland area, hired labor, and household labor. In comparison, we also run an OLS (ordinary least squares) regression without controlling for the treatment effects, in which an interaction term of the two choice variables—participating in contract farming and using modern food distribution channels—is included along with the two choice variables and other explanatory variables (base specification).

The limitation of such an empirical approach, along with many other treatment effects models, is that it exhibits only the conditional mean estimates but not conditional quantile estimates of marginal effects of the treatment effects. A preliminary analysis of the sales revenue for farm households adopting different production and marketing strategies is analyzed through Figure 2. It is demonstrated in Figure 2 that differences in sales revenue corresponding to farmer's choice of production/marketing strategies as well as distributional differences along the revenue distribution exist. Hence, we estimate conditional quantile regressions (CQR) [21] with the above base specification to obtain the impacts estimated on the conditional quantiles instead of on the conditional mean. We also adopt a so-called unconditional quantile regression approach (UQR) proposed by [22] based on the re-centered influence function (RIF) of unconditional quantile on the explanatory variables to estimate the direct effect of increasing the proportion of contract farming and/or with modern distributor partnership on the various quantile of the distribution of sales revenues. Such an approach has also been applied in the literature examining heterogeneous effects of various farming choices or strategies, e.g., organic adoption [23], cooperative membership [24], governmental policy support [25], rural infrastructure [26], and Internet use [27] on farm household performance or well-being. Our UQR estimation is conducted with the module of "uqreg" in STATA, in which point estimates of average treatment effects were provided to capture the "unconditional quantile partial effects (For the details of the STATA module, please refer to [28]).

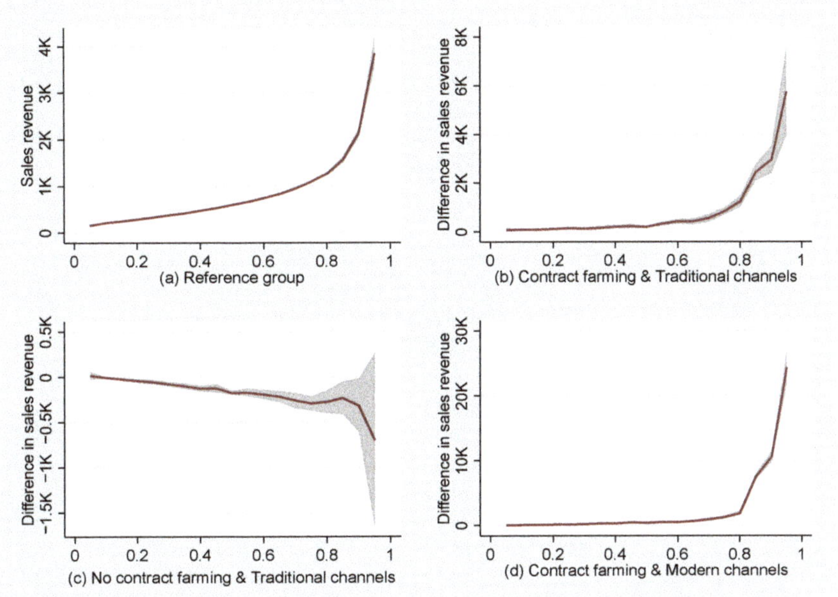

**Figure 2.** Mean revenue differences along the revenue distribution by adoption category: (**a**) choice = 3 (participate in modern channels but not contract farming); (**b**) choice = 2 (participate in contract farming but not modern channels); (**c**) choice = 1 (does not participate in either contract farming or modern marketing channels); (**d**) choice = 4 (participate in both contract farming and modern channels). Source: The primary farm household survey (PFHS).

*2.3. Identification Strategy*

We assume that the principal farm operator's choice of the production/marketing strategies is a rational behavior intended to maximize farm household's expected utility as in previous work (see, for instance, [29]).

Let the vectors of the farm operator's observed characteristics and the corresponding parameters be denoted by, respectively, x and α. The farm operator's choice of production/marketing strategies is indicated by $choice_k$ ($k = 1, 2, 3, 4$). That is, $choice_1$ takes the value of 1 when the farm household participates in contract farming and takes modern food distributors as the major marketing channel and 0 otherwise. The other three dummy variables are similarly defined as $choice_2 = 1$ if the farm household participates in contract farming but not modern channels; $choice_3 = 1$ if the farm household participates in modern channels but not contract farming; and $choice_4 = 1$ if the farm household participates in both contract farming and modern channels.

We assume the farm operator's expected utility associated with $choice_m$ is a linear function of the farm household's and principal operator's characteristics, x. That is, the *i*th farm operator's expected utility is given by:

$$E_i(\pi^m) = x_i \alpha_m + \varepsilon_{im}, \quad m = 1, \ldots, 4. \tag{1}$$

Under the assumption that the disturbance terms, $\varepsilon_{i1}, \ldots, \varepsilon_{i4}$, follow a multinomial logistic distribution, the $i$th farm operator's probability of choosing the $m$th strategy can be expressed as:

$$\text{Prob}[choice_{im} = 1 \,|\, x, \, w, \alpha, \theta] = \frac{exp(x_i \alpha_m + s_i \beta_m)}{1 + \sum_{j=1}^{4} \exp(x_i \alpha_j + s_i \beta_j)} \, , \; m = 1, \ldots, 4. \tag{2}$$

In the above equation, s and $\beta$ are vectors of the latent variables and their corresponding parameter vector. This study then starts out with the outcome equation of the farm operator's crop choices.

The sales revenue is determined by the choice of the production/marketing strategies as the following,

$$R_i = \gamma_0 + \gamma_1 choice_{i1} + \gamma_2 choice_{i2} + \gamma_3 choice_{i3} + \gamma_4 choice_{i4} + x_i \kappa + \mu_i. \tag{3}$$

$R_i$ in the above equation denotes farm household $i$'s sales revenue from rice farming. The parameter vector corresponding to the socioeconomic characteristics that affect farm household's sales revenue is denoted by $\kappa$.

In light of the endogeneity problem associated with the estimation of (3), the outcome equation to be estimated with (2) through the maximum simulated likelihood is what follows:

$$E(R_i \,|\, x, choice, l^*, \kappa, \theta, \lambda) = \theta_0 + x_i \kappa + \theta choice_i + l_i^* \lambda. \tag{4}$$

In the above specification, $l^*$ is the latent factor vector that represent the unobservable characteristics determining both the farm household's sales revenue and the choice of production/marketing strategies based on the underlying preferences (C*), while $\lambda$ denotes the vector of the selectivity correction terms [30].

In conditional quantile regression (CQR) for the $\tau$th-quantile, it is assumed that the $\tau$th-conditional quantile of the dependent variable is given as a linear function of the explanatory variables:

$$q_\tau = Q(R_i \,|\, x, cf, md, \, C^*, \kappa, \beta) = \theta_0 + x_i \kappa + \beta_1 cf_i + \beta_2 md_i + \beta_3 (cf * md)_i, \tag{5}$$

where $cf_i$ takes the value of 1 when the $i$th farm household participates in contract farming and 0 otherwise; $md_i$ takes the value of 1 when choosing modern food distributors as the major marketing channel and 0 otherwise; and their interaction term $(cf * md)_i$, and $q_\tau = Q(\cdot|X)$ denotes the $\tau$th quantile of $Y$ (sales revenue) conditional on $X$. It is easy to see that the various combinations of the above three terms will reproduce the multinomial choice of production/marketing strategies indicated by *choice* in the previous model incorporated with treatment effects.

The approach of unconditional quantile regression (UQR) builds on the concept of the influence function $\text{IF}(Y; q_\tau, F_Y) = (\tau - 1\{Y \le q_\tau\})/f_Y(q_\tau)$. In the so-called recentered influence function (RIF) regression model by [24], the dependent variable for the given quantile is:

$$\text{RIF}(Y; q_\tau, F_Y) = q_\tau + (\tau - 1\{Y \le q_\tau\})/f_Y(q_\tau). \tag{6}$$

It is shown in the study, [22], that the marginal effect of increasing the proportion of a treatment on the $\tau$th-quantile of the distribution of the outcome variable is captured by the UQR estimates rather than CQR estimates. The estimation of the marginal effect on the unconditional quantile is performed by computing the average derivative of the unconditional quantile regression within a small derivation from the specific point in the distribution of covariates, holding everything else constant.

## 3. Results and Discussion

The results are divided into the following subheadings. First, the results from the multinomial treatment–effect model are presented in Section 3.1, with a comparison to the

OLS results. Section 3.2 documents the differential effects at different points of the outcome distribution for conditional and unconditional quantile regression estimation. The three versions of estimation for identifying treatment effects of contract farming and/or modern distributors on sales revenue for rice farm households are compared and further discussed in Section 3.3.

### 3.1. Multinomial Treatment Effects

Table 4 reports the estimation results from the MTE and the OLS models. In the MTE's estimates, the results show that the farm households with $choice_1$ (no contract farming and traditional distributors) received less sales revenues by NTD 115,270 in comparison with the base category—the ones with $choice_3$ (no contract farming and modern distributors), implying that choosing modern distributors as the major marketing channel would produce positive outcomes in sales revenue.

**Table 4.** MTE and OLS regression results.

| Variable | (1)<br>MTE | Variable | (2)<br>OLS |
|---|---|---|---|
| $choice_1$ | −115.27 *** | $choice_1$ | −91.50 *** |
| $choice_2$ | 11.68 | $choice_2$ | −12.13 |
| $choice_4$ | 186.72 *** | $choice_4$ | 144.13 *** |
| Male | −13.25 ** | Male | −9.92 * |
| Age | −8.27 *** | Age | −8.40 *** |
| Elementary | 24.62 ** | Elementary | 26.00 ** |
| Junior high | 14.25 | Junior high | 15.99 |
| Senior high | 6.17 | Senior high | 6.40 |
| University | 3.52 *** | University | 3.59 *** |
| Experience | 106.02 *** | Experience | 102.11 *** |
| Agriculture | −60.48 *** | Agriculture | −58.22 *** |
| 2nd and 3rd industry | −156.51 *** | 2nd and 3rd industry | −155.08 *** |
| Self-employed | 2.03 *** | Self-employed | 2.03 *** |
| Own labor | 84.12 *** | Own labor | 83.81 *** |
| Hired labor | 182.28 *** | Hired labor | 183.15 *** |
| Farmland | 1.62 *** | Farmland | 1.61 *** |
| North | 36.79 * | North | 29.56 |
| South | −27.09 *** | South | −24.22 *** |
| East | 12.92 | East | 36.88 |
| Constant | 276.70 *** | Constant | 181.79 *** |
| Observations | 26,563 | Observations | 26,563 |
| $\lambda_1$ | 27.77 *** | $R^2$ | 0.217 |
| $\lambda_2$ | −26.65 *** | | |
| $\lambda_4$ | −57.35 *** | | |
| sigma | 556.53 | | |

Note: Statistical significance levels *** $p \leq 0.01$; ** $p \leq 0.05$; * $p \leq 0.1$. The base category for MTE regression is choice, the farmers with no contract farming but modern distributors.

This result is consistent with a previous study which also found the positive economic effect of using modern food channels, e.g., [31]. Similarly, we found a positive average treatment effect for contract farming which concurs with the conclusion of a positive effect of contract farming in the systematic review of the economic effects of contract farming in 13 developing countries [2]. Model (1) in Table 4 also shows that the farm households with $choice_4$ (contract farming and modern distributors) received more sales revenues by NTD 186,720 in comparison with the base category; that is, contract farming with modern food distributors brings more revenues to the farm households after taking both treatment effects into account. Relative to contract farming with traditional food distributors, our finding of additional benefits associated with contract farming and modern food distributors concurs with the results found in previous studies, for example [32].

Beyond the choice variables, the farm households with male principal operators, the self-employed, with more own labor or having bigger farmland size, tend to have higher sales revenues, while age and prior experience in agricultural, industrial (second), or service (third) sectors present as negative factors to a farmer's sales revenue. It shows a generally consistent set of estimation results from the two models in terms of the directions and scales of coefficient estimates for those control variables. However, it is worth noting that the estimates of the selectivity coefficients ($\lambda_1$, $\lambda_2$, $\lambda_4$) to the latent factor vector, which represent the unobservable characteristics determining both farm household's sales revenue and the choice of production/marketing strategies based on the underlying preferences, are significantly different from zero. It implies that the MTE estimates are relatively more consistent or unbiased than the OLS estimates.

To further investigate the marginal effects of either choice when holding the other the same, we compute the linear combinations of associated coefficients after the MTE and the OLS regression estimation. As shown in Table 5, the marginal effects estimated from the MTE model are generally greater than the OLS estimates, suggesting that the least square estimates tend to underestimate the effects of choice variables due to the lack of controlling for the endogeneity of treatment choices.

**Table 5.** Marginal effects of contract farming or modern distributors.

| Marginal Effect of Contract Farming | | | |
|---|---|---|---|
| **Traditional Distributors** | | **Modern Distributors** | |
| MTE | OLS | MTE | OLS |
| 126.95 *** | 79.37 * | 186.72 *** | 144.13 *** |
| (43.77) | (43.42) | (34.36) | (34.93) |
| **Marginal Effect of Modern Distributors** | | | |
| **No contract Farming** | | **Contract Farming** | |
| MTE | OLS | MTE | OLS |
| 115.27 *** | 91.50 *** | 175.04 *** | 156.27 *** |
| (9.42) | (7.72) | (61.07) | (59.89) |

Note: Statistical significance levels *** $p \leq 0.01$; ** $p \leq 0.05$; * $p \leq 0.1$.

While controlling for the choice of marketing channels, the marginal effect estimates indicate that contract farming contributes to sales revenue by a greater scale between the ones with modern distributors, in comparison between those with traditional distributors. On the other hand, choosing modern distributors as the major marketing channel shows a positive impact on the sales revenue regardless of whether participating in contract farming but with a bigger scale of impact for those who participated in contract farming than the ones that did not. Take the MTE estimates. The marginal effect of contract farming for the farmers partnered with modern distributors is on average 47% more than the one for farmers with traditional distributors. Between the farmers with and without contract farming, the MTE estimate of marginal effect of partnership with modern distributors for those adopting contract farming is about 52% more compared to those with no contract farming. This presents an interesting picture regarding the economic effects of adopting the two choices, in which choosing modern distributors generates more revenues for those with no contract farming (94% of the analysis sample), and even more for the farmers with contract farming (the rest 6%); additionally, contract farming is more likely to help generate more revenues for those who have taken modern distributors as their major marketing channel—consisting of 74% observations in our analysis sample. These results from both the MTE and the OLS models suggest that contract farming in partnership with modern distributors boosts more sales revenues for rice farmers.

Such results could also correlate to the distributional patterns that we observed in Figure 2, where contract farming makes significantly positive differences in revenues at the higher quantiles of the revenue distribution, while choosing modern distributors reduces the sales revenues towards the higher quantile of the revenue distribution. This calls for further investigations on the conditional (or unconditional) quantiles of the response variable rather than on the conditional means as performed in the above.

### 3.2. Conditional and Direct Effects from Quantile Regressions

We demonstrate the distributional differences in the effects of the determinants on farmer's sales revenue at low (q10), median (q50), and high (q90) quantiles from both conditional and unconditional quantile regressions in Table 6. As illustrated in Section 2.3, the coefficient estimates from the CQR are generally different from the ones from the UQR, especially when the explanatory variable is binary as the choice of contract farming (cf) or modern distributors (md) in our analysis. In our case study, the unconditional quantile regression's coefficient estimates are generally greater in size in comparison to the ones from conditional quantile regression as in Table 6. For instance, farmers choosing modern distributors as their major marketing channel have generated higher sales revenues compared to the ones with traditional distributors, and the effects increase with the quantiles evaluated. In addition, the direct effects estimated from the UQR are larger than the conditional effects from the CQR. Taking the 90th percentile, the UQR's estimate of marginal effect of a partnership with modern distributors (for those who do not adopt contract farming) is 437.32, while the CQR's is only 82.15 more in sales revenue (in thousands of NTD). Between quantiles, there are also significant differences in marginal effects. Take uq90 and uq50 as a comparison; the marginal effect of a partnership with modern distributors for non-contract-farming households is four times the median (50th) at the 90th quantile. Similarly, in comparison to the MTE estimate, the specific UQR's 90-th quantile estimate (437.32) is about 3.78 times of the mean effect (115.27). Such patterns of distributional effects are also observed in most coefficient estimates for most control variables. More sizable marginal effects of other explanatory variables on sales revenue in unconditional quantile estimates than in conditional quantile ones are also observed. Generally, we observe negative associations of age and prior experiences with farmer's sales revenue throughout the distribution, while holding other variables constant.

**Table 6.** Conditional and unconditional quantile regression results.

| | (1) | (2) | (3) | (4) | (5) | (6) |
|---|---|---|---|---|---|---|
| | **Conditional Quantile Regression** | | | **Unconditional Quantile Regression** | | |
| Variable | cq10 | cq50 | cq90 | uq10 | uq50 | uq90 |
| contract farming (cf) | 54.69 *** | 24.21 ** | −125.76 | 47.99 | 80.38 | 7.64 |
| modern distributors (md) | 5.52 *** | 20.46 *** | 82.15 *** | 16.71 | 106.94 *** | 437.32 *** |
| interaction (cf * md) | 28.85 *** | 212.17 *** | 343.31 *** | 26.18 | 72.97 | 619.98 |
| Male | −49.96 *** | −42.55 *** | 51.20 *** | 26.70 | 34.42 | 464.56 * |
| Age | −1.17 *** | −3.64 *** | −9.48 *** | −0.58 | −4.36 *** | −28.48 *** |
| Elementary | −46.07 *** | −18.77 *** | 49.22 ** | 17.91 | 31.34 | 102.48 |
| Junior high | −0.68 | 0.48 | 44.87 * | 19.74 | 45.51 | 183.02 |
| Senior high | −22.76 *** | −13.5 | −45.16 | 34.23 | −25.19 | −439.15 |
| University | −0.07 | 0.98 *** | 5.67 *** | 0.89 | 2.88 ** | 0.52 |
| Experience | −37.17 *** | −39.20 *** | −43.27 | −12.62 | 8.14 | −397.07 |
| Agriculture | 0.89 | 12.94 *** | −29.00 | −50.49 *** | −47.57 * | −677.59 *** |
| 2nd and 3rd industry | −63.89 *** | −37.70 *** | −93.58 *** | −64.12 *** | −131.42 *** | −915.56 *** |
| Self−employed | 0.01 | 0.86 *** | 4.31 *** | −0.05 | 1.00 *** | 6.45 *** |
| Own labor | −7.59 *** | 10.49 *** | 13.82 | 5.97 | 52.65 *** | 406.53 *** |
| Hired labor | −4.62 | 72.02 | 3077.43 | −29.73 *** | −29.51 * | 118.03 |
| Farmland | 0.97 *** | 1.43 *** | 1.91 *** | 0.20 *** | 0.63 *** | 2.81 *** |
| North | −44.10 *** | −71.77 *** | 128.69 *** | −70.04 *** | −42.17 | 648.11 *** |
| South | −16.83 *** | −10.45 *** | −9.80 | 38.18 *** | 70.26 *** | 54.83 |
| East | −78.08 *** | −78.72 *** | 234.61 *** | −40.35 * | −7.82 | 229.90 |
| Constant | 305.57 *** | 306.03 *** | 250.03 *** | | | |
| Observations | 26,563 | 26,563 | 26,563 | 26,563 | 26,563 | 26,563 |

Note: Statistical significance levels *** $p \leq 0.01$; ** $p \leq 0.05$; * $p \leq 0.1$.

To further decompose the treatment effects, we compute the marginal effects of contract farming or modern distributor partnership when keeping the other constant by linear combinations of coefficient estimates from the CQR and the UQR at the quantiles between the 5th–95th percentiles. Figure 3 compares the associated marginal effects from the OLS, multinomial treatment effects (MTE), conditional (CQR), and unconditional quantile regressions (UQR). Chart (a) shows no statistically significant effect of contract farming on sales revenue between the farmers partnered mainly with traditional distributors.

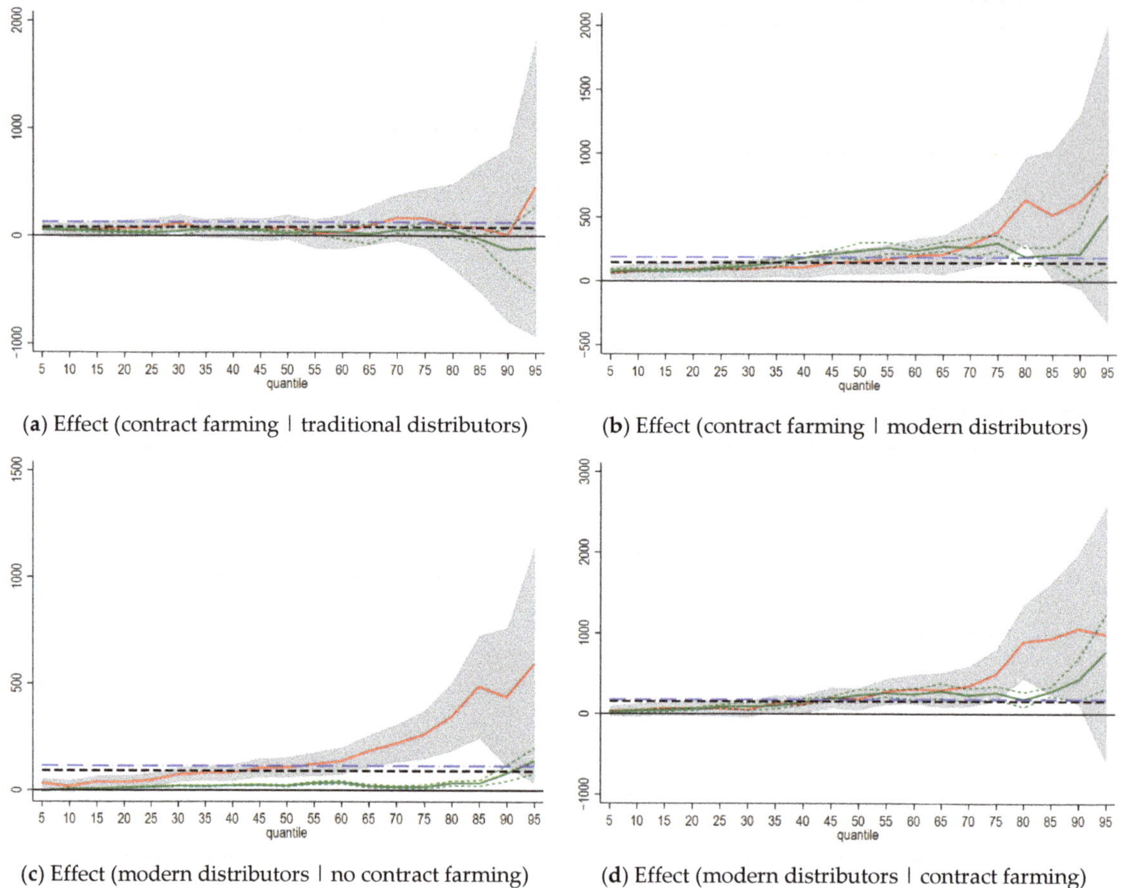

**(a)** Effect (contract farming | traditional distributors)

**(b)** Effect (contract farming | modern distributors)

**(c)** Effect (modern distributors | no contract farming)

**(d)** Effect (modern distributors | contract farming)

**Figure 3.** Marginal effects of contract farming or channel choice (modern/traditional) on revenue. Note: The associated marginal effects are computed by linear combinations of coefficient estimates from various models. The black dash (− −) line represents the OLS estimates, the blue dash-dot (— · —) line represents the estimates from the multinomial treatment–effect model, the green solid line (—) represents the CQR estimates with 95% confidence intervals in green dot lines ( . . . ), and the red solid line (—) represents the UQR estimates with 95% confidence intervals in gray shaded areas at various quantiles between the 5th and 95th percentiles.

As shown in Figure 3, we observe generally larger effects of contract farming between the farmers partnered mainly with modern distributors, and both the CQR estimates and the UQR estimates of marginal effects increase as quantile increases, even though the variation becomes larger too (chart (b)). The marginal effects of partnering with modern distributors on sales revenue are generally positive and increasing as the quantile increases between non-contract-farming farmers (chart (c)) or between contract-farming farmers (chart (d)). There are noticeably greater marginal effects of partnership with modern distributors at higher quantiles (75th or above) regardless of the farmer's contract farming choice.

In sum, the effects estimated from the conditional or unconditional quantile regression provide a vivid image on how the changes in the quantiles of the marginal distribution of outcome variable, i.e., sales revenue in our analysis. The marginal treatment effects vary by the distribution of sales revenue, generally in an increasing trend as the quantile increases.

### 3.3. A Remark on the Determinants and Distributional Effects on Profit

In addition, we also conduct analyses on the determinants and distributional effects of contract farming and channel choice on the profit of rice farm households in Taiwan. The results are generally similar to the ones for sales revenue of rice farm households. The estimates based on the conditional means, including the MTE and the OLS models, resemble the results from the revenue equations. One of the noticeable differences between the profit and sales revenue estimation results is a downturn in marginal effects of the two production/marketing choice variables at the high quantiles of the distribution. As shown in Appendix A, the marginal (direct) effects of either choice on profit from the unconditional quantile regression are positive and increasing as the quantile increases but become insignificantly different from zero or even go into the negative territory at the high quantiles (85th–95th percentiles) of profit distribution, especially for the cases when farmers already adopt one of the two choices (contract farming or modern distributors). It implies that adopting one additional production or marketing strategy may generate more sales revenue on one hand but could hinder a farmer's profit for those in the higher quantiles of the distribution likely associated with additional costs corresponding to such dual partnerships.

## 4. Conclusions

We provide a comprehensive exploration of the economic effects of farm households' production and marketing strategies. Specifically, this study focuses on investigating the effect of contract farming on the sales revenue of rice producers in Taiwan. In light of the important role of supermarkets, hypermarkets, and convenience stores in Taiwanese food purchases and the emerging trend of farmers' contract farming with modern food distribution channels, we also examine if contract farming with modern food retailers can increase farmers' sales revenues.

There have been mixed empirical results on the economic impacts of contract farming on farm income, especially for the smallholder farmers in the existing literature, even though contract farming has been thought beneficial to small-scale farmers for being able to provide access to higher-end markets. Such mixed results may be due to biased estimates due to the design of the empirical approach. In our present study, we apply both the multinomial treatment effects model and quantile regression models to assess both mean treatment effects and marginal effects over the quantiles in a distribution of sales revenue. We believe that such methodological efforts are proven to bring more understanding and insights on how contract farming as well as partnership with modern distributors impact farmers' profitability at various scales. It resolves the puzzle of mixed results being observed in the previous studies, as the marginal effects of contract farming vary by their profitability scale.

Our major findings are summarized as the following. First, the marginal effect estimates from the MTE model indicate that contract farming contributes positively to sales revenue regardless of their channel choice, but more between the ones with modern distributors than between those with traditional distributors. This result is consistent with previous studies' findings that farmers participating in contract farming with modern food distributors outperform those who do not. Second, the results from the conditional and unconditional quantile regression provide a vivid image that treatment effect changes in the quantiles of the marginal distribution of sales revenue—the marginal treatment effects vary by the distribution of sales revenue, generally in an increasing trend as the quantile increases. Although it is shown that the partnership of contract farming with modern food distributors could boost more sales revenues for rice farmers based on the results from both MTE and quantile regressions, it may limit farmers' ability to gain profit for those in the higher quantiles of the distribution from the quantile estimations.

While the ratio of partnership with modern retailers has been high (with an average of 74%), the contract farming rate accounts for only 6% among rice farmers in Taiwan. Through quantile regressions, especially the UQR approach, we show that the marginal effects of contract farming are positive and higher for those who have partnerships with modern distributors, especially at higher quantiles. It implies that contract farming paired with modern distributors has the potential to boost farmer's revenues by risk sharing with the contractors and extending markets with modern distributors. It is an important feature to encourage more sustainable agriculture as the risk factor plays a key role in farmers' decision making when facing a greater degree of uncertainty from global warming, especially for an agriculture with smallholders in Taiwan or some South Asian countries and many developing countries. The more income security, the more sustainable agriculture will be.

There are two major research limitations of the present study. Our first research limitation concerns the data availability. Although the PFHS is a dated set of data, it is the most-up-to-date available and a representative set of data that fits for the present subject of interest. For one thing, the data are lack of the proportion of contracted farm produce in the farm households' total sales. Therefore, how the treatment effect varies with farmers' degree of contract farming participation was not examined. Another shortcoming of the data is that some details of contract farming were not recorded. One of the unobserved information is the type of contract, and different contracts may vary in their economic impacts on farming outcomes. Therefore, where data permit, future research may seek to identify and compare the economic effects of different types of contract farming, including spot contracts, contracted production, contracted sale, and management contacts in vertical integration.

**Author Contributions:** Conceptualization, Y.-H.L.; data curation, M.-F.H. and Y.-H.L.; formal analysis, M.-F.H. and Y.-H.L.; funding acquisition, Y.-H.L.; investigation, M.-F.H. and Y.-H.L.; methodology, M.-F.H. and Y.-H.L.; project administration, Y.-H.L.; resources, Y.-H.L.; software, M.-F.H. and Y.-H.L.; supervision, Y.-H.L.; validation, Y.-H.L.; visualization, M.-F.H. and Y.-H.L.; writing—original draft, M.-F.H. and Y.-H.L.; writing—review and editing, M.-F.H. and Y.-H.L. All authors have read and agreed to the published version of the manuscript.

**Funding:** Part of this research is funded by the National Science and Technology Council (previously, Ministry of Science of Technology) in Taiwan under project number MOST-106-2410-H-002-018.

**Institutional Review Board Statement:** Not applicable.

**Informed Consent Statement:** Not applicable.

**Data Availability Statement:** The data used in this study are available from the Survey Research Data Archive, Center for Survey Research, Academia Sinica, through application.

**Conflicts of Interest:** The authors declare no conflict of interest.

## Appendix A. Marginal Effects of Contract Farming or Channel Choice (Modern/Traditional) on Profits

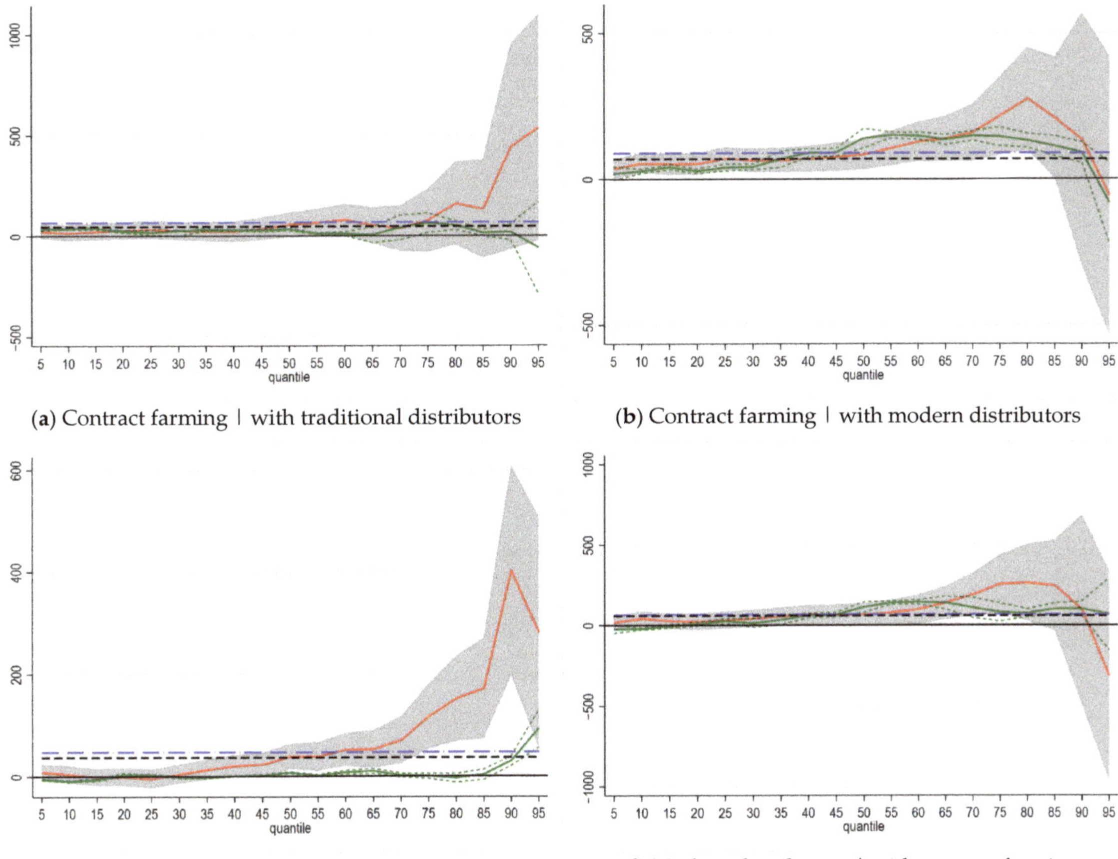

(**a**) Contract farming | with traditional distributors

(**b**) Contract farming | with modern distributors

(**c**) Modern distributors | no contract farming

(**d**) Modern distributors | with contract farming

**Figure A1.** Marginal effects of contract farming or channel choice (modern/traditional) on profits. Note: The associated marginal effects are computed by linear combinations of coefficient estimates from various models. The black dash (− −) line represents the OLS estimates, the blue dash-dot (— · —) line represents the estimates from the multinomial treatment effects model, the green solid line (—) represents the CQR estimates with 95% confidence intervals in green dot lines ( . . . ), and the red solid line (—) represents the UQR estimates with 95% confidence intervals in gray shaded areas at various quantiles between 5th and 95th percentiles.

## References


1. Bellemare, M.F.; Bloem, J.R. Does contract farming improve welfare? A review. *World Dev.* **2018**, *112*, 259–271. [CrossRef]
2. Ton, G.; Vellema, W.; Desiere, S.; Weituschat, S.; D'Haese, M. Contract farming for improving smallholder incomes: What can we learn from effectiveness studies? *World Dev.* **2018**, *104*, 46–64. [CrossRef]
3. Gramzow, A.; Batt, P.J.; Afari-Sefa, V.; Petrick, M.; Roothaert, R. Linking smallholder vegetable producers to markets-A comparison of a vegetable producer group and a contract-farming arrangement in the Lushoto District of Tanzania. *J. Rural Stud.* **2018**, *63*, 168–179. [CrossRef]
4. Nguyen, H.K.; Chiong, R.; Chica, M.; Middleton, R.H.; Pham, D.T.K. Contract farming in the Mekong Delta's rice supply chain: Insights from an agent-based modeling study. *J. Artif. Soc. Soc. Simul.* **2019**, *22*, 1. [CrossRef]

5.  Zeller, B.; Langa, L. Contract farming: Global standards or market forces? The case of the Australian dairy industry. *Unif. Law Rev.* **2018**, *23*, 282–297. [CrossRef]
6.  Luh, Y.-H. Inclusiveness of contract farming along the modern food supply chain: Empirical evidence from Taiwan. *Agriculture* **2020**, *10*, 187. [CrossRef]
7.  Nguyen, A.T.; Dzator, J.; Nadolny, A. Does contract farming improve productivity and income of farmers? A review of theory and evidence. *J. Dev. Areas* **2015**, *49*, 531–538. [CrossRef]
8.  Wang, H.H.; Wang, Y.; Delgado, M.S. The transition to modern agriculture: Contract farming in developing economies. *Am. J. Agric. Econ.* **2014**, *96*, 1257–1271. [CrossRef]
9.  Liu, Y.; Minot, N.; Wang, M. Improving market access for smallholders. In *The Sustainable Intensification of Smallholder Farming Systems*, 1st ed.; Robinson, M., Klauser, D., Eds.; Burleigh Dodds Science Publishing: London, UK, 2020; pp. 339–360.
10. Ogutu, S.O.; Ochieng, D.O.; Qaim, M. Supermarket contracts and smallholder farmers: Implications for income and multidimensional poverty. *Food Policy* **2020**, *95*, 101940. [CrossRef]
11. Ochieng, D.O.; Veettil, P.C.; Qaim, M. Farmers' preferences for supermarket contracts in Kenya. *Food Policy* **2017**, *68*, 100–111. [CrossRef]
12. Custodio, M.C.; Demont, M.; Laborte, A.; Velasco, M. Rice Quality Defined by Urban Consumers and Value Chain Actors in South and Southeast Asia. In Proceedings of the Asian Society of Agricultural Economists 9th International Conference, Bangkok, Thailand, 11–13 January 2017.
13. Blandon, J.; Henson, S.; Cranfield, J. Small-scale farmer participation in new agri-food supply chains: Case of the supermarket supply chain for fruit and vegetables in Honduras. *J. Int. Dev. J. Dev. Stud. Assoc.* **2009**, *21*, 971–984. [CrossRef]
14. Otsuka, K.; Nakano, Y.; Takahashi, K. Contract farming in developed and developing countries. *Annu. Rev. Resour. Econ.* **2016**, *8*, 353–376. [CrossRef]
15. van Landbouw, M. Structure, Land Use and Profitability of Farming in Japan. Market Report 2020 (105-0011), Department of Agriculture, Embassy of the Kindom of the Netherlands, Tokyo. Available online: https://www.agroberichtenbuitenland.nl/landeninformatie/japan/achtergrond/marktstudies/large-agricultural-corporations (accessed on 12 August 2022).
16. Reardon, T.; Timmer, C.P.; Minten, B. Supermarket revolution in Asia and emerging development strategies to include small farmers. *Proc. Natl. Acad. Sci. USA* **2012**, *109*, 12332–12337. [CrossRef]
17. Lin, Y.-T. *Where Do You Get Your Food and Groceries? The Physical Channel Quanlian Takes the Lead While Line Shopping Is the Fastest Growing.* Available online: https://www.foodnext.net/column/columnist/paper/5616516067 (accessed on 12 August 2022). (In Chinese).
18. Moustier, P.; Tam, P.T.G.; Anh, D.T.; Binh, V.T.; Loc, N.T.T. The role of farmer organizations in supplying supermarkets with quality food in Vietnam. *Food Policy* **2010**, *35*, 69–78. [CrossRef]
19. Chang, Y.-C.; Wei, M.-F.; Luh, Y.-H. Choice of modern food distribution channels and its welfare effects: Empirical evidence from Taiwan. *Agriculture* **2021**, *11*, 499. [CrossRef]
20. Cariappa, A.G.; Sinha, M. Choice of paddy marketing channel and its impact: Evidence from Indian farm households. *Agric. Econ. Res. Rev.* **2020**, *33*, 191–204. [CrossRef]
21. Koenker, R. *Quantile Regression*; Cambridge University Press: New York, NY, USA, 2005.
22. Firpo, S.; Fortin, N.M.; Lemieux, T. Unconditional quantile regressions. *Econometrica* **2009**, *77*, 953–973.
23. Khanal, A.R.; Mishra, S.K.; Honey, U. Certified organic food production, financial performance, and farm size: An unconditional quantile regression approach. *Land Use Policy* **2018**, *78*, 367–376. [CrossRef]
24. Ma, W.; Zheng, H.; Zhu, Y.; Qi, J. Effects of cooperative membership on financial performance of banana farmers in China: A heterogeneous analysis. *Ann. Public Coop. Econ.* **2022**, *93*, 5–27. [CrossRef]
25. Kostov, P.; Davidova, S.; Gjokaj, E. Does policy support really help farmers' incomes: The case of Kosovo. In Proceedings of the International Association of Agricultural Economists, Virtual, 17–31 August 2021.
26. Wu, Q.; Guan, X.; Zhang, J.; Xu, Y. The role of rural infrastructure in reducing production costs and promoting resource-conserving agriculture. *Int. J. Environ. Res. Public Health* **2019**, *16*, 3493. [CrossRef]
27. Zheng, H.; Ma, W.; Wang, F.; Li, G. Does internet use improve technical efficiency of banana production in China? Evidence from a selectivity-corrected analysis. *Food Policy* **2021**, *102*, 102044. [CrossRef]
28. Rios-Avila, F. Recentered influence functions (RIFs) in Stata: RIF regression and RIF decomposition. *Stata J.* **2020**, *20*, 51–94. [CrossRef]
29. Luh, Y.-H.; Tsai, M.-H.; Fang, C.L. Do first-movers in the organic market stand to gain? Implications for promoting cleaner production alternatives. *J. Clean. Prod.* **2020**, *262*, 121156. [CrossRef]
30. Di Paolo, A. (Endogenous) occupational choices and job satisfaction among recent Spanish PhD recipients. *Int. J. Manpow.* **2016**, *37*, 511–535. [CrossRef]
31. Slamet, A.S.; Nakayasu, A.; Ichikawa, M. Small-scale vegetable farmers' participation in modern retail market channels in Indonesia: The determinants of and effects on their income. *Agriculture* **2017**, *7*, 11. [CrossRef]
32. Miyata, S.; Minot, N.; Hu, D. Impact of contract farming on income: Linking small farmers, packers, and supermarkets in China. *World Dev.* **2009**, *37*, 1781–1790. [CrossRef]

 *sustainability*

*Article*

# Chemometric Screening of Oregano Essential Oil Composition and Properties for the Identification of Specific Markers for Geographical Differentiation of Cultivated Greek Oregano

**Eleftheria S. Tsoumani, Ioanna S. Kosma * and Anastasia V. Badeka ***

Laboratory of Food Chemistry, Department of Chemistry, University of Ioannina, GR-45110 Ioannina, Greece
* Correspondence: i.kosma@uoi.gr (I.S.K.); abadeka@uoi.gr (A.V.B.)

**Abstract:** The present study investigated the potential interconnection between the place of cultivation of Greek oregano samples and the composition and properties of their essential oils (EOs). In addition, it attempted to identify characteristic chemical features that could differentiate between geographical origins with the use of chemometric tools. To this end, a total of 142 samples of commercially available Greek oregano (*Origanum vulgare* ssp. *hirtum*) plants harvested during the calendar years 2017–2018 were obtained for this study. The samples came from five different geographical areas of Greece and represented twelve localities. After appropriate processing, the oregano samples were subjected to hydrodistillation (HD), and the resulting EOs were analyzed for their total phenolic content (TPC), antioxidant activity, and chemical composition. The acquired data were subjected to the chemometric methods of multivariate analysis of variance (MANOVA) and linear discriminant analysis (LDA) to investigate the potential of classifying the oregano samples in terms of geographical origin. In addition, stepwise LDA (SLDA) was used as a final step to narrow down the number of variables and identify those wielding the highest discriminatory power (marker compounds). Carvacrol was identified as the most abundant component in the majority of samples, with a content ranging from 28.74% to 68.79%, followed by thymol, with a content ranging from 7.39% to 35.22%. The TPC values, as well as the Trolox equivalent antioxidant capacity (TEAC) values, showed no significant variations among the samples, ranging from 74.49 ± 3.57 mg GAE/g EO to 89.03 ± 4.76 mg GAE/g EO, and from 306.83 ± 5.01 µmol TE/g EO to 461.32 ± 7.27 µmol TE/g EO, respectively. The application of the cross-validation method resulted in high correct classification rates in both geographical groups studied (93.3% and 82.7%, respectively), attesting to a strong correlation between location and oregano EO composition.

**Keywords:** Greek oregano; essential oils; geographical differentiation; chemometrics

**Citation:** Tsoumani, E.S.; Kosma, I.S.; Badeka, A.V. Chemometric Screening of Oregano Essential Oil Composition and Properties for the Identification of Specific Markers for Geographical Differentiation of Cultivated Greek Oregano. *Sustainability* 2022, 14, 14762. https://doi.org/10.3390/su142214762

Academic Editor: Alessandra Durazzo

Received: 1 October 2022
Accepted: 4 November 2022
Published: 9 November 2022

**Publisher's Note:** MDPI stays neutral with regard to jurisdictional claims in published maps and institutional affiliations.



## 1. Introduction

A trend toward healthier lifestyles has recently emerged among consumers, leading to increasing demand for herbal medicines, nutraceuticals, and natural foods worldwide. Medicinal and aromatic plants (MAPs), one of the wealthiest bioresources of drugs in traditional and modern medicine and an abundant source of fragrances, condiments, decoctions, and essential oils (EOs), have become a burgeoning area of research due to their treasured active ingredients [1]. An extensively employed herb that enjoys wide industrial, pharmaceutical, and traditional usage worldwide is oregano. Oregano, apart from its proven biological (antimicrobial, fungicidal, and antioxidant) properties, has a unique aroma that distinguishes it from other plants [2–7]. The term "oregano", attributed to more than 60 species globally, is mainly associated with the genus *Origanum* of the Lamiaceae family, which is for the most part spread throughout the Mediterranean [8]. *Origanum* presents excellent morphological and chemical diversity and is assorted into 49 taxa and 42 species. In most European countries, *Origanum vulgare* L. is the most predominant species of the genus [9,10].

Greek oregano, or *Origanum vulgare* L. ssp. *hirtum* (Link) Ietswaart, is regarded as one of the best varieties in the world in terms of quality due to its EO composition and high EO yield [11–14]. Previously published data have revealed significant variability in the chemical composition of the EOs of oregano and their yield, even within species [14–16]. The main constituents of the EOs of Greek oregano include four biosynthetically related monoterpene compounds: $\gamma$-terpinene, *p*-cymene, and either thymol or carvacrol, depending on the chemotype [8,17]. The chemotypes of aromatic plants are generally defined by the predominant compound of their EO. In the case of Greek oregano, the types that prevail are the carvacrol type; the thymol type; and the carvacrol/thymol type, wherein carvacrol and thymol are present in almost equal amounts. As a rule, the carvacrol chemotype designates a condiment as oregano; however, the amount of carvacrol may vary significantly among *O. vulgare* plants (from traces to over 90%) depending on the region, season, and subspecies [18]. While studying autumnal Greek oregano plants from several parts of Greece, Kokkini et al. (1997) [19] recorded noticeable differences in their total EO content and the concentration of their four main components: the $\gamma$-terpinene content ranged from 0.6 to 3.6% of the total EO, *p*-cymene from 17.3 to 51.3%, thymol from 0.2 to 42.8%, and carvacrol from 1.7% to 69.6%. Additionally, when comparing these data to those obtained from plants collected from the same localities in the mid-summer, the authors found that the carvacrol ratio was much higher in the summer, while in the autumn, *p*-cymene predominated. Likewise, Russo et al. (1998) [20] reported significant quantitative and quality variations when studying the chemical composition of wild populations of *Origanum vulgare* ssp. *hirtum* in Calabria, Italy.

The analysis of the active ingredients of Greek oregano EOs can be challenging due to the aforementioned chemical diversity and variability. Such challenges can be met using chemometrics, a discipline that integrates mathematics, statistics, and formal logic and provides helpful information through processing multivariate chemical data [21]. Qualitative chemometric models are widely used in food analysis to determine authentication, trace geographical or genetic origins, and detect impurities. In contrast, quantitative models are mainly used to estimate concentrations of food ingredients [22]. Chromatographic methods such as gas chromatography (GC), liquid chromatography (LC), high-performance liquid chromatography (HPLC), and high-temperature gas chromatography (HTGC) are often coupled with chemometrics to identify unique marker compounds that could indicate differentiation with respect to the place of origin [23]. Most studies using these methods have focused on honey [24–26] and dairy products [27–29], whereas fewer have included spices such as saffron [30], paprika [31], and oregano [14]. Even though studies on Greek oregano and its many properties are abundant in the literature, Vokou et al. [14] were the first, and to the best of the authors' knowledge the only, authors who attempted to correlate the chemical properties of wild *O. vulgare* ssp. *hirtum* to its geographical origin. However, rather than attempting to correlate the sampling regions to the composition and yield of the oregano EOs, they employed multifactor ANOVA to process the geographical and climatic characteristics of the areas, aiming to identify their effect on the attributes of oregano. In particular, they assessed six factors—altitude, distance from the sea, moisture index, summer water deficiency, thermal efficiency (TE), and summer concentration of TE—in relation to the EO yield; the concentration sum of thymol and carvacrol; and the concentration sum of thymol, carvacrol, $\gamma$-terpinene, and *p*-cymene. They observed that four out of the six factors (altitude, summer water deficiency, TE, and summer concentration of TE) significantly affected the yield, whereas only thermal efficiency appeared to influence the compound concentrations.

Despite the sharp increase in consumption and the significant commercial value of Greek oregano, coordinated efforts to domesticate and systematically cultivate oregano in Greece have only begun in recent decades [32–34]. Because oregano growers more often than not use oregano populations without any appropriate plant material selection, a wide array of products of varying quality, particularly in terms of composition, are commercially produced. Apart from generating inconsistency, such tactics may leave room for acts of profiteering and fraud due to adulteration. Additionally, the individual morpho- and ontogenetic variability and the ecological and environmental effects add to the vast heterogeneity of the species [8,14,34–38]. Therefore, a comparative study of Greek oregano EOs from different regions of Greece is a valuable tool to explore the cultivated species' chemical diversity and realize their actual commercial value.

Thus, the aims of this study were: (i) to determine the main constituents of the essential oil of Greek oregano plants collected from cultivated populations from twelve different localities of Greece with the use of gas chromatography–mass spectrometry (GC-MS); (ii) to assess the total phenolic content and antioxidant activity of these essential oils; and finally (iii) to combine the acquired data with chemometric methods in an attempt to identify characteristic chemical attributes, also known as marker compounds, that potentially signify a differentiation of geographical origin. To achieve this, the acquired data were initially treated with MANOVA; the geographical origin was set as the independent variable, while the experimental data were appointed as the dependent variables. After establishing the significant dependent variables for geographical differentiation, LDA was then applied to these designated variables in order to explore the possibility of classifying the oregano samples according to their geographical origin. The combination of multiple analytical parameters resulted in a greater aggregation of the oregano samples in the respective regions. Since the number of significant variables ($p < 0.05$) resulting from MANOVA was quite large, stepwise LDA (SLDA) was subsequently employed so as to reduce the parameters to those considered as the best set of authenticity predictors/markers in relation to the herein-studied regions.

This study is the first to focus exclusively on cultivated Greek oregano samples from areas throughout Greece. In addition, the combination of analytical parameters with the specific cultivation origin of the Greek oregano samples constitutes the novelty of the present work.

## 2. Materials and Methods

### 2.1. Oregano Samples

A total of 142 Greek oregano samples were obtained at the stage of optimum maturity from professional oregano growers in late June–early July of 2017 and 2018, respectively. The plants originated from five geographical areas of Greece (Epirus, Thessaly, Northern Greece, the Peloponnese, and Crete) and twelve localities in total (Figure 1). The samples, consisting of dried oregano aerial parts, were all processed by hand to ascertain homo- geneity. The resulting rubbed oregano samples were refrigerated within air-tight glass containers until the analyses.

All oregano growers procured the initial oregano seedlings from the Hellenic Agricul- tural Organization, Elgo-Dimitra, as certified O. vulgare ssp. hirtum material. Moreover, the selected oregano plants were of similar cultivation characteristics: organic, open-field, mostly non-irrigated, and beyond their second year of cultivation.

### 2.2. Extraction of EOs

The EOs were extracted from the rubbed oregano samples using hydrodistillation (HD), a method chosen because it is devoid of organic solvents and is extensively used in the food industry. Exactly 20 g of each oregano sample and 300 mL of distilled water were placed into a 500 mL round-bottomed flask, which was then connected to a Clevenger apparatus (Auxilab, Spain). A heating mantle was used as a heating medium, and the samples were subjected to HD for 3 h [39]. After being dried over anhydrous sodium sulfate, the obtained EOs were filtered and stored in 3 mL amber glass bottles at 4 °C until use.

Northern Greece
(Central & Western Macedonia)
1. Thessaloniki
2. Katerini
3. Kilkis
4. Kozani

Thessaly
5. Volos
6. Kalambaka

Epirus
7. Ioannina
8. Preveza

Peloponnese
9. Ileia
10. Achaea

Crete
11. Rethymno
12. Heraklion

**Figure 1.** Oregano cultivation and sampling sites.

*2.3. Gas Chromatography–Mass Spectrometry (GC/MS) Instrumentation and Analysis Conditions*

The GC detector used in this study was an Agilent 7890A coupled with an Agilent 5975C inert XL MSD mass selective detector (Agilent, Wilmington, DE, USA). The GC fused silica capillary column was a BP20 (WAX) polar column, 25 m × 0.32 mm × 0.25 mm (J & W Scientific, Folsom, CA, USA.), and the carrier gas used was ultra-high purity helium at a flow rate of 1.5 mL/min. The injector operated in split mode (30:1 split ratio), and its temperature was kept at 260 °C.

An aliquot of 50 µL of each essential oil sample and 200 µL of a 4-methyl-2-pentanol internal standard solution were placed in a 5 mL volumetric flask. The flask was filled with hexane to the mark, and 1 µL of the final solution was then injected into the GC inlet port.

The temperature program used was as follows: The oven temperature was initially maintained at 40 °C for 4 min, increased to 120 °C at a rate of 20 °C/min, maintained for 2 min, raised again to 200 °C at 8 °C/min, and increased anew to 230 °C at a rate of 15 °C/min. The final temperature was maintained for 1 min, while a solvent delay was also set at 1.5 min. The acquisition was performed in the MS, operating with electron impact ionization (EI, 200 eV) and 2.92 scans/s in a 35–300 ($m/z$) mass range, while the transfer line temperature was set at 230 °C. Peak identification was performed by comparing the eluting compounds' retention times and mass spectra to the Wiley Library database [40].

Retention indices (RI) of the EO compounds were calculated using appropriate n-alkane (C8–C20) standards (Fluka, Buchs, Switzerland). All determinations were conducted thrice.

### 2.4. Determination of Total Phenolic Content (TPC) and Antioxidant Activity

The process followed regarding the methanolic extracts was based on liquid–liquid extractions and, more specifically, olive oil assays [41,42]. Precisely 0.1 g of each EO sample was mixed with 2 mL of hexane and 3 mL of MeOH/$H_2O$ (60:40). The mixture was initially vortexed for 2 min and then centrifuged at 4000 rpm for 10 min at 4 °C to ascertain the separation of the two phases. The methanol phase was separated, and the process was repeated. The combined methanolic extracts were collected in a 10 mL volumetric flask, which was then filled with MeOH/$H_2O$ (60:40) up to the mark. This methanolic sample solution was used in the following determinations.

TPC was spectrophotometrically estimated according to the Folin–Ciocalteu colorimetric method [43]. The reaction mixture was prepared in a 100 mL volumetric flask by mixing 0.2 mL of the methanolic sample solution, 0.25 mL of Folin–Ciocalteu reagent, and 2.3 mL of $H_2O$. After 3 min, 0.5 mL of $Na_2CO_3$ 20% was added to the mixture, which was then supplemented with water up to the mark. The samples were incubated at room temperature for 30 min in the dark, and the absorbance was measured at λmax = 725 nm against a blank using a Perkin Elmer Lambda 25 UV/VIS Spectrophotometer. The TPC concentrations were estimated using a calibration curve obtained over the range of 50–200 mg/kg of gallic acid. The results were expressed as mg of gallic acid equivalents (GAE)/g of oregano EO.

The antioxidant activity of the samples was assessed according to the DPPH (2,2-diphenyl-1-picrylhydrazyl) free-radical scavenging method [44,45]: 2.9 mL of DPPH solution was mixed with 0.1 mL of the methanolic sample solution and kept at room temperature for 30 min in the dark. The control solution consisted of methanol and DPPH, and the absorbance was measured at 517 nm. A calibration curve was created in the range of 10–175 mg/kg of Trolox, and the DPPH radical scavenging activity was expressed as μmol of Trolox equivalents per EO g (μmol TE/g EO sample).

All determinations were carried out in triplicate, and the results are presented as the mean average.

### 2.5. Statistical Analysis

IBM SPSS 25.0 [46] was used for all statistical analyses in this study. The acquired data were subjected to MANOVA to determine the significant variables for the geographical differentiation of oregano. Geographical origin was set as the independent variable, while several analytical parameters were selected as the dependent variables (essential oil composition, total phenolics, antioxidant capacity, and combinations thereof). After that, linear discriminant analysis (LDA) was applied using the same parameters to identify characteristic chemical attributes that potentially differentiated between geographical origins. The original and leave-one-out cross-validation methods were implemented to evaluate the prediction classification ability. The procedure was repeated for all the parameters of the samples. Box's M test was conducted to assess the homogeneity of variability in this study [47,48].

Finally, stepwise LDA (SLDA) was applied as the ultimate classification method to distinguish the most significant variables through a stepwise process in order to optimize the discrimination. The classification evaluation of SLDA was conducted by leave-one-out cross-validation [49,50].

## 3. Results and Discussion

### 3.1. Essential Oil Chemical Composition

The composition of the essential oil of each oregano sample, as determined by GC and combined GC-MS, is shown in Table 1. In total, 35 compounds and two chemotypes were identified. The following fifteen compounds were present in all samples: α-pinene, β-myrcene, α-terpinene, γ-terpinene, *p*-cymene, 1-octen-3-ol, *cis*-sabinene hydrate, *trans*-sabinene hydrate, caryophyllene, 4-terpineol, borneol, $\beta$-bisabolene, caryophyllene oxide, thymol, and carvacrol. However, these were present in varying proportions, and carvacrol was the predominant constituent in all samples apart from the Ileia oregano. The fundamental components of the EOs were primarily oxygenated monoterpenes (65.67–83.98%) and monoterpene hydrocarbons (10.80–30.43%). Sesquiterpene hydrocarbons, oxygenated sesquiterpenes, and miscellaneous compounds followed at lower rates: 2.44–3.97%, 0.40–1.92%, and 0.10–1.29%, respectively.

The most abundant regions in terms of compounds were Kozani and Ioannina, each recording 33 in total, followed closely by Preveza, Ileia, and Achaea with 31, 28, and 27 compounds, respectively. Contrarily, the region of Thessaloniki presented the fewest constituents, featuring only 16 in total, while the rest of the studied locations recorded around 22 compounds each.

Most oregano samples pertained to the carvacrol chemotype, with carvacrol rates ranging from 46.19% (Rethymno) to 68.79% (Thessaloniki). In contrast, only two belonged to the carvacrol/thymol chemotype, with both sampling regions situated in the Peloponnese: Ileia, with 28.74% carvacrol and 35.22% thymol, and Achaea, with 34.77% carvacrol and 34.68% thymol. This finding contradicted the results previously reported by Vokou et al. [14], who found that the oregano essential oil samples from the Peloponnese were primarily composed of carvacrol. In a similar study on *Origanum vulgare* L. subsp. *hirtum* cultivated in Turkey, Esen et al. [39] reported contents of carvacrol and thymol that varied significantly from 5.3 to 85.4% and from 0.3 to 68.0%, respectively. Nevertheless, these diverging results further support the assertion that the species boasts considerable variability, so such fluctuations are to be expected [3,14].

Similar to previous reports, the four principal components present in considerable amounts in all samples were the aromatic monoterpenes carvacrol, thymol, *p*-cymene, and $\gamma$-terpinene [15,17,19,39,51–56]. Despite the quantitative variations in these main components, their sum content appeared almost equivalent in the EOs of different regions and represented more than 80% of the total oil, specifically ranging between 85.12% and 89.51%. These results aligned with the findings of past studies. In particular, Kokkini et al. and Vokou et al. [3,14] reported a similar range (85.0 to 96.8%) when studying the EO composition of *O. vulgare* ssp. *hirtum* of different geographic origins as well as over different seasons. Thymol, the second most abundant ingredient, recorded its highest values in the oregano of Ileia and Achaea (35.22% and 34.68%, respectively), both located in the Peloponnese, and its lowest in the Heraklion sample (7.39%). Additionally, the lowest percentages of $\gamma$-terpinene and *p*-cymene were recorded in the Thessaloniki sample (2.72% and 6.60%, respectively), whereas the highest rates were registered for $\gamma$-terpinene in the Ileia samples (10.49%) and *p*-cymene in the Heraklion samples (12.53%).

Delving deeper into the EO composition, six additional compounds—four monoterpenes ($\beta$-myrcene, $\alpha$-terpinene, 4-terpineol, and borneol) and two sesquiterpenes (caryophyllene and $\beta$-bisabolene)—exhibited relatively high content rates in all samples. Caryophyllene, a common bicyclic sesquiterpene, was the fifth most abundant constituent in 8 out of the 12 regions examined, with contents ranging from 1.72% (Ioannina) to 2.43% (Katerini). As for the remaining four regions, the fifth most abundant component was the monoterpene $\alpha$-terpinene in the Ileia and Achaea samples (2.48% and 1.63%, respectively); $\beta$-myrcene in the Rethymno samples (1.65%); and the sesquiterpene $\beta$-bisabolene in the Heraklion samples (2.07%). In addition to these compounds, the oxygenated monoterpenes 4-terpineol and borneol were also notably present in all samples. The lowest content of 4-terpineol was recorded in the sample originating from the region of Ileia (0.17%), and the highest was registered in the sample from Kozani (1.28%). Likewise, borneol reached its highest content rate in the oregano sample from Thessaloniki (1.41%) and its lowest in the samples from Ileia and Rethymno (both 0.51%).

**Table 1.** Essential oil chemical composition (%) for each of the twelve regions studied. Main compounds and their values are marked in bold blue.

| S/N [a] | Compounds | RI$_{EXP}$ | RI$_{LIT}$ | THESSALONIKI | KATERINI | KILKIS | KOZANI | VOLOS | KALAMBAKA | IOANNINA | PREVEZA | ILEIA | ACHAEA | RETHYMNO | HERAKLION |
|---|---|---|---|---|---|---|---|---|---|---|---|---|---|---|---|
| 1 | α-pinene | 1012 | 1012 | 0.26 | 0.62 | 0.60 | 0.75 | 0.53 | 0.85 | 0.62 | 0.73 | 0.90 | 0.54 | 0.71 | 0.80 |
| 2 | α-thujene | 1019 | 1017 | | 0.09 | 0.56 | 0.90 | 0.45 | 0.59 | 0.70 | 0.96 | 1.03 | 0.68 | 0.67 | 0.06 |
| 3 | camphene | 1054 | 1055 | | | | 0.16 | tr [b] | | 0.13 | 0.12 | 0.16 | 0.07 | | |
| 4 | β-pinene | 1096 | 1100 | | | | 0.15 | | | 0.11 | 0.12 | 0.17 | 0.07 | | 0.06 |
| 5 | δ-3-carene | 1143 | 1138 | | | | | | | 0.06 | | 0.09 | | | |
| 6 | α-phellandrene | 1166 | 1166 | | | 0.16 | 0.19 | 0.05 | | 0.18 | 0.22 | 0.29 | 0.22 | | 0.07 |
| 7 | β-myrcene | 1175 | 1175 | 0.55 | 0.91 | 1.39 | 1.46 | 1.20 | 1.32 | 1.51 | 1.51 | 2.01 | 1.47 | 1.65 | 1.15 |
| 8 | α-terpinene | 1191 | 1201 | 0.67 | 1.09 | 1.45 | 1.63 | 1.22 | 1.57 | 1.46 | 1.58 | 2.48 | 1.63 | 1.56 | 1.26 |
| 9 | limonene | 1199 | 1208 | | 0.14 | 0.16 | 0.25 | 0.12 | 0.22 | 0.24 | 0.26 | 0.38 | 0.29 | 0.29 | 0.19 |
| 10 | β-phellandrene | 1221 | 1228 | | 0.10 | 0.18 | 0.29 | 0.08 | 0.11 | 0.28 | 0.28 | 0.29 | 0.28 | 0.24 | 0.07 |
| 11 | γ-terpinene | 1243 | 1246 | 2.72 | 3.13 | 5.76 | 7.05 | 5.40 | 6.67 | 6.96 | 7.00 | 10.49 | 8.18 | 7.28 | 4.13 |
| 12 | 3-octanone | 1255 | 1252 | | 0.09 | 0.21 | 0.14 | 0.22 | 0.24 | 0.22 | 0.23 | 0.24 | | | |
| 13 | p-cymene | 1270 | 1269 | 6.60 | 11.10 | 8.32 | 9.77 | 8.93 | 12.26 | 8.47 | 10.45 | 11.98 | 9.98 | 12.03 | 12.53 |
| 14 | α-terpinolene | 1280 | 1282 | | | tr | 0.14 | tr | | 0.15 | 0.15 | 0.16 | 0.12 | 0.11 | tr |
| 15 | 1-octen-3-ol | 1448 | 1447 | 0.41 | 0.57 | 0.50 | 0.34 | 0.52 | 1.07 | 0.54 | 0.45 | 0.32 | 0.41 | 0.10 | 0.43 |
| 16 | cis-sabinene hydrate | 1462 | 1471 | 0.49 | 0.51 | 0.67 | 0.58 | 0.56 | 0.36 | 0.50 | 0.55 | 0.49 | 0.68 | 0.43 | 0.52 |
| 17 | trans-sabinene hydrate | 1546 | 1556 | 0.39 | 0.43 | 0.50 | 0.46 | 0.50 | 0.44 | 0.37 | 0.39 | 0.34 | 0.47 | 0.38 | 0.54 |
| 18 | bornyl acetate | 1561 | 1566 | | | | | | | tr | | | | | |
| 19 | caryophyllene | 1598 | 1590 | 1.75 | 2.43 | 2.08 | 1.92 | 1.77 | 1.96 | 1.72 | 1.74 | 1.50 | 1.13 | 0.74 | 1.45 |
| 20 | 4-terpineol | 1604 | 1605 | 1.22 | 0.85 | 0.82 | 1.28 | 1.20 | 0.33 | 1.00 | 0.88 | 0.17 | 0.89 | 0.80 | 1.25 |
| 21 | carvacrol methyl ether | 1610 | 1601 | 0.79 | 0.88 | 0.61 | 0.43 | 0.41 | 0.64 | 0.20 | 0.27 | | 0.27 | | 0.70 |
| 22 | cis-dihydrocarvone | 1629 | - | | | | 0.05 | | | 0.16 | 0.05 | | | | |
| 23 | α-humulene | 1668 | 1677 | | 0.09 | 0.10 | 0.27 | 0.05 | | 0.23 | 0.21 | 0.10 | 0.09 | | |
| 24 | α-terpineol | 1698 | 1698 | | | | tr | | | | | | | | |
| 25 | borneol | 1702 | 1717 | 1.41 | 1.16 | 1.06 | 0.90 | 1.20 | 1.03 | 0.80 | 0.75 | 0.51 | 0.65 | 0.51 | 0.84 |
| 26 | β-bisabolene | 1726 | 1722 | 1.33 | 1.45 | 0.87 | 1.24 | 1.19 | 1.53 | 1.06 | 1.10 | 1.32 | 1.22 | 1.45 | 2.07 |
| 27 | δ-cadinene | 1733 | 1756 | | | | 0.25 | | | 0.13 | 0.17 | | | | |
| 28 | p-cymen-8-ol | 1851 | 1865 | | | | 0.06 | | | | tr | | | tr | |
| 29 | carvacryl acetate | 1875 | 1880 | | | | tr | | | tr | | | | | |
| 30 | caryophyllene oxide | 1973 | 1994 | 1.72 | 1.31 | 0.96 | 0.68 | 1.00 | 0.72 | 0.66 | 0.68 | 0.37 | 0.52 | 0.40 | 0.98 |
| 31 | spathulenol | 2122 | 2136 | | | | 0.08 | | | tr | tr | 0.06 | | | |
| 32 | 4-isopropyl-m-cresol | 2155 | - | | | | 0.08 | | | tr | 0.14 | 0.20 | 0.17 | 0.13 | |
| 33 | thymol | 2173 | 2186 | 10.89 | 10.22 | 12.85 | 8.41 | 13.03 | 16.45 | 10.64 | 13.65 | 35.22 | 34.68 | 24.01 | 7.39 |
| 34 | 5-isopropyl-m-cresol | 2208 | - | | | | 0.10 | | tr | 0.10 | 0.18 | | 0.26 | 0.14 | |
| 35 | carvacrol | 2211 | 2212 | 68.79 | 62.95 | 60.28 | 59.89 | 60.04 | 51.62 | 60.05 | 55.09 | 28.74 | 34.77 | 46.19 | 63.55 |
| | Main compounds [c] | | | 89.00 | 87.40 | 87.21 | 85.12 | 87.40 | 87.00 | 86.12 | 86.19 | 86.43 | 87.61 | 89.51 | 87.60 |
| | Monoterpene hydrocarbons | | | 10.80 | 17.18 | 18.58 | 22.74 | 17.98 | 23.59 | 20.87 | 23.38 | 30.43 | 23.53 | 24.60 | 20.26 |
| | Oxygenated Monoterpenes | | | 83.98 | 77.00 | 76.79 | 72.24 | 76.94 | 70.87 | 73.82 | 71.95 | 65.67 | 72.84 | 72.59 | 74.79 |
| | Sesquiterpene hydrocarbons | | | 3.08 | 3.97 | 3.05 | 3.68 | 3.01 | 3.49 | 3.14 | 3.22 | 2.92 | 2.44 | 2.19 | 3.52 |
| | Oxygenated Sesquiterpenes | | | 1.72 | 1.31 | 0.96 | 0.76 | 1.00 | 0.72 | 0.66 | 0.68 | 0.43 | 0.52 | 0.40 | 0.98 |
| | Miscellaneous | | | 0.41 | 0.57 | 0.59 | 0.55 | 0.66 | 1.29 | 0.78 | 0.67 | 0.55 | 0.65 | 0.10 | 0.43 |

[a] Compounds listed in order of elution from a BP-20 capillary column; [b] concentrations below 0.05% are marked as tr (traces); [c] total percentage of the 4 main compounds (carvacrol, thymol, γ-terpinene, and p-cymene) in the essential oil samples; RIEXP: experimentally determined retention indices; RILIT: retention indices from literature (NIST MS search).

### 3.2. Total Phenolic Content and Antioxidant Activity

As stated in the relevant literature, the primary phenolic constituents found in plants of the Lamiaceae family include phenolic compounds, such as hydroxycinnamic acids, along with flavonoids in the form of esters and glycosides [57,58]. Apart from being influenced by genotype, environmental and handling conditions can also affect the total phenolic content in plants. For this reason, it was essential to determine the actual content of these compounds in the oregano EOs of different geographical origins.

The total phenolic content and antioxidant activity determined for the oregano EO samples are presented in Table 2.

**Table 2.** Mean values and standard deviation (SD) of TPC and antioxidant activity of the oregano sample methanolic extracts for each region studied.

| Location | TPC (mg GAE/g EO) | TEAC (µmol TE/g EO) |
|---|---|---|
| Thessaloniki | 88.25 ± 7.92 | 395.84 ± 12.03 |
| Katerini | 83.20 ± 9.24 | 387.79 ± 13.65 |
| Kilkis | 86.48 ± 5.15 | 382.51 ± 11.37 |
| Kozani | 79.92 ± 7.39 | 457.00 ± 7.42 |
| Volos | 83.92 ± 6.98 | 375.81 ± 9.37 |
| Kalambaka | 85.92 ± 6.41 | 321.89 ± 7.53 |
| Ioannina | 84.17 ± 6.93 | 410.71 ± 10.95 |
| Preveza | 81.58 ± 7.28 | 397.06 ± 6.71 |
| Ileia | 85.38 ± 3.32 | 382.29 ± 20.33 |
| Achaea | 89.03 ± 4.76 | 306.83 ± 5.01 |
| Rethymno | 75.27 ± 3.31 | 461.32 ± 7.27 |
| Heraklion | 74.49 ± 3.57 | 361.43 ± 16.06 |

TPC: total phenolic content; TEAC: Trolox equivalent antioxidant capacity.

All studied samples contained high levels of phenolics without exhibiting any significant fluctuations. The TPC values of all specimens ranged from 74.49 ± 3.57 mg GAE/g EO (Heraklion) to 89.03 ± 4.76 mg GAE/g EO (Achaea). These results were comparable to earlier reports. More specifically, Pasias et al. [59], while investigating the chemical composition of the EOs of aromatic and medicinal herbs cultivated in Greece, reported a TPC value for *Origanum vulgare* L. of 42.6 ± 3.9 mg GAE/g EO. Similarly, Semiz et al. [60], while studying four different *Origanum* species, documented TPC rates ranging from 3.81 to 47.54 mg GAE/g extract. According to Oniga et al. [61], *O. vulgare* ssp. *vulgare* rendered a TPC value of 94.69 ± 4.03 mg GAE/g extract. Moreover, Spiridon et al. [62] compared the TPC values of oregano (*Origanum vulgare*), lavender (*Lavandula angustifolia*), and lemon balm (*Melissa officinalis*) extracts from Romania and found that *O. vulgare* yielded the highest rates of the three, reaching 67.8 ± 3.41 mg GAE/g.

As a general rule, the antioxidant potential of EOs is determined by their chemical composition. Secondary metabolites, such as phenolic compounds, have the ability to bind with double bonds and subsequently exhibit substantial antioxidant activity [63]. Apart from safely preventing food deterioration, natural antioxidants have been reported to help prevent health conditions such as cancer and coronary heart disease [64].

Greek oregano's highly valued antioxidant and antimicrobial activity is strongly associated with the prevalence of the phenols carvacrol and thymol in its essential oil, followed by the abundance of phenolic constituents such as rosmarinic acid and its derivatives within the nonvolatile fraction. A synergistic effect of oxygen-containing compounds has also been proposed [65,66]. Moreover, Al-Mansori et al. [67] investigated potential synergistic effects while examining the antioxidant activity of thymol and carvacrol. It was reported that even though the two phenols exhibited high antioxidant activity, there was no synergistic effect at play.

The Trolox equivalent antioxidant capacity (TEAC) values of the studied samples ranged between $306.83 \pm 5.01$ µmol TE/g EO (Achaea) and $461.32 \pm 7.27$ µmol TE/g EO (Rethymno). In their study, Kosakowska et al. [65] found similar results when comparing the antioxidant activity of the EOs and ethanolic extracts of Greek oregano to those of common oregano. Using the DPPH method, the EO of *O. vulgare* ssp. *hirtum* attained a value of $220.29 \pm 2.83$ µmol Trolox/g EO, whereas the EO of *O. vulgare* ssp. *vulgare* attained a value of $218.78 \pm 2.68$ µmol Trolox/g EO. Even though one would expect Greek oregano to prevail due to its higher levels of carvacrol and thymol, the amount of oxygenated and hydrocarbon monoterpenes in the common oregano proposedly narrowed the gap. In addition, Rostro-Alanis et al. [68] reported considerable variations while investigating the biological activities of Mexican oregano EOs; the application of the DPPH method produced values ranging from 2.91 to 22,129.54 µmol TE/g EO, depending on the corresponding analyzed fraction.

In tune with the TPC values, no noticeable fluctuations were recorded. No apparent correlation between the antioxidant activity and the total phenolic content of the samples was observed, since high TPC values did not correspond to high TEAC values; the oregano from Achaea attained the highest phenolic content while contrarily yielding the lowest TEAC rates. Simirgiotis et al. [69] also reported a lack of correlation between TPC and TEAC, suggesting that the FC method performed to determine the total phenolic content has certain limitations. Additionally, no association was noted between the chemical composition of the EOs and the TPC or TEAC results. In their study on the antioxidant capacity variation of oregano, Yan et al. [70] reached the same conclusion; no association was established between the oxygen radical absorbance capacity (ORAC) value and EO content of the 352 oregano samples investigated, while only a weak correlation was reported between TPC rates and EO content.

*3.3. Geographical Differentiation of Greek Oregano Based on EO Composition, TPC, and Antioxidant Capacity*

One of the most frequently committed types of fraud in the agricultural market, according to Katerinopoulou et al. [71], is when second-rate agrarian products are promoted as "local". Thus, tools such as the certification of geographic origin must be put into use to safeguard valuable, high-quality products both internationally and within a nation's boundaries.

This study, contrary to that of Vokou et al. [14], investigated the interconnection between the geographical origin of oregano and the composition and properties of its EOs. All 142 samples were initially subjected to MANOVA. However, as several regions overlapped, the samples were divided into two geographical groups: Group A, consisting of Ileia, Heraklion, Kalambaka, Thessaloniki, Kilkis, and Preveza; and Group B, consisting of Rethymno, Volos, Kozani, Katerini, Achaea, and Ioannina.

Sixty oregano samples in Group A and eighty-two samples in Group B were subjected to MANOVA to determine the significant parameters eligible for geographical differentiation. In both groups, dependent variables included essential oil composition, total phenolics, and antioxidant capacity, while geographical origin was set as the independent variable. The significant variables are presented in Table 3.

Thirty-four parameters were detected as significant in Group A ($p < 0.05$) and were further analyzed using LDA. Three statistically significant discriminant functions were generated: Wilks' $\lambda = 0.000$, $X^2 = 661.668$, df = 150 with $p$-value = 0.000 < 0.05 for the first; Wilks' $\lambda = 0.000$, $X^2 = 469.112$, df = 116 with $p$-value = 0.000 < 0.05 for the second; and Wilks' $\lambda = 0.001$, $X^2 = 306.040$, df = 84 with $p$-value = 0.001 < 0.05 for the third, respectively. The variance homogeneity test (Box's M) was insignificant at the 5% significance level (169.516, with F = 1.702, $p$-value = 0.059), indicating the homogeneity of the sample variations for each region. The first discriminant function interpreted 53.5% of the total dispersion with normal distribution $R^2 = 0.995$, the second 25.8% with normal distribution $R^2 = 0.991$, and the third 13.3% with normal distribution $R^2 = 0.982$.

**Table 3.** Statistically significant values in Groups A and B (F-ratio and $p < 0.05$).

| Dependent Variables | Group A | | Group B | |
|---|---|---|---|---|
| | *F* | *Sig.* | *F* | *Sig.* |
| $\alpha$-pinene | 3.747 | 0.006 | 0.707 | 0.620 |
| $\alpha$-thujene | 16.479 | 0.000 | 7.269 | 0.000 |
| camphene | 19.979 | 0.000 | 6.349 | 0.000 |
| $\beta$-pinene | 43.232 | 0.000 | 12.184 | 0.000 |
| $\delta$-3-carene | 19.328 | 0.000 | 7.860 | 0.000 |
| $\alpha$-phellandrene | 16.598 | 0.000 | 16.736 | 0.000 |
| $\beta$-myrcene | 7.314 | 0.000 | 3.602 | 0.006 |
| $\alpha$-terpinene | 12.833 | 0.000 | 2.341 | 0.050 |
| limonene | 12.281 | 0.000 | 4.912 | 0.001 |
| $\beta$-phellandrene | 8.004 | 0.000 | 12.216 | 0.000 |
| $\gamma$-terpinene | 17.102 | 0.000 | 8.733 | 0.000 |
| 3-octanone | 9.956 | 0.000 | 10.266 | 0.000 |
| $p$-cymene | 12.625 | 0.000 | 4.350 | 0.002 |
| $\alpha$-terpinolene | 11.586 | 0.000 | 23.839 | 0.000 |
| 1-octen-3-ol | 13.330 | 0.000 | 10.677 | 0.000 |
| *cis*-sabinene hydrate | 3.399 | 0.010 | 2.573 | 0.033 |
| *trans*-sabinene hydrate | 1.948 | 0.101 | 2.545 | 0.035 |
| bornyl acetate | | | 1.049 | 0.395 |
| caryophyllene | 4.710 | 0.001 | 7.956 | 0.000 |
| 4-terpineol | 14.092 | 0.000 | 1.743 | 0.135 |
| carvacrol methyl ether | 5.480 | 0.000 | 8.447 | 0.000 |
| *cis*-dihydrocarvone | 2.623 | 0.034 | 9.498 | 0.000 |
| $\alpha$-humulene | 8.666 | 0.000 | 16.284 | 0.000 |
| $\alpha$-terpineol | | | 1.253 | 0.293 |
| borneol | 17.003 | 0.000 | 18.737 | 0.000 |
| $\beta$-bisabolene | 40.163 | 0.000 | 2.540 | 0.035 |
| $\delta$-Cadinene | 11.646 | 0.000 | 46.986 | 0.000 |
| $p$-cymen-8-ol | 10.009 | 0.000 | 9.047 | 0.000 |
| carvacryl acetate | | | 2.416 | 0.044 |
| caryophyllene oxide | 20.672 | 0.000 | 13.091 | 0.000 |
| spathulenol | 3.750 | 0.005 | 5.293 | 0.000 |
| 4-isopropyl-m-cresol | 64.496 | 0.000 | 20.953 | 0.000 |
| thymol | 19.504 | 0.000 | 40.470 | 0.000 |
| 5-isopropyl-m-cresol | 28.198 | 0.000 | 22.509 | 0.000 |
| carvacrol | 59.721 | 0.000 | 33.509 | 0.000 |
| TPC | 6.587 | 0.000 | 4.570 | 0.001 |
| TEAC | 2.726 | 0.029 | 26.044 | 0.000 |

The overall interpreted percentage accounted for 92.7% of the total variance, which was highly satisfactory. The values of the group centroids were the average values of the variables as defined by the discriminant functions (Figure 2a). For Ilia, the values were (15.246, −3.748); for Heraklion (−15.280, 1.156); for Kalambaka (−0.247, −8.277); for Thessaloniki (−7.958, 2.675); for Kilkis (1.195, −3.456); and for Preveza (7.853, 13.864). Figure 2a demonstrates the complete separation of the Group A oregano regions.

Thirty-two parameters were found to be significant in Group B ($p < 0.05$) and were further analyzed using LDA. Three statistically significant discriminant functions were generated: Wilks' $\lambda = 0.000$, $X^2 = 671.051$, df = 160 with $p$-value = 0.000 < 0.05 for the first; Wilks' $\lambda = 0.001$, $X^2 = 448.496$, df = 124 with $p$-value = 0.000 < 0.05 for the second; and Wilks' $\lambda = 0.009$, $X^2 = 287.835$, df = 90 with $p$-value = 0.001 < 0.05 for the third, respectively. The variance homogeneity test (Box's M) was insignificant at the 5% significance level (174.858, with F = 1.885, $p$-value = 0.052), indicating the homogeneity of the sample variations of each region. The first discriminant function interpreted 58.5% of the total dispersion with normal distribution $R^2 = 0.987$, the second 20.2% with $R^2 = 0.963$, and the third 12.2% with $R^2 = 0.942$. The overall interpreted percentage accounted for 91.0% of the total variance, which was very satisfactory. The values of the group centroids were the average values of the parameters. For Achaia, the values were (−0.586, 3.384); for Rethymno (1.464, 7.866); for Volos (−5.917, −1.583); for Katerini (−8.250, −2.060); for Kozani (8.297, −2.618); and for Ioannina (1.457, 0.777). Figure 2b shows the adequate differentiation of the regions of Group B.

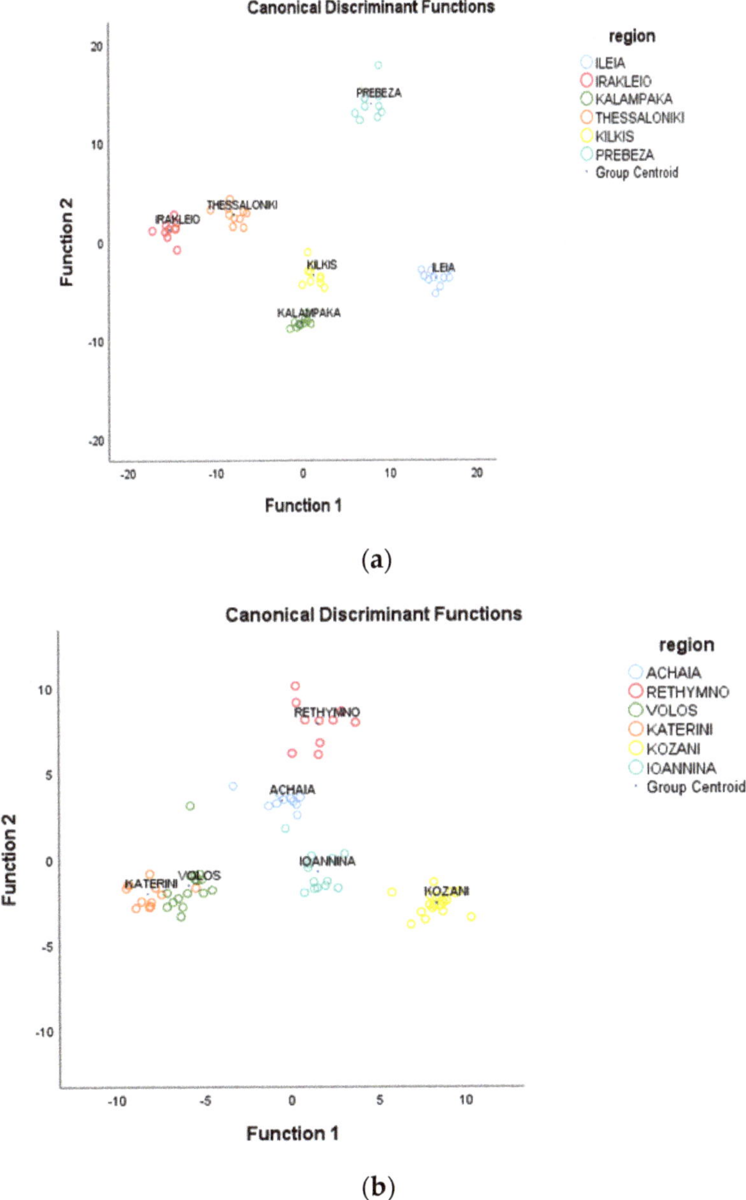


**Figure 2.** (**a**) Group A: oregano geographical differentiation based on essential oil content, total phenolics, and antioxidant capacity. (**b**) Group B: oregano geographical differentiation based on essential oil content, total phenolics, and antioxidant capacity.

In Group A, when using the original method, 100% of all grouped cases were correctly classified; this percentage decreased to 93.3% after the application of the cross-validation method. Regarding Group B, the results of the original method attained a correct classification rate of 98.8% of all grouped cases, while the rate fell to 82.7% with the use of the cross-validation method.

For Groups A and B, the statistical analysis of the EO composition and the TPC and TEAC variables, each separately, provided the results listed in Table 4. The EO composition of Group A presented a very satisfactory separation rate (cross-validation = 93.3%), while the TPC and TEAC values showed inefficient results. Nonetheless, combining the above variables did not seem to affect the results, as the areas' differentiation percentage remained the same after using the cross-validation method. This stability demonstrated the strong effect of the EO composition on the geographical differentiation of oregano. The EO composition of Group B presented a reasonably satisfactory separation rate (cross-validation = 81.5%); nevertheless, the resulting distribution diagram (Figure 2b) seemed to indicate an overlap between the samples of the Rethymno–Achaia and Volos–Katerini areas. Once more, the TPC and TEAC results were unsatisfactory, as shown in Table 4. However, the combination of the above parameters produced good results, and by incorporating more analyses, a better distribution of the samples was attained in the diagram.

**Table 4.** Classification rates of Groups A and B using the original and cross-validation methods (including individual and combinations).

| | Discriminant Function | Original Method (%) | Cross-Validation (%) |
|---|---|---|---|
| GROUP A | EO compounds | 100 | 93.3 |
| | TPC and TEAC | 38.3 | 33.3 |
| GROUP B | EO compounds | 96.3% | 81.5% |
| | TPC and TEAC | 43.9 | 39.0 |

Stepwise LDA was performed as the final step of the statistical data analysis in order to determine the variables with the highest discriminant ability in Groups A and B. Sixteen out of the thirty-four significant EO compounds exhibited a higher discriminant ability in Group A, while in Group B, fourteen out of the thirty-two stood out; in both cases, three statistically significant discriminant functions were formed. As shown in Figure 3, all samples included in Group A were significantly differentiated, while those included in Group B were well differentiated. More specifically, in Group A, the samples from Ileia, Preveza, Heraklion, and Thessaloniki were clearly distinct from all others; however, a small number of those from Kilkis appeared to overlap with those of Kalampaka. On the other hand, in Group B, the specimens originating from Kozani and Achaea were very well differentiated from the rest, contrary to those from Rethymno, which partly overlapped those of Ioannina, and those from Katerini, which overlapped slightly with those from Volos. The overall correct classification rate of Group A was 100% for the original and 100% for the cross-validation method. Respectively, the overall correct classification rate of Group B was 93.9% for the original and 87.8% for the cross-validation method. All in all, 100% correct geographical classification was attained for the regions of Ileia, Preveza, Heraklion, Thessaloniki, and Achaea, followed closely by Kozani (95%) and Rethymno (90%).

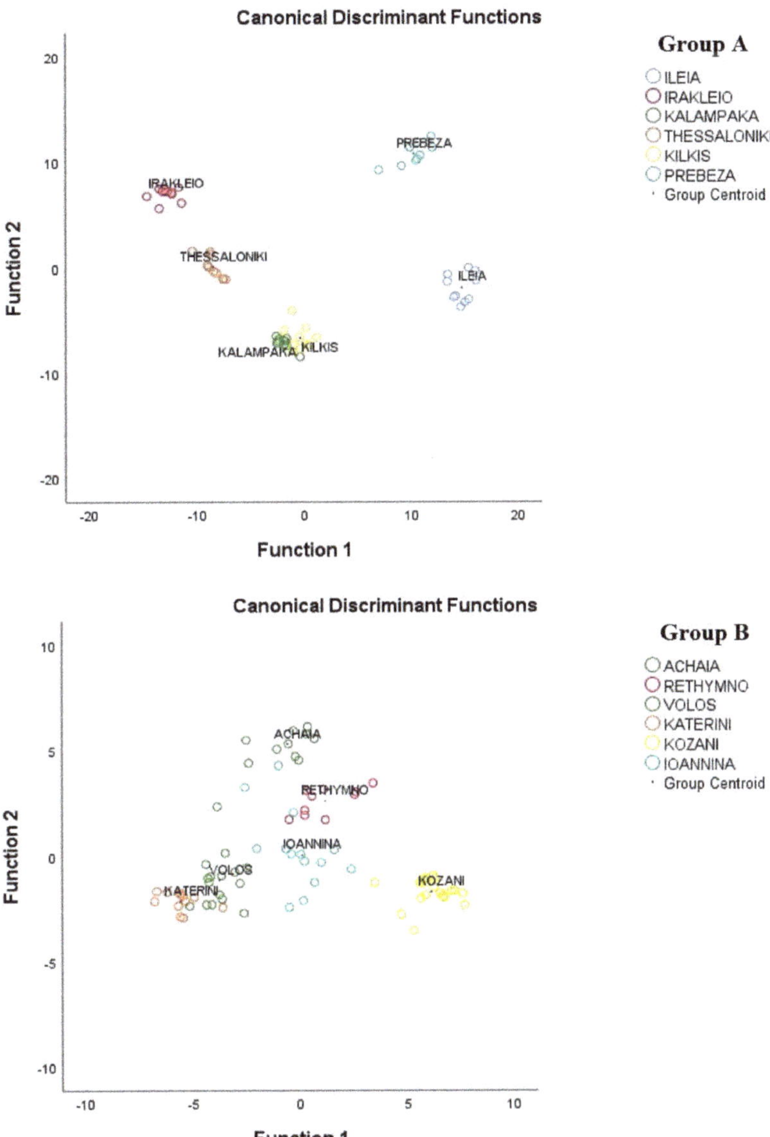


**Figure 3.** Oregano geographical differentiation based on EO composition, total phenolics, and antioxidant capacity. Scatter plot from SLDA analysis (Group A: 100.0% original, 100.0% cross-validation; Group B: 93.9% original, 87.8% cross-validation).

## 4. Conclusions

The chromatographic analysis of Greek oregano EO samples obtained from 12 different locations in Greece indicated remarkable variability in terms of composition, even though the overall content of the four main components (carvacrol, thymol, $p$-cymene, and $\gamma$-terpinene) was relatively stable and higher than 85% in all cases.

The statistical treatment of the acquired data yielded satisfactory correct classification rates for both oregano cultivation groups regarding geographical origin. The EO composition was found to be the most significant discriminant parameter (Group A, correct classification rate 93.3% using the cross-validation method; Group B, correct classification rate 81.5% using the cross-validation method), while TPC and TEAC variables displayed no substantial effect on the geographical differentiation of the samples.

Overall, the acquired results provide preliminary evidence that the chromatographic profile of EOs extracted from Greek oregano samples can act as a powerful tool for geographical origin discrimination purposes, ratifying that GC/MS profiling constitutes an effective approach toward food traceability. Moreover, the implementation of chemometric tools in our study further enabled us to identify which chemical features (markers) are explicitly associated with geographical origin. As the herb market is quite susceptible to fraud and adulteration, a fast and easy methodology such as that described in the current paper could represent a valuable asset for testing authenticity. In addition, apart from their authentication and traceability applications, the techniques used in this work may also act as a tentative guide characterizing the traits and olfactory profiles of commercially available Greek oregano samples originating from different areas of Greece. This guide could be further utilized by the relevant trade authorities when promoting Greek oregano by establishing relative quality standards based on its origin of cultivation.

Nevertheless, the findings of this study must be viewed in light of some limitations. More specifically, the sample size could have been more adequate, and more cultivation regions could have been covered. Even though the results are satisfactory, an even larger sample pool could validate our hypothesis further and eliminate any potential statistical errors. Notwithstanding the preliminary nature of our research, the methods considered in this work represent a promising solution for Greek oregano traceability, thus deserving further investigation.

**Author Contributions:** Conceptualization, A.V.B.; data curation, E.S.T. and I.S.K.; investigation, E.S.T.; formal analysis, E.S.T.; methodology, A.V.B., I.S.K. and E.S.T.; supervision, A.V.B.; writing—original draft, E.S.T. and I.S.K.; writing—review and editing, A.V.B., I.S.K. and E.S.T.; project administration, A.V.B. All authors have read and agreed to the published version of the manuscript.

**Funding:** This research received no external funding.

**Institutional Review Board Statement:** Not applicable.

**Data Availability Statement:** Not applicable.

**Acknowledgments:** We sincerely thank all the growers who provided us with the free oregano samples used in this study.

**Conflicts of Interest:** The authors declare no conflict of interest.

## References

1. Atanasov, A.G.; Waltenberger, B.; Pferschy-Wenzig, E.M.; Linder, T.; Wawrosch, C.; Uhrin, P.; Temml, V.; Wang, L.; Schwaiger, S.; Heiss, E.H.; et al. Discovery and Resupply of Pharmacologically Active Plant-Derived Natural Products: A Review. *Biotechnol. Adv.* **2015**, *33*, 1582–1614. [CrossRef] [PubMed]
2. Karakaya, S.; El, S.N.; Karagözlü, N.; Şahin, S. Antioxidant and Antimicrobial Activities of Essential Oils Obtained from Oregano (*Origanum vulgare* ssp. *hirtum*) by Using Different Extraction Methods . *J. Med. Food* **2011**, *14*, 645–652. [CrossRef] [PubMed]
3. Kokkini, S. Taxonomy, Diversity and Distribution of *Origanum* Species. In *Proceedgs of the IPGRI International Workshop on Oregano, Valenzano, Italy, 8–12 May 1996*; Padulosi, S., Ed.; IPGRI: Bari, Italy, 1996; pp. 2–12.
4. Skoula, M.; Harborne, J. The Taxonomy and Chemistry of *Origanum*. In *Oregano: The Genera Origanum and Lippia*; Kintzios, S.E., Ed.; Taylor & Francis: London, UK, 2002; pp. 67–108.
5. Stefanakis, M.K.; Touloupakis, E.; Anastasopoulos, E.; Ghanotakis, D.; Katerinopoulos, H.E.; Makridis, P. Antibacterial Activity of Essential Oils from Plants of the Genus *Origanum*. *Food Control* **2013**, *34*, 539–546. [CrossRef]
6. Bakkali, F.; Averbeck, S.; Averbeck, D.; Idaomar, M. Biological Effects of Essential Oils-A Review. *Food Chem. Toxicol.* **2008**, *46*, 446–475. [CrossRef] [PubMed]
7. Sakkas, H.; Papadopoulou, C. Antimicrobial Activity of Basil, Oregano, and Thyme Essential Oils. *J. Microbiol. Biotechnol.* **2017**, *27*, 429–438. [CrossRef]
8. Kintzios, S.E. Profile of the multifaceted prince of the herbs. In *Oregano: The Genera Origanum and Lippia*; Kintzios, S.E., Ed.; Taylor & Francis: London, UK, 2002; pp. 3–8.

9.  Kokkini, S.; Vokou, D.D.; Karousou, R.R. Morphological and Chemical Variation of *Origanum vulgare* L. in Greece. *Bot. Chron.* **1991**, *10*, 337–346.

10. Tucker, A.O.; Maciarello, M.J. Oregano: Botany, Chemistry, and Cultivation. *Dev. Food Sci.* **1994**, *34*, 439–456.

11. Soltani, S.; Shakeri, A.; Iranshahi, M.; Boozari, M. A Review of the Phytochemistry and Antimicrobial Properties of *Origanum vulgare* L. and Subspecies. *Iran. J. Pharm. Res.* **2021**, *20*, 268–285. [CrossRef]

12. Węglarz, Z.; Kosakowska, O.; Przybył, J.L.; Pióro-Jabrucka, E.; Bączek, K. The Quality of Greek Oregano (*O. vulgare* L. subsp. *Hirtum* (Link) Ietswaart) and Common Oregano (*O. vulgare* L. subsp. *Vulgare*) Cultivated in the Temperate Climate of Central Europe. *Foods* **2020**, *9*, 1671. [CrossRef]

13. Skoufogianni, E.; Solomou, A.D.; Danalatos, N.G. View of Ecology, Cultivation and Utilization of the Aromatic Greek Oregano (*Origanum vulgare* L.): A Review. *Not. Bot. Horti Agrobot. Cluj-Napoca* **2019**, *47*, 545–552. [CrossRef]

14. Vokou, D.; Kokkini, S.; Bessiere, J.-M.M.; Kokkinit, S.; Bessiere, J.-M.M.; Kokkini, S.; Bessiere, J.-M.M. Geographic Variation of Greek Oregano (*Origanum vulgare* ssp. *Hirtum*) Essential Oils . *Biochem. Syst. Ecol.* **1993**, *21*, 287–295. [CrossRef]

15. Kokkini, S.; Karousou, R.; Vokou, D. Pattern of Geographic Variations of *Origanum vulgare* Trichomes and Essential Oil Content in Greece. *Biochem. Syst. Ecol.* **1994**, *22*, 517–528. [CrossRef]

16. Wogiatzi, E.; Gougoulias, N.; Papachatzis, A.; Vagelas, I.; Chouliaras, N. Greek Oregano Essential Oils Production, Phytotoxicity, and Antifungal Activity. *Biotechnol. Biotechnol. Equip.* **2009**, *23*, 1150–1152. [CrossRef]

17. Skoula, M.; Gotsiou, P.; Naxakis, G.; Johnson, C.B. A Chemosystematic Investigation on the Mono- and Sesquiterpenoids in the Genus *Origanum* (Labiatae). *Phytochemistry* **1999**, *52*, 649–657. [CrossRef]

18. Stefanaki, A.; van Andel, T. Mediterranean Aromatic Herbs and Their Culinary Use. In *Aromatic Herbs in Food. Bioactive Compounds, Processing, and Applications*, 1st ed.; Galanakis, C.M., Ed.; Academic Press: London, UK, 2021; Chapter 3; pp. 93–121.

19. Kokkini, S.; Karousou, R.; Dardioti, A.; Krigas, N.; Lanaras, T. Autumn Essential Oils of Greek Oregano. *Phytochemistry* **1997**, *44*, 883–886. [CrossRef]

20. Russo, M.; Galletti, G.C.; Bocchini, P.; Carnacini, A. Essential Oil Chemical Composition of Wild Populations of Italian Oregano Spice (*Origanum vulgare* ssp. *Hirtum* (Link) Ietswaart): A Preliminary Evaluation of Their Use in Chemotaxonomy by Cluster Analysis. 1. Inflorescences. *J. Agric. Food Chem.* **1998**, *46*, 3741–3746. [CrossRef]

21. Jing, D.; Deguang, W.; Linfang, H.; Shilin, C.; Minjian, Q. Application of Chemometrics in Quality Evaluation of Medicinal Plants. *J. Med. Plants Res.* **2011**, *5*, 4001–4008.

22. Andre, C.M.; Soukoulis, C. Food Quality Assessed by Chemometrics. *Foods* **2020**, *9*, 897. [CrossRef]

23. Dimitrakopoulou, M.E.; Vantarakis, A. Does Traceability Lead to Food Authentication? A Systematic Review from A European Perspective. *Food Rev. Int.* **2021**. [CrossRef]

24. Nousias, P.; Karabagias, I.K.; Kontakos, S.; Riganakos, K.A. Characterization and Differentiation of Greek Commercial Thyme Honeys According to Geographical Origin Based on Quality and Some Bioactivity Parameters Using Chemometrics. *J. Food Process. Preserv.* **2017**, *41*, e13061. [CrossRef]

25. Karabagias, I.K.; Badeka, A.; Kontakos, S.; Karabournioti, S.; Kontominas, M.G. Characterization and Classification of Thymus Capitatus (L.) Honey According to Geo-graphical Origin Based on Volatile Compounds, Physicochemical Parameters and Chemometrics. *Food Res. Int.* **2014**, *55*, 363–372. [CrossRef]

26. Senyuva, H.Z.; Gilbert, J.; Silici, S.; Charlton, A.; Dal, C.; Gürel, N.; Cimen, D. Profiling Turkish Honeys to Determine Authenticity Using Physical and Chemical Characteristics. *J. Agric. Food Chem.* **2009**, *57*, 3911–3919. [CrossRef] [PubMed]

27. Vatavali, K.A.; Kosma, I.S.; Louppis, A.P.; Badeka, A.V.; Kontominas, M.G. Analyses in Combination with Chemometrics for the Discrimination of the Geographical Origin of Mean Values and SDs of the Percentages of Fatty Acids (% Fatty Acids) for the Graviera Cheese Samples Tested. Cheeses. *Molecules* **2020**, *25*, 3507. [CrossRef] [PubMed]

28. Danezis, G.; Theodorou, C.; Massouras, T.; Zoidis, E.; Hadjigeorgiou, I.; Georgiou, C.A. Greek Graviera Cheese Assessment through Elemental Metabolomics—Implications for Authentication, Safety and Nutrition. *Molecules* **2019**, *24*, 670. [CrossRef] [PubMed]

29. Rutkowska, J.; Bialek, M.; Adamska, A.; Zbikowska, A. Differentiation of Geographical Origin of Cream Products in Poland According to Their Fatty Acid Profile. *Food Chem.* **2015**, *178*, 26–31. [CrossRef]

30. Anastasaki, E.; Kanakis, C.; Pappas, C.; Maggi, L.; del Campo, C.P.; Carmona, M.; Alonso, G.L.; Polissiou, M.G. Geographical Differentiation of Saffron by GC-MS/FID and Chemometrics. *Eur. Food Res. Technol.* **2009**, *229*, 899–905. [CrossRef]

31. Barbosa, S.; Campmajó, G.; Saurina, J.; Puignou, L.; Núñez, O. Determination of Phenolic Compounds in Paprika by Ultrahigh Performance Liquid Chromatography-Tandem Mass Spectrometry: Application to Product Designation of Origin Authentication by Chemometrics. *J. Agric. Food Chem.* **2020**, *68*, 591–602. [CrossRef]

32. Bernáth, J. Some Scientific and Practical Aspects of Production and Utilization of Oregano in Central Europe. In Proceedings of the IPGRI International Workshop on Oregano, Valenzano, Italy, 8–12 May 1996; Padulosi, S., Ed.; IPGRI: Bari, Italy, 1996; pp. 75–92.

33. Goliaris, A.H.; Chatzopoulou, P.S.; Katsiotis, S.T. Production of New Greek Oregano Clones and Analysis of Their Essential Oils. *J. Herbs Spices Med. Plants* **2003**, *10*, 29–35. [CrossRef]

34. Sarrou, E.; Tsivelika, N.; Chatzopoulou, P.; Tsakalidis, G.; Menexes, G.; Mavromatis, A. Conventional Breeding of Greek Oregano (*Origanum vulgare* ssp. *hirtum*) and Development of Improved Cultivars for Yield Potential and Essential Oil Quality. *Euphytica* **2017**, *213*, 104. [CrossRef]

35. Farías, G.; Brutti, O.; Grau, R.; Di Leo Lira, P.; Retta, D.; van Baren, C.; Vento, S.; Bandoni, A.L. Morphological, Yielding and Quality Descriptors of Four Clones of *Origanum* spp. (Lamiaceae) from the Argentine Littoral Region Germplasm Bank. *Ind. Crops Prod.* **2010**, *32*, 472–480. [CrossRef]

36. Torres, L.E.; Brunetti, P.C.; Baglio, C.; Bauzá, P.G.; Chaves, A.G.; Massuh, Y.; Ocaño, S.F.; Ojeda, M.S. Field Evaluation of Twelve Clones of Oregano Grown in the Main Production Areas of Argentina: Identification of Quantitative Trait with the Highest Discriminant Value. *ISRN Agron.* **2012**, *2012*, 349565. [CrossRef]

37. Azizi, A.; Wagner, C.; Honermeier, B.; Friedt, W. Intraspecific Diversity and Relationship between Subspecies of *Origanum vulgare* Revealed by Comparative AFLP and SAMPL Marker Analysis. *Plant Syst. Evol.* **2009**, *281*, 151–160. [CrossRef]

38. Khan, M.; Khan, S.T.; Khan, M.; Mousa, A.A.; Mahmood, A.; Alkhathlan, H.Z. Chemical Diversity in Leaf and Stem Essential Oils of *Origanum vulgare* L. and Their Effects on Microbicidal Activities. *AMB Express* **2019**, *9*, 176. [CrossRef] [PubMed]

39. Esen, G.; Azaz, A.D.; Kurkcuoglu, M.; Baser, K.H.C.; Tinmaz, A. Essential Oil and Antimicrobial Activity of Wild and Cultivated *Origanum vulgare* L. subsp. *hirtum* (Link) Ietswaart from the Marmara Region, Turkey. *Flavour Fragr. J.* **2007**, *22*, 371–376. [CrossRef]

40. NIST. *National Institute of Standards and Technology*; J. Wiley & Sons Ltd.: West Sussex, UK, 2005.

41. Oil IOC Olive Oil Extraction 2009.

42. Kosma, I.; Badeka, A.; Vatavali, K.; Kontakos, S.; Kontominas, M. Differentiation of Greek Extra Virgin Olive Oils According to Cultivar Based on Volatile Compound Analysis and Fatty Acid Composition. *Eur. J. Lipid Sci. Technol.* **2016**, *118*, 849–861. [CrossRef]

43. Singleton, V.L.; Orthofer, R.; Lamuela-Raventós, R.M. Analysis of Total Phenols and Other Oxidation Substrates and Antioxidants by Means of Folin-Ciocalteu Reagent. *Methods Enzymol.* **1999**, *299*, 152–178. [CrossRef]

44. Payum, T.; Das, A.K.; Shankar, R.; Tamuly, C.; Hazarika, M. Antioxidant Potential of *Solanum spirale* Shoot and Berry: A Medicinal Food Plant Used in Arunachal Pradesh. *Am. J. Pharmtech Res.* **2015**, *5*, 307–314.

45. Bondet, V.; Brand-Williams, W.; Berset, C. Kinetics and Mechanisms of Antioxidant Activity Using the DPPH• Free Radical Method. *LWT Food Sci. Technol.* **1997**, *30*, 609–615. [CrossRef]

46. IBM SPSS Statistics 25. Available online: https://www.ibm.com/support/pages/downloading-ibm-spss-statistics-25 (accessed on 25 January 2021).

47. Friendly, M.; Sigal, M. Visualizing Tests for Equality of Covariance Matrices. *Am. Stat.* **2020**, *74*, 144–155. [CrossRef]

48. Field, A. *Discovering Statistics Using SPSS ISM (London, England) Introducing Statistical Methods Series*; SAGE Publications Ltd.: New York, NY, USA, 2009; Volume 2, ISBN1 1847879071, ISBN2 9781847879073.

49. Pizarro, C.; Rodríguez-Tecedor, S.; Pérez-del-Notario, N.; González-Sáiz, J.M. Recognition of Volatile Compounds as Markers in Geographical Discrimination of Spanish Extra Virgin Olive Oils by Chemometric Analysis of Non-Specific Chromatography Volatile Profiles. *J. Chromatogr. A* **2011**, *1218*, 518–523. [CrossRef]

50. Kosma, I.S.; Kontominas, M.G.; Badeka, A.V. The Application of Chemometrics to Volatile Compound Analysis for the Recognition of Specific Markers for Cultivar Differentiation of Greek Virgin Olive Oil Samples. *Foods* **2020**, *9*, 1672. [CrossRef] [PubMed]

51. Khan, M.; Khan, S.T.; Khan, N.A.; Mahmood, A.; Al-Kedhairy, A.A.; Alkhathlan, H.Z. The Composition of the Essential Oil and Aqueous Distillate of *Origanum vulgare* L. Growing in Saudi Arabia and Evaluation of Their Antibacterial Activity. *Arab. J. Chem.* **2018**, *11*, 1189–1200. [CrossRef]

52. Kokkini, S.; Karousou, R.; Hanlidou, E.; Lanaras, T. Essential Oil Composition of Greek (*Origanum vulgare* ssp. *hirtum*) and Turkish (*O. onites*) Oregano: A Tool for Their Distinction. *J. Essent. Oil Res.* **2004**, *16*, 334–338. [CrossRef]

53. Bosabalidis, A.; Kokkini, S. Infraspecific Variation of Leaf Anatomy in *Origanum vulgare* Grown Wild in Greece. *Bot. J. Linn. Soc.* **1997**, *123*, 353–362. [CrossRef]

54. Karousou, R. Taxonomic Studies on the Cretan Labiatae: Distribution, Morphology and Essential Oils. In Proceedings of the X OPTIMA Meeting, Palermo, Italy, 13–19 September 2001; University of Thessaloniki: Thessaloniki, Greek, 1996.

55. Lagouri, V.; Blekas, G.; Tsimidou, M.; Kokkini, S.; Boskou, D. Composition and Antioxidant Activity of Essential Oils from Oregano Plants Grown Wild in Greece. *Z Leb. Unters Forsch* **1993**, *197*, 20–23. [CrossRef]

56. Sivropoulou, A.; Papanikolaou, E.; Nikolaou, C.; Kokkini, S.; Lanaras, T.; Arsenakis, M. Antimicrobial and Cytotoxic Activities of *Origanum* Essential Oils. *J. Agric. Food Chem.* **1996**, *44*, 1202–1205. [CrossRef]

57. Martins, N.; Barros, L.; Santos-Buelga, C.; Henriques, M.; Silva, S.; Ferreira, I.C.F.R. Decoction, Infusion and Hydroalcoholic Extract of *Origanum vulgare* L.: Different Performances Regarding Bioactivity and Phenolic Compounds. *Food Chem.* **2014**, *158*, 73–80. [CrossRef]

58. Modnicki, D.; Balcerek, M. Estimation of Total Polyphenols Contents in *Ocimum basilicum* L., *Origanum vulgare* L. and *Thymus vulgaris* L. Commercial Samples. *Herba Pol.* **2009**, *55*, 35–42.

59. Pasias, I.N.; Ntakoulas, D.D.; Raptopoulou, K.; Gardeli, C.; Proestos, C. Chemical Composition of Essential Oils of Aromatic and Medicinal Herbs Cultivated in Greece—Benefits and Drawbacks. *Foods* **2021**, *10*, 2354. [CrossRef]

60. Semiz, G.; Semiz, A.; Mercan-Doğan, N. International Journal of Food Properties Essential Oil Composition, Total Phenolic Content, Antioxidant and Antibiofilm Activities of Four Origanum Species from Southeastern Turkey Essential Oil Composition, Total Phenolic Content, Antioxidant and Antibiofilm Activities of Four Origanum Species from Southeastern Turkey. *Int. J. Food Prop.* **2018**, *21*, 194–204. [CrossRef]

61. Oniga, I.; Pus, C.; Silaghi-Dumitrescu, R.; Olah, N.K.; Sevastre, B.; Marica, R.; Marcus, I.; Sevastre-Berghian, A.C.; Benedec, D.; Pop, C.E.; et al. *Origanum vulgare* ssp. *vulgare*: Chemical Composition and Biological Studies. *Molecules* **2018**, *23*, 2077. [CrossRef] [PubMed]

62. Spiridon, I.; Colceru, S.; Anghel, N.; Teaca, C.A.; Bodirlau, R.; Armatu, A. Antioxidant Capacity and Total Phenolic Contents of Oregano (*Origanum vulgare*), Lavender (*Lavandula Angustifolia*) and Lemon Balm (*Melissa Officinalis*) from Romania. *Nat. Prod. Res.* **2011**, *25*, 1657–1661. [CrossRef] [PubMed]

63. Bhavaniramya, S.; Vishnupriya, S.; Al-Aboody, M.S.; Vijayakumar, R.; Baskaran, D. Role of Essential Oils in Food Safety: Antimicrobial and Antioxidant Applications. *Grain Oil Sci. Technol.* **2019**, *2*, 49–55. [CrossRef]

64. Tungmunnithum, D.; Thongboonyou, A.; Pholboon, A.; Yangsabai, A. Flavonoids and Other Phenolic Compounds from Medicinal Plants for Pharmaceutical and Medical Aspects: An Overview. *Medicines* **2018**, *5*, 93. [CrossRef]

65. Kosakowska, O.; Zenon, W.; Pióro-Jabrucka, E.; Przybył, J.L.; Krasniewska, K.; Gniewosz, M.; Baczek, K. Antioxidant and Antibacterial Activity of Essential Oils and Hydroethanolic Extracts of Greek Oregano. *Molecules* **2021**, *26*, 15. [CrossRef]

66. Kulisic, T.; Radonic, A.; Katalinic, V.; Milos, M. Use of Different Methods for Testing Antioxidative Activity of Oregano Essential Oil. *Food Chem.* **2004**, *85*, 633–640. [CrossRef]

67. Al-Mansori, B.; El-Ageeli, W.H.; Alsagheer, S.H.; Ben-Khayal, F.A.F. Antioxidant Activity-Synergistic Effects of Thymol and Carvacrol. *Al-Mukhtar J. Sci.* **2020**, *35*, 185–194. [CrossRef]

68. Rostro-Alanis, M.; Báez-González, J.; Torres-Alvarez, C.; Parra-Saldívar, R.; Rodriguez-Rodriguez, J.; Castillo, S. Chemical Composition and Biological Activities of Oregano Essential Oil and Its Fractions Obtained by Vacuum Distillation. *Molecules* **2019**, *24*, 1904. [CrossRef]

69. Simirgiotis, M.J.; Burton, D.; Parra, F.; López, J.; Muñoz, P.; Escobar, H.; Parra, C. Antioxidant and Antibacterial Capacities of *Origanum vulgare* L. Essential Oil from the Arid Andean Region of Chile and Its Chemical Characterization by Gc-Ms. *Metabolites* **2020**, *10*, 414. [CrossRef]

70. Yan, F.; Azizi, A.; Janke, S.; Schwarz, M.; Zeller, S.; Honermeier, B. Antioxidant Capacity Variation in the Oregano (*Origanum vulgare* L.) Collection of the German National Genebank. *Ind. Crops Prod.* **2016**, *92*, 19–25. [CrossRef]

71. Katerinopoulou, K.; Kontogeorgos, A.; Salmas, C.E.; Patakas, A.; Ladavos, A. Geographical Origin Authentication of Agri-Food Products: A Review. *Foods* **2020**, *9*, 489. [CrossRef] [PubMed]

 *sustainability*

*Article*

# Bioactivity of Grape Skin from Small-Berry Muscat and Augustiatis of Samos: A Circular Economy Perspective for Sustainability

Afroditi Michalaki [1,*], Elpida Niki Iliopoulou [1], Angeliki Douvika [1], Constantina Nasopoulou [1], Dimitris Skalkos [2] and Haralabos Christos Karantonis [1,*]

[1] Laboratory of Food Chemistry, Biochemistry and Technology, Department of Food Science and Nutrition, School of the Environment, University of the Aegean, Metropolitan Ioakeim 2, 81400 Myrina, Greece
[2] Laboratory of Food Chemistry, Department of Chemistry, University of Ioannina, 45110 Ioannina, Greece
* Correspondence: fnsd20011@fns.aegean.gr (A.M.); chkarantonis@aegean.gr (H.C.K.); Tel.: +30-225408311 (H.C.K.)

**Abstract:** Consumer interest in health-promoting foods has prompted researchers to use wine by-products to increase food's functional characteristics. This research aims to examine the skin bioactivities of Samos white (small-berry Muscat) and red (Augustiatis) grape skin extracts (M-GSkE, A-GSkE). Total phenolic content, antiradical activity, the inhibition of plasma oxidation and platelet aggregation, and the phenolic profile were examined. A-GSkE and M-GSkE showed high total phenolics (1.19 ± 0.13 vs. 2.12 ± 0.23 mM GAE), antiradical activity (7.7 ± 0.4 vs. 6.6 ± 0.3 µM GAE for ABTS; 31.12 ± 0.8 vs. 26.4 ± 1.0 µM GAE for DPPH), resistance to plasma oxidation (5.7 ± 0.4 vs. 1.1 ± 0.2 µM GAE), and antithrombotic activity (19.7 ± 0.1 vs. 26.6 ± 0.2 µM GAE). Ferulic (41.3 ± 0.1 > 13.2 ± 0.1 µg/g DM), vanillic (26.3 ± 1.7 > 12.2 ± 1.2 µg/g DM), and gallic (16.6 ± 0.1 > 8.4 ± 2.9 µg/g DM) acids along with ε-viniferin (3.6 ± 0.4 > 2.8 ± 0.3 µg/g DM) were identified in higher content in A-GSkE. Catechin (59.8 ± 1.5 µg/g DM), chlorogenic acid (43.8 ± 0.9 µg/g DM), and resveratrol (0.83 ± 0.13 µg/g DM) were identified only in M-GSkE, while caffeic acid 19.8 ± 0.4 µg/g DM) and daidzein (16.8 ± 0.1 µg/g DM) were identified only in A-GSkE. The specialized bioactivities researched in two previously unexplored Samos' wine grape skin extracts give them added value. The valorization of such by-products promises a sustainable future in the food sector of local communities and an improvement in local public health.

**Keywords:** grape skin; wine making by-products; phenolics; antiradical activity; antioxidant activity; antiplatelet activity

**Citation:** Michalaki, A.; Iliopoulou, E.N.; Douvika, A.; Nasopoulou, C.; Skalkos, D.; Karantonis, H.C. Bioactivity of Grape Skin from Small-Berry Muscat and Augustiatis of Samos: A Circular Economy Perspective for Sustainability. *Sustainability* **2022**, *14*, 14576. https://doi.org/10.3390/su142114576

Academic Editor: Marian Rizov

Received: 21 October 2022
Accepted: 4 November 2022
Published: 6 November 2022

**Publisher's Note:** MDPI stays neutral with regard to jurisdictional claims in published maps and institutional affiliations.

## 1. Introduction

Because of the significant quantity of by-products generated in the food sector, economic and environmental issues have escalated in recent decades. One of the major concerns in this area involves the redefining of such by-products as raw materials that may be processed, with the objective of limiting their negative environment impacts while gaining high value-added foods. This action could lead in the future to a more sustainable food sector.

The wine industry produces a lot of by-products, which are usually utilized as organic fertilizer or animal feed. Moreover, viticulture has seen tremendous expansion in the last decade [1], and wine production, along with its by-products, have been increased dramatically.

Viticulture plays a vital role in the European economy. Greece ranks seventh in the EU [2] in terms of grape volume produced. The result of the processing of these grapes during wine production is a substantial quantity of grape skin in the form of grape pomace.

Alternative uses for wine industry by-products are being researched, having in mind factors such as environmental benefits, financial savings, and new potential for industrial development. Recent studies have suggested that nonextracted items, such as grape skin bioactive chemicals, may be of significant interest [3–5].

The demand for functional foods has been increasing, and one solution that might be considered for satisfying this need is the use of by-products, such as the grape skin derived from grape pomace during wine production [6–8].

Applying this reasoning to areas with many small islands, such as that of the northern Aegean in Greece, the goal of promoting innovation and entrepreneurship in the context of the agri-food sector is served.

The exploitation of the grape skin by-product, which is abundant in these islands, could lead to the production of new functional foods or the further development of already existing traditional products. Increased consumer demand for such foods will in turn lead to more sustainable conditions in the future for local communities in these areas.

Grape skin has been recognized for its large quantities of bioactive compounds [9,10] that add value to these by-products owing to their various potential applications in the food and pharmaceutical sectors [11,12]. Previous studies have referred to the antioxidant, antibacterial, anti-inflammatory, antiobesity, and anticancer properties of the biomolecules that are obtained from grape skin [3–5,13–17].

Other studies have shown that the oxidation of plasma lipoproteins has been demonstrated to initiate atherosclerosis at the molecular level. This oxidation procedure results in the production of thrombotic and inflammatory lipid mediators such as platelet-activating factor (PAF) and PAF-like oxidized phospholipids, which mediate the early stages of inflammation on the aortic endothelium as well as thrombosis and free radical production [18–20]. In addition, minor bioactive substances in plant-originating foods that exhibit antioxidant and/or PAF-inhibitory effects have been shown in studies to be important in the prevention of cardiovascular disease [18,21].

The aim of this study was to highlight the bioactivities of grape skins from two unexplored winemaking grape varieties that are cultivated in the Greek island of Samos in Northern Aegean from the perspective of the nutraceutical value related to antiatherogenic activities such as free radical scavenging and the inhibition of plasma oxidation and PAF-induced platelet aggregation. Total phenolic content (TPC), radical scavenging, and the inhibition of platelet aggregation and plasma oxidation were evaluated in grape skin methanolic extracts.

## 2. Materials and Methods

### 2.1. Chemicals and Reagents

Folin–Ciocalteu and di-sodium hydrogen phosphate dihydrate was supplied by Merck (Darmstadt, Germany). Anhydrous sodium carbonate was purchased from SDS (Peypin, France). The reagents of gallic acid; 1,1-Diphenyl-2-picryl-hydrazyl (DPPH); fatty acid-free bovine serum albumin, beta-acetyl-O-hexadecyl-L-phosphatidylcholine (PAF), as well as formic acid and solvents of methanol and water for HPLC analysis, were obtained from Sigma–Aldrich Co. (St. Louis, MO, USA). Trolox was supplied by Acros Organics (Waltham, MA, USA). Sodium chloride and sodium dihydrogen phosphate dehydrate were purchased from Penta (CZ Ltd., Chrudim, Czech Republic). The 2,2′-Azino-bis(3-ethylbezothiazoline-6-sulphonic acid (ABTS) reagent was acquired from Applichem (Darmstadt, Germany). Chem-Lab was the supplier of potassium persulfate (Zedelgem, Belgium). The purchase of copper sulphate pentahydrate was from Alfa Aesar (Ward Hill, MA, USA).

### 2.2. Material for Analysis

#### 2.2.1. Preparation of Grape Skin Samples

Grapes cultivated in the Greek island of Samos from small-berry Muscat and Augustiatis were harvested at technological maturation in August 2020. The grapes were then pressed to obtain the juice for the production of wine. The by-product of the processing consisted of grape skin, grape seeds, and grape stems and was used to manually obtain the grape skin samples. To reduce the humidity of the samples, grape skins were stored in a deep freeze (DW-HL388, Zhongke Meiling Cryogenics corp., Hefei, China) at $-86\,^{\circ}C$ for 1 day and then lyophilized for 48 h under vacuum (5.0 Pascal) at $-60\,^{\circ}C$ using a freeze dryer BK-FD10PT (Biobase Biodustry Co., Ltd., Jinan, China). The dry grape skin samples were then processed for one minute in a laboratory grinder IKA A 10 basic (IKA Works, Wilmington, NC, USA) to produce a sample of fine powder.

### 2.2.2. Extraction of Grape Skin Samples

The selection of the solvent for the extraction of phenolics from the grape skin samples was based on an analysis which utilized three distinct methanol/water solvent mixtures with ratios 80/20, 70/30, and 60/40 (*v/v*) and a ratio of (solvent mixture)/(grape skin sample) equal to 100 (*v/w*). The grape skin samples with the various solvent mixtures were placed in polystyrene test tubes with a stopper and extracted through pulsed ultrasound assistance at 37 kHz and 220 W ultrasonic power using an elmasonic P 70 H ultrasonic device (Elmasonic P; Elma, Singen, Germany) at $50\,^{\circ}C$ for 30 min. The samples were then stored at $-86\,^{\circ}C$ for 24 h, followed by centrifugation at $20\,^{\circ}C$ for 15 min at $20,000\times\ g$. Until further analysis, the supernatant was aliquoted in 2.0 mL portions and kept in polypropylene microvials at $-86\,^{\circ}C$. The extract with the methanol/water ratio that resulted in the strongest antiradical activity based on the ABTS assay was adopted for further study of the samples.

### 2.3. Determination of Phenolic Compounds

The total phenolic content was measured in triplicate in grape skin methanolic extracts using a modified version of Singleton and Rossi's technique [22]. The experiment was carried out by combining 0.01 to 0.001 mL of extract with 1.8 mL of distilled water and 0.1 mL of Folin–Ciocalteu reagent. The materials were then rapidly mixed and incubated in the dark for two minutes. After adding 0.3 mL of 20% (*w/v*) aqueous Na2CO3, the samples were rapidly agitated and incubated at $40\,^{\circ}C$ in a water bath for 30 min. Absorbance was measured spectrophotometrically at 765 nm using a Spectrophotometer Lambda 25 (Perkin Elmer, Norwalk, CT, USA). Gallic acid was used to develop a standard curve. The final findings were expressed as equivalent concentrations of gallic acid (mM GAE).

### 2.4. Radical Scavenging Properties Evaluation

### 2.4.1. ABTS Assay

The ABTS radical scavenging activity of extracts was determined using a modified version of the technique of Re et al. [23]. The $ABTS^{+\bullet}$ was generated by reacting a 7 mmol/L stock solution of ABTS with a 2.45 mmol/L final concentration of potassium persulphate $(K_2S_2O_8)$. The $ABTS^{+\bullet}$ solution was diluted to an absorbance of $0.700\pm0.050$ at 734 nm with distilled water. Aliquots of grape skin extracts or suitable volumes of Trolox, as positive reference compound, were combined with 1.0 mL $ABTS^{+\bullet}$. The absorbance at 734 nm was determined spectrophotometrically after vigorous stirring and a 15-min incubation of samples in the dark at room temperature. The capacity of the extracts to scavenge the ABTS free cationic radical was examined compared to a control sample containing distilled water instead of each amount of the extracts tested. The results were expressed as concentration of GAE in $\mu M$ able for 50% scavenging of $ABTS^{+\bullet}$ $(IC_{50\text{-}ABTS})$ and as a Trolox-equivalent amount. Each sample was evaluated in triplicate.

### 2.4.2. DPPH Assay

The ability of grape skin extracts to scavenge the DPPH free radical was evaluated using a modified version of Abe, Murata, and Hirota's technique [24]. An aliquot of the extracts or a suitable standard solution of Trolox, as a positive reference compound, was diluted to a volume of 0.9 mL with methanol. Then, 0.1 mL of DPPH reagent in methanol at a concentration of 6.0 mM was added, followed by vigorous stirring. After 15 min in the dark, the absorbance was measured spectrophotometrically at 515 nm against a reference sample containing methanol in place of each volume of grape skin extracts examined. The findings were presented as concentration of GAE in $\mu$M able for 50% scavenging of DPPH ($IC_{50\text{-}DPPH}$) and as a Trolox-equivalent amount. Each sample was examined in triplicate.

### 2.5. Plasma Oxidation

The human plasma oxidation inhibition experiment was performed according to Schnitzer et al. [25] with minimal modifications utilizing a Lambda 25 Spectrophotometer (Perkin–Elmer, Norwalk, CT, USA) equipped with an eight-position thermostatic sample changer. Grape skin extracts were deposited in UV-transparent disposable cuvettes (Brand, Wertheim, Germany). Then, 880 $\mu$L of phosphate buffer solution (PBS), pH 7.4, 146 mM in NaCl, and 20 $\mu$L of human plasma were added. After moderate shaking and incubation at room temperature for 1 min, the samples were placed in the photometer's thermostatic chamber and incubated for 10 min at 37 °C. The oxidation process was initiated by adding 100 $\mu$L of 1 mM $CuSO_4 \cdot 5H_2O$. The absorbance was measured continuously for 3.0 h at 245 nm and at a constant temperature of 37 °C. The duration of plasma's resistance to oxidation in the presence of grape skin extracts, PBS, or Trolox was assessed by the absence of a rise in absorbance at 245 nm. The prevention of in vitro plasma oxidation generated by the extracts was assessed by comparing the plasma oxidation resistance time of each sample containing the extract to that of a reference sample containing PBS instead of each tested volume of grape skin sample. Trolox was used as a positive reference compound. The findings were represented as concentration of GAE in $\mu$M able to produce a 50% increase in plasma oxidation lag time ($LTIC_{50\text{-}POX}$) and as a Trolox-equivalent amount.

### 2.6. Platelet Aggregation Assay

Grape skin extracts were evaluated for their in vitro antithrombotic effectiveness using the PAF-induced thrombosis inhibition assay in platelet-rich plasma (PRP). The experiment was performed using a Chrono-Log 500-Ca aggregometer (Chrono-Log Co., Havertown, PA, USA) linked to a computer (Aggro/Link software; Chrono-Log, Hawerstown, PA, USA) [26]. Under a stream of nitrogen, aliquots of grape skin extracts and PAF ethanolic solution were evaporated and reconstituted in bovine serum albumin (BSA) (2.5 mg/mL saline). Then, 250 $\mu$L aliquots of PRP and stir bars were placed in siliconized glass cuvettes and incubated for 15 min at 37 °C in the incubation wells of the aggregometer. Next, the platelet response caused by a final concentration of 0.27 $\mu$M PAF in PRP was measured. The resulting curves were recorded before (assumed to be 0% inhibition) and after the addition of different quantities of the studied extracts in the presence of PAF while stirring at 1200 rpm. The quantity of extracts necessary to inhibit PAF activity by 50 percent was estimated utilizing the area of 20 to 80 percent inhibition against varying quantities of grape skin extracts and was expressed as concentration of GAE in $\mu$M capable of producing 50% inhibition of the PAF-induced PRP aggregation ($IC_{50\text{-}PAF}$).

### 2.7. HPLC-DAD Analysis of Phenolics in Grape Skin Extracts

For the analysis of phenolic compounds in grape skin extracts, a Shimadzu LC-2030 C prominence-i system equipped with a binary pump, degasser, autosampler, column heater, and PDA detector was used (Shimadzu, Kyoto, Japan). For the separation of the phenolic compounds under analysis, the analytical column of Luna C18(2), 5 $\mu$m, 4.6 mm $\times$ 250 mm from Phenomenex (Aschaffenburg, Germany) was used. The elution was carried out using 0.2% ($v/v$) formic acid-acidified water (mobile phase A) and methanol (mobile phase B). The following is how the chosen elution gradient scheme was implemented: 5% mobile phase B at 0 min; 5% mobile phase B at 2 min; 95% mobile phase B at 20 min; 95% mobile phase B at 25 min; 5% mobile phase B at 25.01 min; 5% mobile phase B at 28 min. The volume of injection was 20 $\mu$L. UV–vis spectra were recorded between 190 and 800 nm, and chromatograms were obtained at 280 nm. Gallic acid, ferulic acid, vanillic acid, daidzein, chlorogenic acid, caffeic acid, (+)-catechin, protocatechic acid, tyrosol, resveratrol, and viniferin were employed as standards.

*2.8. Statistical Analysis*

Data are presented as mean $\pm$ standard deviation (SD) after statistical analysis by ANOVA with Tukey's post hoc, Student's *t*-test, and Pearson's correlation analysis using SPSS (version 28.0; SPSS Inc., Chicago, IL, USA).

## 3. Results

The study of grape skin methanolic extracts from small-berry Muscat and Augustiatis cultivated in the Greek island of Samos were carried out by estimating the following parameters: (1) extractive capacity of antioxidants by aqueous methanol solutions, (2) total phenolic determination and radical scavenging activity, (3) antiplatelet and plasma antioxidant activities, and (4) free phenolic profile by HPLC-DAD analysis.

*3.1. Extractive Capacity of Antioxidants by Aqueous Methanol Solutions*

The extractive capacity of antioxidants by aqueous methanol solutions was evaluated using ABTS radical cation scavenging activity. The results, as assessed by the ABTS assay, are presented in Table 1. Grape skin extracts from both samples exerted higher radical scavenging activity when extracted through methanol/water 60/40 (*v/v*). For this reason, the ratio of 60/40 for methanol/water was adopted as the solvent to produce extracts for further study.

**Table 1.** ABTS radical cation scavenging amount of grape skin extracts obtained from different solvent mixtures.

| | [1]M-GSkE | [2]A-GSkE |
|---|---|---|
| Solvent mixture | [3]$SA_{50\text{-ABTS}}$ (µL) | $SA_{50\text{-ABTS}}$ (µL) |
| [4]M/W: 80/20 (*v/v*) | $10.3 \pm 0.4$ [a] | $4.9 \pm 0.3$ [b] |
| M/W: 70/30 (*v/v*) | $9.8 \pm 0.3$ [a] | $4.7 \pm 0.2$ [b] |
| M/W: 60/40 (*v/v*) | $6.5 \pm 0.3$ [c] | $3.1 \pm 0.1$ [d] |

[1]M-GSkE: small-berry Muscat grape skin extract; [2]A-GSkE: Augustiatis grape skin extract; [3]$SA_{50}$: Amount for 50% scavenging of the ABTS radical cation; [4]M/W: Methanol/Water. Different letters (a–d) in rows and columns denote values of statistically significant difference. Results are expressed as mean $\pm$ SD in microliters of extract. Extraction was performed using 100 mL of solvent per 1 g of dry powder matter of each grape skin sample.

*3.2. Total Phenolic Content and Radical Scavenging Activity Determination*

The TPC of grape skin methanolic extracts are reported in Table 2 as mM of gallic acid equivalent (GAE). The total phenolic content in Augustiatis grape skin extract (A-GSkE) was higher than that from the small-berry Muscat grape skin extract (M-GSkE) ($2.12 \pm 0.23 > 1.19 \pm 0.13$ mM GAE; $p < 0.05$). The radical scavenging activity of the two extracts, as evaluated by the ABTS and DPPH assays, was expressed as µM GAE and showed higher antiradical activity for A-GSkE compared to M-GSkE ($p < 0.05$).

**Table 2.** Total phenolic content and radical scavenging activities.

| Parameters | [1]M-GSkE | [2]A-GSkE |
|---|---|---|
| [3]TPC (mM GAE) | $1.19 \pm 0.13$ [a] | $2.12 \pm 0.23$ |
| [4]$IC_{50\text{-ABTS}}$ (µM GAE) | $7.7 \pm 0.4$ [a] | $6.6 \pm 0.3$ |
| $IC_{50\text{-DPPH}}$ (µM GAE) | $31.2 \pm 0.8$ [a] | $26.4 \pm 1.0$ |

[1]M-GSkE: small-berry Muscat grape skin extract; [2]A-GSkE: Augustiatis grape skin extract; [3]TPC: Total phenolic content; [4]$IC_{50}$: Concentration for 50% scavenging of the ABTS radical cation or DPPH radical. Letter [a] in rows denote values of statistically significant difference. Results are expressed as mean $\pm$ SD in µM GAE.

*3.3. Antiplatelet Activity and Plasma Oxidation Inhibition*

Antiplatelet activity of M-GSkE was higher than A-GSkE (Table 3) as a lower concentration of µM GAE was required to inhibit PAF-induced platelet aggregation ($19.7 \pm 0.1 < 26.6 \pm 0.2$; $p < 0.05$). On the other hand, a concentration such as the µM GAE of A-GSkE was required to increase the lag phase time of the plasma oxidation curves by 50% was lower than M-GSkE ($1.1 \pm 0.2 < 5.7 \pm 0.3$; $p < 0.05$), showing higher antioxidant activity of A-GSkE compared to M-GSkE ($p < 0.05$) toward copper-induced plasma oxidation (Table 3).

**Table 3.** Antiplatelet activity and plasma oxidation Inhibition.

| Bioactivity | [1]M-GSkE | [2]A-GSkE |
|---|---|---|
| [3]IA$_{50\text{-PAF}}$ ($\mu$M GAE) | $19.7 \pm 0.1$ [a] | $26.6 \pm 0.2$ |
| [4]LTIC$_{50\text{-POX}}$ ($\mu$M GAE) | $5.7 \pm 0.4$ [a] | $1.1 \pm 0.2$ |

[1]M-GSkE: small-berry Muscat grape skin extract; [2]A-GSkE: Augustiatis grape skin extract; [3]IC$_{50}$: concentration for 50% inhibition. [4]LTIC$_{50\text{-POX}}$: concentration for 50% lag time increase for plasma oxidation. Letter [a] in rows denote values of statistically significant difference. Results are expressed as mean $\pm$ SD in $\mu$M GAE.

### 3.4. Free Phenolic Profile by HPLC-DAD Analysis

The results from the HPLC-DAD phenolic analysis are presented in Table 4. Gallic acid, vanillic acid, ferulic acid, and $\varepsilon$-viniferin were detected in both extracts. The content of those four phenolics was higher in A-GSkE compared to M-GSkE ($p < 0.05$). Moreover, catechin, chlorogenic acid, and resveratrol were identified only in M-GSkE, while caffeic acid and Daidzein were identified only in A-GSkE.

**Table 4.** Free phenolic profile in methanolic extracts of grape skin samples.

| Standard Phenolic Compounds | [1]M-GSkE Phenolic Compounds ($\mu$g/g) [3] | [2]A-GSkE Phenolic Compounds ($\mu$g/g) |
|---|---|---|
| Ferulic acid | $13.2 \pm 0.1$ [a] | $41.3 \pm 0.1$ |
| Vanillic acid | $12.2 \pm 1.2$ [a] | $26.3 \pm 1.7$ |
| Gallic acid | $8.4 \pm 2.9$ [a] | $16.6 \pm 0.1$ |
| $\varepsilon$-viniferin | $2.8 \pm 0.3$ [a] | $3.6 \pm 0.4$ |
| Catechin | $59.8 \pm 1.5$ | n.d. |
| Chlorogenic acid | $43.8 \pm 0.9$ | n.d. |
| Resveratrol | $0.83 \pm 0.13$ | n.d. |
| Caffeic acid | n.d. | $19.8 \pm 0.4$ |
| Daidzein | n.d. | $16.8 \pm 0.1$ |
| Tyrosol | n.d. | n.d. |

[1]M-GSkE: small-berry Muscat grape skin extract; [2]A-GSkE: Augustiatis grape skin extract; [3]results are presented as the mean value of two independent analysis in $\mu$g of each phenolic per g of dried grape skin before extraction. Letter [a] in rows denote values of statistically significant difference.

Representative HPLC chromatographs are reported in Figure 1.

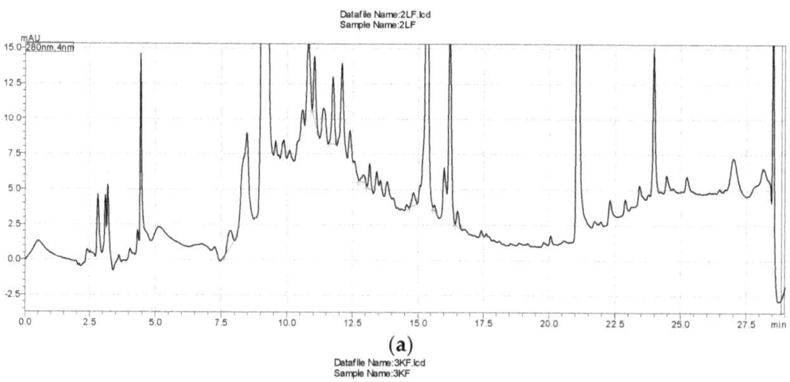


(**a**)

**Figure 1.** *Cont.*

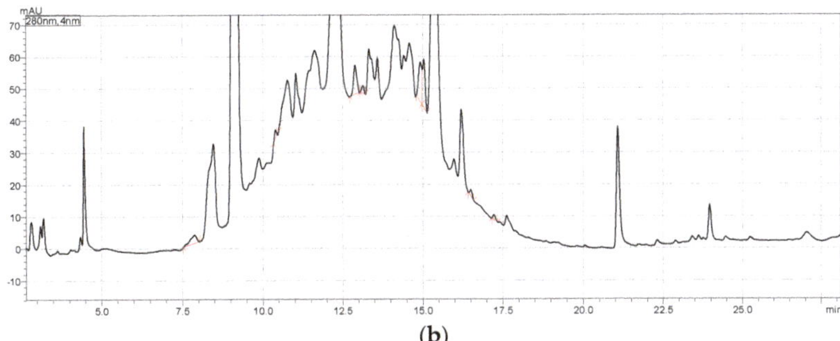

**Figure 1.** Representative HPLC chromatographs at 280 nm of (**a**) M-GSkE and (**b**) A-GSkE along with chemical structures of identified compounds. Retention times in min were gallic acid, 7.8; catechin, 11.0; chlorogenic acid, 11.8; tyrosol, 12.2; caffeic acid, 12.9; vanillic acid, 13.1; resveratrol, 14.5; ferulic acid, 14.8; ε-viniferin, 16.4; daidzein, 17.3.

## 4. Discussion

In the context of the circular economy, researchers have tried to exploit wine by-products with the final goal of producing human health-promoting foods.

This fact is an opportunity for local communities that have unique varieties and species that can be used as raw materials. The exploitation of such raw materials can lead to unique food products or to the improvement of traditional foods of such regions. In this approach, the effort to exploit winemaking by-products, such as wine grape skins, in the North Aegean region offers the Greek islands of the North Aegean a tool for both economic development and public health prevention. This is consistent with recent research presenting that food value chains are widely realized as more fair alternatives to conventional supply networks [27–30].

Under this working hypothesis, the bioactivity of methanolic grape skin extracts of a white (small-berry Muscat) and a red (Augustiatis) winemaking grape species cultivated in Samos was investigated. We showed that methanol/water 60/40 ($v/v$) was a good choice in solvent mixture to produce extracts with higher antiradical activity based on the ABTS assay compared to other aqueous methanolic solvents (Table 1). Methanol is a solvent that has been used in various ratios with water to extract grape skin antioxidants [31]. Our results are in accordance with the work of Ćurko et al. that showed that the 62.7% of methanol in water acted as a solvent for optimized total phenolic extraction from grape skin pomaces through microwave-assisted extraction [32].

Both extracts exerted high total phenolic content. Augustiatis had 1.8-times higher phenolic content compared to small-berry Muscat ($p < 0.05$) (Table 2). More specifically, we found 2.12 ± 0.23 mM GAE in A-GSkE and 1.19 ± 0.13 mM GAE in M-GSkE. These values are equivalent to 36. 06 ± 3.9 and 20.24 ± 2.21 mg GAE/g, respectively. The results are consistent with other studies that have referred values of TPC ranging from 12.74 to 47.72 mg GAE/g [33–35]. Differences in the content of the total phenolics are attributed to the different grape varieties studied as well as to the differences in the methodologies used to produce the extracts.

Augustiatis also exerted 1.15- and 1.18-times higher antiradical activity based on ABTS and DPPH assays, respectively. The results showed $IC_{50\text{-}ABTS}$ values of 6.6 ± 0.3 and 7.7 ± 0.4 μM GAE for A-GSkE and M-GSkE, respectively. These values are equivalent to 213.54 ± 4.27 and 101.84 ± 8.56 μmol Trolox/g for A-GSkE and M-GSkE, respectively. The results are within the range 42.07 ± 5.93 to 447.27 ± 10.49 μmol Trolox/g of values reported in previous studies on grape skins of other varieties [33,36]. The $IC_{50\text{-}DPPH}$ values of 26.4 ± 1.0 and 31.2 ± 0.8 μM GAE for A-GSkE and M-GSkE, respectively, are equivalent to 224.37 ± 8.50 and 119.13 ± 3.05 μmol Trolox/g. $IC_{50\text{-}DPPH}$ values of our study are also in agreement with previous studies on the grape skins of other varieties, which presented ranges from $79.71 \pm 1.13$ to $390.0 \pm 4.3$ µmol of Trolox/g [33,36].

Our results show that radical scavenging activity is corelated to the phenolic content ($p < 0.05$ for both ABTS and DPPH results) that is consistent with previous studies [31,37].

Both extracts exerted inhibitory bioactivity against in vitro copper induced plasma oxidation; however, Augustiatis was five-times more bioactive compared to small-berry Muscat ($p < 0.05$; Table 3). This result is in accordance with our results concerning the total phenolic content and antiradical activities in the two samples. According to our knowledge the inhibition of copper-induced plasma oxidation from grape skin samples has not been studied previously. Nonetheless, our results are supported by other studies in wine samples showing that the extent of LDL oxidation inhibition is directly related to the total phenolic content in the wine samples [38].

Concerning antiplatelet activity, small-berry Muscat exerted 1.34-times higher antiplatelet activity compared to Augustiatis ($p < 0.05$; Table 3). Antiplatelet activity against the thrombotic and inflammatory lipid mediator of PAF from the grape skin samples has not been previously examined, as far as we are aware. Nevertheless, our results are supported by studies in wine samples showing that the protective effect of a wine is independent of its color but is strongly associated with its microconstituent phenolic profile [39].

Indeed, the phenolics identified in the two extracts are different. Ferulic, vanillic, and gallic acid along with ε-viniferin were identified in both Augustiatis and small-berry Muscat. In those common phenolic compounds, Augustiatis presents higher contents ($p < 0.05$; Table 4). Catechin, chlorogenic, and resveratrol were identified only in small-berry Muscat, while caffeic acid and daidzein were identified only in Augustiatis.

Other researchers have also posited that the polyphenolic composition of grape skin extracts depend on the grape variety [40]. Although many phenolic compounds have been presented to have antiplatelet activities [41–45], our results show clearly that the type and concentration of phenolics existing in the extracts determine which of the samples will have higher antiplatelet activity.

## 5. Conclusions

There are several factors that must be considered before the commercial implementation of the recovery of value-added chemicals from food by-products can be considered. Improving the value of by-products from the wine industry will help cut down on expenses and recoverable materials. This is in line with European rules about the management of food by-products, which stress the need to reduce the quantity of by-products while simultaneously increasing their value.

Methanolic extracts of grape skin samples were tested for their total phenolic content (TPC) and their ability to scavenge free radicals, to prevent platelet aggregation, and to reduce plasma oxidation. The phenolic profile of methanolic grape skin extracts was noteworthy. Both extracts contained abundant phenolic components, including ferulic acid, vanillic acid, gallic acid, ε-viniferin, catechin, chlorogenic acid, resveratrol, caffeic acid, and daidzein.

This is the first time that small berry Muscut and Augustiatis grape varieties have been studied for their skin bioactivities. Methanol:water 60:40 ($v/v$) yielded extracts from the grape skins studied with higher antiradical activities based on the ABTS assay compared to methanol:water 70:30 or 80:20 ($v/v$). The total phenolic content and antiradical activities based on the ABTS and DPPH assays were higher for the skin sample extract of Augustiatis (A-GSkE) compared to that of small berry Muscut (M-GSkE). A-GSkE was also more active compared to M-GSkE toward the inhibition of copper-induced plasma oxidation. On the other hand, M-GSkE was more potent compared to A-GSkE toward platelet aggregation induced by the thrombotic and inflammatory lipid mediator of the platelet-activating factor (PAF). These bioactivities are determined by bioactive molecules in the obtained extracts, and the differences between A-GSkE and M-GSkE concerning the quantity and quality of bioactive phenolics may explain the obtained results. These findings demonstrate that each variety excels in certain bioactivities. This information should be considered while planning the future valorization of food by-products such as grape skins. The present study highlights the nutraceutical potential of the grape skins of two unexplored winemaking grape varieties.

The considerable grape skin by-product presents an opportunity for the creation of novel functional foods or the refinement of current traditional products with superiority in consumer health protection. The results indicate that grape skins of small berry Muscut and Augustiatis could be the subject of a mixture design for the formulation of new enriched healthy animal or plant food products such as meat products, dairy products, bakery snacks, traditional pasta, spread products, beverages, or even wine with increased antioxidant and antiplatelet activities.

Increased consumer demand for such products would assist local economies in these locations by helping them establish a more solid economic basis for the future, and research in this area will continue to provide support to this endeavor.

**Author Contributions:** Conceptualization, methodology, software, validation, data curation, visualization, supervision H.C.K. and A.M.; formal analysis, A.D. and E.N.I.; investigation, H.C.K., A.M., C.N., A.D. and E.N.I.; resources, H.C.K.; writing—original draft preparation, H.C.K.; writing—review and editing, H.C.K., A.M., C.N. and D.S.; project administration, H.C.K.; funding acquisition, H.C.K. All authors have read and agreed to the published version of the manuscript.

**Funding:** This research received partial funding by the "nutriwine" ERDF-North Aegean region funded program (BAP2-0060608) of the United Winemaking Agricultural Cooperative (UWAC) of Samos.

**Institutional Review Board Statement:** Not applicable.

**Informed Consent Statement:** Not applicable.

**Data Availability Statement:** Not applicable.

**Acknowledgments:** We would like to thank United Winemaking agriculture Cooperative of Samos (UAWC of Samos), Samos, Greece, for the supply of the grape pomace samples used in the present study.

**Conflicts of Interest:** The authors declare no conflict of interest. The funders had no role in the design of the study; in the collection, analyses, or interpretation of data; in the writing of the manuscript; or in the decision to publish the results.

## References

1. Food and Agriculture Organization of the United Nations (FAO). FAOSTAT Crop InformationGrapes. 2020. Available online: https://www.fao.org/statistics/en/ (accessed on 10 September 2022).
2. Vlachos, V.A. A Macroeconomic Estimation of Wine Production in Greece. *Wine Econ. Policy* **2017**, *6*, 3–13. [CrossRef]
3. Fernández-Fernández, A.M.; Dellacassa, E.; Nardin, T.; Larcher, R.; Ibañez, C.; Terán, D.; Gámbaro, A.; Medrano-Fernandez, A.; Del Castillo, M.D. Tannat Grape Skin: A Feasible Ingredient for the Formulation of Snacks with Potential for Reducing the Risk of Diabetes. *Nutrients* **2022**, *14*, 419. [CrossRef] [PubMed]
4. Gorrasi, G.; Viscusi, G.; Gerardi, C.; Lamberti, E.; Giovinazzo, G. Physicochemical and Antioxidant Properties of White (Fiano Cv) and Red (Negroamaro Cv) Grape Pomace Skin Based Films. *J. Polym. Environ.* **2022**, *30*, 3609–3621. [CrossRef]
5. Mokrani, M.; Charradi, K.; Limam, F.; Aouani, E.; Urdaci, M.C. Grape Seed and Skin Extract, a Potential Prebiotic with Anti-Obesity Effect through Gut Microbiota Modulation. *Gut Pathog.* **2022**, *14*, 30. [CrossRef] [PubMed]
6. Plasek, B.; Temesi, Á. The Credibility of the Effects of Functional Food Products and Consumers' Willingness to Purchase/Willingness to Pay—Review. *Appetite* **2019**, *143*, 104398. [CrossRef] [PubMed]
7. Bakshi, A.; Chhabra, S.; Kaur, R. Consumers' Attitudes Toward Functional Foods: A Review. *Curr. Top. Nutraceutical Res.* **2020**, *18*, 343–347. [CrossRef]
8. Topolska, K.; Florkiewicz, A.; Filipiak-Florkiewicz, A. Functional Food—Consumer Motivations and Expectations. *Int. J. Environ. Res. Public Health* **2021**, *18*, 5327. [CrossRef]
9. de Andrade, R.B.; Machado, B.A.S.; Barreto, G.A.; Nascimento, R.Q.; Corrêa, L.C.; Leal, I.L.; Tavares, P.P.L.G.; Ferreira, E.S.; Umsza-Guez, M.A. Syrah Grape Skin Residues Has Potential as Source of Antioxidant and Anti-Microbial Bioactive Compounds. *Biology* **2021**, *10*, 1262. [CrossRef]

10. Serea, D.; Râpeanu, G.; Constantin, O.E.; Bahrim, G.E.; Stănciuc, N.; Croitoru, C. Ultrasound and Enzymatic Assisted Extractions of Bioactive Compounds Found in Red Grape Skins Băbească Neagră (*Vitis vinifera*) Variety. *Ann. Univ. Dunarea Jos Galati Fascicle VI Food Technol.* **2021**, *45*, 9–25. [CrossRef]

11. Felice, F.; Zambito, Y.; Di Colo, G.; D'Onofrio, C.; Fausto, C.; Balbarini, A.; Di Stefano, R. Red Grape Skin and Seeds Polyphenols: Evidence of Their Protective Effects on Endothelial Progenitor Cells and Improvement of Their Intestinal Absorption. *Eur. J. Pharm. Biopharm.* **2012**, *80*, 176–184. [CrossRef]

12. Soares De Moura, R.; Da Costa, G.F.; Moreira, A.S.B.; Queiroz, E.F.; Moreira, D.D.C.; Garcia-Souza, E.P.; Resende, A.C.; Moura, A.S.; Teixeira, M.T. *Vitis vinifera* L. Grape Skin Extract Activates the Insulin-Signalling Cascade and Reduces Hyperglycaemia in Alloxan-Induced Diabetic Mice. *J. Pharm. Pharmacol.* **2012**, *64*, 268–276. [CrossRef] [PubMed]

13. Morré, D.M.; Morré, D.J. Anticancer Activity of Grape and Grape Skin Extracts Alone and Combined with Green Tea Infusions. *Cancer Lett.* **2006**, *238*, 202–209. [CrossRef] [PubMed]

14. Oueslati, N.; Charradi, K.; Bedhiafi, T.; Limam, F.; Aouani, E. Protective Effect of Grape Seed and Skin Extract against Diabetes-Induced Oxidative Stress and Renal Dysfunction in Virgin and Pregnant Rat. *Biomed. Pharmacother.* **2016**, *83*, 584–592. [CrossRef] [PubMed]

15. Bomfim, G.H.S.; Musial, D.C.; Miranda-Ferreira, R.; Nascimento, S.R.; Jurkiewicz, A.; Jurkiewicz, N.H.; de Moura, R.S. Antihypertensive Effects of the *Vitis vinifera* Grape Skin (ACH09) Extract Consumption Elicited by Functional Improvement of P1 ($A_1$) and P2 (P2X1) Purinergic Receptors in Diabetic and Hypertensive Rats. *PharmaNutrition* **2019**, *8*, 100146. [CrossRef]

16. Sochorova, L.; Prusova, B.; Cebova, M.; Jurikova, T.; Mlcek, J.; Adamkova, A.; Nedomova, S.; Baron, M.; Sochor, J. Health Effects of Grape Seed and Skin Extracts and Their Influence on Biochemical Markers. *Molecules* **2020**, *25*, 5311. [CrossRef]

17. De Costa, G.F.; Ognibene, D.T.; Da Costa, C.A.; Teixeira, M.T.; Da Silva Cristino Cordeiro, V.; De Bem, G.F.; Moura, A.S.; De Castro Resende, A.; De Moura, R.S. *Vitis vinifera* L. Grape Skin Extract Prevents Development of Hypretension and Altered Lipid Profile in Sponteneously Hypertensive Tats: Role of Oxidative Stress. *Prev. Nutr. Food Sci.* **2020**, *25*, 25–31. [CrossRef]

18. Demopoulos, C.A.; Karantonis, H.C.; Antonopoulou, S. Platelet Activating Factor - A Molecular Link between Atherosclerosis Theories. *Eur. J. Lipid Sci. Technol.* **2003**, *105*, 705–716. [CrossRef]

19. Pégorier, S.; Stengel, D.; Durand, H.; Croset, M.; Ninio, E. Oxidized Phospholipid: POVPC Binds to Platelet-Activating-Factor Receptor on Human Macrophages. Implications in Atherosclerosis. *Atherosclerosis* **2006**, *188*, 433–443. [CrossRef]

20. Boffa, M.B.; Koschinsky, M.L. Oxidized Phospholipids as a Unifying Theory for Lipoprotein(a) and Cardiovascular Disease. *Nat. Rev. Cardiol.* **2019**, *16*, 305–318. [CrossRef]

21. Nomikos, T.; Fragopoulou, E.; Antonopoulou, S. Food Ingredients and Lipid Mediators. *Curr. Nutr. Food Sci.* **2007**, *3*, 255–276. [CrossRef]

22. Singleton, V.L.; Rossi, J.A. Colorimetry of Total Phenolics with Phosphomolybdic-Phosphotungstic Acid Reagents. *Am. J. Enol. Vitic.* **1965**, *16*, 144–158.

23. Re, R.; Pellegrini, N.; Proteggente, A.; Pannala, A.; Yang, M.; Rice-Evans, C. Antioxidant Activity Applying an Improved ABTS Radical Cation Decolorization Assay. *Free Radic. Biol. Med.* **1999**, *26*, 1231–1237. [CrossRef]

24. Abe, N.; Murata, T.; Hirota, A. Novel DPPH Radical Scavengers, Bisorbicillinol and Demethyltrichodimerol, from a Fungus. *Biosci. Biotechnol. Biochem.* **1998**, *62*, 661–666. [CrossRef] [PubMed]

25. Schnitzer, E.; Pinchuk, I.; Bor, A.; Fainaru, M.; Samuni, A.; Lichtenberg, D. Lipid Oxidation in Unfractionated Serum and Plasma. *Chem. Phys. Lipids* **1998**, *92*, 151–170. [CrossRef]

26. Karantonis, H.C.; Antonopoulou, S.; Perrea, D.N.; Sokolis, D.P.; Theocharis, S.E.; Kavantzas, N.; Iliopoulos, D.G.; Demopoulos, C.A. In Vivo Antiatherogenic Properties of Olive Oil and Its Constituent Lipid Classes in Hyperlipidemic Rabbits. *Nutr. Metab. Cardiovasc. Dis.* **2006**, *16*, 174–185. [CrossRef]

27. Erälinna, L.; Szymoniuk, B. Managing a Circular Food System in Sustainable Urban Farming. Experimental Research at the Turku University Campus (Finland). *Sustainability* **2021**, *13*, 6231. [CrossRef]

28. Krzyzanowski Guerra, K.; Hanks, A.S.; Plakias, Z.T.; Huser, S.; Redfern, T.; Garner, J.A. Local Value Chain Models of Healthy Food Access: A Qualitative Study of Two Approaches. *Nutrients* **2021**, *13*, 4145. [CrossRef]

29. Mariutti, L.R.B.; Rebelo, K.S.; Bisconsin-Junior, A.; de Morais, J.S.; Magnani, M.; Maldonade, I.R.; Madeira, N.R.; Tiengo, A.; Maróstica, M.R.; Cazarin, C.B.B. The Use of Alternative Food Sources to Improve Health and Guarantee Access and Food Intake. *Food Res. Int.* **2021**, *149*, 110709. [CrossRef]

30. Oakden, L.; Bridge, G.; Armstrong, B.; Reynolds, C.; Wang, C.; Panzone, L.; Rivera, X.S.; Kause, A.; Ffoulkes, C.; Krawczyk, C.; et al. The Importance of Citizen Scientists in the Move Towards Sustainable Diets and a Sustainable Food System. *Front. Sustain. Food Syst.* **2021**, *5*, 596594. [CrossRef]

31. Milinčić, D.D.; Stanisavljević, N.S.; Kostić, A.Ž.; Soković Bajić, S.; Kojić, M.O.; Gašić, U.M.; Barać, M.B.; Stanojević, S.P.; Tešić, L.; Pešić, M.B. Phenolic Compounds and Biopotential of Grape Pomace Extracts from Prokupac Red Grape Variety. *LWT* **2021**, *138*, 110739. [CrossRef]

32. Ćurko, N.; Kelšin, K.; Dragović-Uzelac, V.; Valinger, D.; Tomašević, M.; Kovačević Ganić, K. Microwave-Assisted Extraction of Different Groups of Phenolic Compounds from Grape Skin Pomaces: Modeling and Optimization. *Polish J. Food Nutr. Sci.* **2019**, *69*, 235–246. [CrossRef]

33. Xu, Y.; Sismour, E.; Abraha-Eyob, Z.; McKinney, A.; Jackson, S. Physicochemical, Microstructural, and Antioxidant Properties of Skins from Pomaces of Five Virginia-Grown Grape Varieties and Their Response to High Hydrostatic Pressure Processing. *J. Food Meas. Charact.* **2021**, *15*, 5547–5555. [CrossRef]
34. Di Stefano, V.; Buzzanca, C.; Melilli, M.G.; Indelicato, S.; Mauro, M.; Vazzana, M.; Arizza, V.; Lucarini, M.; Durazzo, A.; Bongiorno, D. Polyphenol Characterization and Antioxidant Activity of Grape Seeds and Skins from Sicily: A Preliminary Study. *Sustainability* **2022**, *14*, 6702. [CrossRef]
35. Serea, D.; Horincar, G.; Constantin, O.E.; Aprodu, I.; Stănciuc, N.; Bahrim, G.E.; Stanciu, S.; Rapeanu, G. Value-Added White Beer: Influence of Red Grape Skin Extract on the Chemical Composition, Sensory and Antioxidant Properties. *Sustainability* **2022**, *14*, 9040. [CrossRef]
36. Karaman, H.T.; Küskü, D.Y.; Söylemezoğlu, G. Phenolic Compounds and Antioxidant Capacities in Grape Berry Skin, Seed and Stems of Six Wine Grape Varieties Grown in Turkey. *Acta Sci. Pol. Hortorum Cultus* **2021**, *20*, 15–25. [CrossRef]
37. Du, Y.; Li, X.; Xiong, X.; Cai, X.; Ren, X.; Kong, Q. An Investigation on Polyphenol Composition and Content in Skin of Grape (*Vitis vinifera* L. Cv. Hutai No.8) Fruit during Ripening by UHPLC-MS2 Technology Combined with Multivariate Statistical Analysis. *Food Biosci.* **2021**, *43*, 101276. [CrossRef]
38. Fuhrman, B.; Volkova, N.; Suraski, A.; Aviram, M. White Wine with Red Wine-like Properties: Increased Extraction of Grape Skin Polyphenols Improves the Antioxidant Capacity of the Derived White Wine. *J. Agric. Food Chem.* **2001**, *49*, 3164–3168. [CrossRef]
39. Fragopoulou, E.; Petsini, F.; Choleva, M.; Detopoulou, M.; S Arvaniti, O.; Kallinikou, E.; Sakantani, E.; Tsolou, A.; Nomikos, T.; Samaras, Y. Evaluation of Anti-Inflammatory, Anti-Platelet and Anti-Oxidant Activity of Wine Extracts Prepared from Ten Different Grape Varieties. *Molecules* **2020**, *25*, 5054. [CrossRef]
40. Katalinić, V.; Možina, S.S.; Skroza, D.; Generalić, I.; Abramovič, H.; Miloš, M.; Ljubenkov, I.; Piskernik, S.; Pezo, I.; Terpinc, P.; et al. Polyphenolic Profile, Antioxidant Properties and Antimicrobial Activity of Grape Skin Extracts of 14 *Vitis vinifera* Varieties Grown in Dalmatia (Croatia). *Food Chem.* **2010**, *119*, 715–723. [CrossRef]
41. He, X.; Guo, X.; Ma, Z.; Li, Y.; Kang, J.; Zhang, G.; Gao, Y.; Liu, M.; Chen, H.; Kang, X. Grape Seed Proanthocyanidins Protect PC12 Cells from Hydrogen Peroxide-Induced Damage via the PI3K/AKT Signaling Pathway. *Neurosci. Lett.* **2021**, *750*, 135793. [CrossRef]
42. Wang, Z.; Huang, Y.; Zou, J.; Cao, K.; Xu, Y.; Wu, J.M. Effects of Red Wine and Wine Polyphenol Resveratrol on Platelet Aggregation in Vivo and in Vitro. *Int. J. Mol. Med.* **2002**, *9*, 77–79. [CrossRef] [PubMed]
43. Cho, H.-J.; Kang, H.-J.; Kim, Y.-J.; Lee, D.-H.; Kwon, H.-W.; Kim, Y.-Y.; Park, H.-J. Inhibition of Platelet Aggregation by Chlorogenic Acid via CAMP and CGMP-Dependent Manner. *Blood Coagul. Fibrinolysis* **2012**, *23*, 629–635. [CrossRef] [PubMed]
44. Amin, R.P.; Kunaparaju, N.; Kumar, S.; Taldone, T.; Barletta, M.A.; Zito, S.W. Structure Elucidation and Inhibitory Effects on Human Platelet Aggregation of Chlorogenic Acid from Wrightia Tinctoria. *J. Complement. Integr. Med.* **2013**, *10*, 97–104. [CrossRef] [PubMed]
45. Zhang, Y.; Wang, X.; Lu, B.; Gao, Y.; Zhang, Y.; Li, Y.; Niu, H.; Fan, L.; Pang, Z.; Qiao, Y. Functional and Binding Studies of Gallic Acid Showing Platelet Aggregation Inhibitory Effect as a Thrombin Inhibitor. *Chinese Herb. Med.* **2022**, *14*, 303–309. [CrossRef] [PubMed]

*Review*

# Exploring the Impact of COVID-19 Pandemic on Food Choice Motives: A Systematic Review

**Dimitris Skalkos * and Zoi C. Kalyva**

Laboratory of Food Chemistry, Department of Chemistry, University of Ioannina, 45110 Ioannina, Greece
* Correspondence: dskalkos@uoi.gr; Tel.: +30-2651008345

**Abstract:** The economic crisis caused by the COVID-19 pandemic has effected the global economy, with the main changes expected to affect human life in the future, including food consumption. However, could this pandemic be assumed as a threshold for the suspension of the usual rules behind food choices? This review highlights the changes in food choice motivations before, during, and after the pandemic that have been reported in the literature to date to answer the research question on the changes in food choice motives caused by the pandemic to consumers worldwide. The review comes up with ten key food motives important for consumers, namely health, convenience, sensory appeal, nutritional quality, moral concerns, weight control, mood and anxiety, familiarity, price, and shopping frequency behavior; these motives continue to be significant in the post-pandemic era. Our findings indicate that it is too premature to give definite answers as to what food choice motives in the post-COVID-19 era will be like. Consumers' perceptions and attitudes toward food in the new era are contradictory, depending on the country of the study, the average age, and the sex of the study group. These controversial results illustrate that, for food consumption, motives depend on the population being searched, with changes identified occurring in two directions. The definite answers will be given in three to five years when the new conditions will be clear and a number of studies will have been published. Even though it is too early to fully understand the definite food choice motive changes, defining a "new" index of consumer satisfaction is necessary since it can alter the food sale strategies of retail managers, food companies, and the other parties involved in the agri-food chain.

**Keywords:** food consumption; food choice motives (FCM); convenience; health; sensory appeal; nutritional quality; ethical concern; weight control; mood and stress; familiarity; price

**Citation:** Skalkos, D.; Kalyva, Z.C. Exploring the Impact of COVID-19 Pandemic on Food Choice Motives: A Systematic Review. *Sustainability* **2023**, *15*, 1606. https://doi.org/10.3390/su15021606

Academic Editor: Colin Michael Hall

Received: 17 December 2022
Revised: 9 January 2023
Accepted: 12 January 2023
Published: 13 January 2023



## 1. Introduction

The COVID-19 crisis forced a significant percentage of the world's population to suddenly confine themselves at home, with limited social contacts, exposure to repeated information on the numbers of infections and deaths, and changes in daily habits and emotional well-being [1]. Daily routines were disrupted by isolation and remote works [2], with decreased physical activity level [3] and increased sedentary behavior [4,5], as well as increased meal and snacking frequency [5]. Consumers are informed about the new situation and choose their food based on the main food choice motives (FCM) of health, convenience, sensory appeal, nutritional quality, moral concerns, weight control, mood and anxiety, familiarity, price, and shopping frequency. FCM are critical parameters for consumers to choose food which include social, cultural, aesthetic, political, and contextual factors, as well as food values [6].

According to Salari et al. [7], a better health status will bring variations in food consumption. The impact of the pandemic on mood, mental health, and emotional well-being can also affect food intake and choices. Aoun et al. [8] reported that unbalanced eating behaviors are frequent in people with emotional disorders, depression, and/or anxiety. Contradictory results are recorded in terms of the influence of the pandemic on diet, with some studies reporting a positive influence while other studies reporting a negative influence or no influence. A study in France revealed improvement in diet quality in some cases, while in others, diet quality worsened or there was no change [9]. A Canadian study indicated a slight improvement in diet quality during the early lockdown [10]. In contrast, a Saudi Arabian cross-sectional study with adults showed that food quality deteriorated during the pandemic [11]. The scoping review by Bennett et al. on the pandemic's impact on food quality also showed the contradiction of the results recorded [5]. A limitation of these studies is that a change in diet quality is a result of a change in FCM, and this latter is not elucidated widely. This concept is crucial since it provides a basis to influence diet quality efficiently and in a long-lasting manner. Furthermore, people consider food not only as a means to meet caloric intake and body needs, but also as a means of satisfaction (e.g., appearance, lifestyle, image, and health). Given the controversy in diet behaviors and associated body weight changes caused by the pandemic, there is a need to better understand the motives for specific food choices and their changes as result of the COVID-19 pandemic restrictions. COVID-19-related motivations for consumer food choice can be interpreted into informational codes and advertising campaigns by actors and food chain participants to reach more consumers and vulnerable groups [12,13].

Reports on the motives for food during the pandemic and beyond are still rare, while FCM are addressed only partially and not thoroughly enough. However, more and more papers from different countries are published on the subject on a monthly basis, indicating an increased interest in FCM on the global market. In this paper, we review the reported data exploring changes in FCM caused by the pandemic. So far, there appears to be a significant increase in online shopping, an increase in prices, and a more conservative household management toward buying quality foods. In contrast, familiarity, convenience, and sensory appeal are not significantly affected by the pandemic.

## 2. Methodology and Literature Search

This review followed the guidelines for systematic reviews and meta-analysis extension for scoping reviews (*PRISMA-ScR*) and was in line with the *JBI Manual for Evidence Synthesis*, which is based on the first methodological guide for such reviews reported by Arksey and O'Malley, who noticed and responded to the early appearance in the literature noting similarities and the lack of uniformity [14]. This review also followed the improved methodologies sometimes referred to as "mapping review" or "scoping study" [15–18] and the latest update [19].

A search of PubMed, Google Scholar, and Science Direct was performed for studies published in 2020, 2021, and 2022 and studies that were published before this period using pre-defined terminology. The search terms (COVID-19) and (Food Choice Motives) and (lockdown); (Food Habits) and (Lockdown); (Dietary change) and (COVID-19) and (lockdown); and (COVID-19) and (nutrition or diet) were used initially in the three databases to obtain an understanding of the current research on this topic area. Following this, an alternative phrasing search in relevant publications and a guidance on the search strategy were finalized. The search terms were then finalized with the 10 search terms (research themes) used in this systematic review. This search took place in September and October of 2022. No restrictions and filters were used to avoid excluding any papers of interest. The results were evaluated for eligibility based on the title, abstract, and full text. Two researchers independently screened the articles for eligibility (DS and ZCK) following these inclusion criteria:

- Limitation to papers published in the years 2020, 2021, 2022 (including prior papers for the definition of terminologies).
- Studies investigating the connection of the pandemic and FCM.
- Studies in English only.

The search was broad to identify all studies fitting the review's aim. No authors were contacted for further information.

The limitations of the review process included the following factors:

- Only full-text publications in English were considered, which might have led to selection bias.
- As with most nutritional research studies, dietary intake was assessed through self-reported data, where misreporting, or underreporting, was possible
- The majority of studies were cross-sectional in design and, therefore, the risk of bias and the quality of each study were difficult to assess due to nature of this review and the included studies.
- It was impossible to evaluate quality compared to longer-term cohort/cause–effect research.

## 3. Results

After reviewing all eligible papers and exploring changes in FCM caused by the pandemic, ten research themes were extracted from each publication for evaluation (Table 1). These included health, convenience, sensory appeal, nutritional quality, moral concerns, weight control, mood and anxiety, familiarity, price, and shopping frequency behavior.

### 3.1. Food Consumption and Health

Food consumption and consumer health have always been one of the main issues that all countries have to address in the new post-COVID-19 era [20]. In recent decades, trends in food consumption have been linked to an increase in chronic food-related diseases, such as obesity, cancer, and coronary heart disease [21]. Experts have focused on promoting medical rules about nutrient intake and proper consumption, while avoiding targeting foods. They, thus, issued guidelines for a balanced diet that does not exclude the consumption of specific food products [22]. Consumption of foods high in fat, sugar, and sodium, and low in fiber, are considered to be risk factors for hypertension, cardiovascular disease, diabetes, breast cancer, colon cancer, rectal cancer, prostate cancer, and obesity [23].

Studies that occurred during the pandemic have shown that energy intake exceeding energy expenditure is a major risk factor for a wide range of medical conditions, ranging from diabetes and cancer to musculoskeletal disorders [24,25]. Health attitudes have changed during the COVID-19 era. Due to long periods of limited mobility, consumers were more prone to unhealthy lifestyles, such as reduced or no physical activity and excessive sedentary behavior, which had negative effects on eating habits as well as on body composition [25,26].

The post-COVID-19 era seems to have altered the lives of people, leading to significant changes in various health behaviors. In particular, according to Drieskens et al. [27], increased consumption of sweet or salty snacks and less physical activity have led to an increase in body weight during pandemic-related confinement among adults in Belgium, and more measures are needed to support individuals to achieve healthier behaviors to tackle overweight and obesity. Furthermore, Martínez-de-Quel et al. [28] showed that pandemic-related confinement caused a drawback on the levels of physical activity and sleeping on Spanish citizens, while body weight and self-perceived well-being were also adversely affected, indicating that those with an active life were more susceptible to such disruptions. Robinson et al. [29] reported perceived negative changes in weight-related eating behaviors and physical activity and perceived negative changes in the barriers that adults living in the UK faced in the management of their weight (e.g., motivation problems and control around eating), compared to pre-lockdown. A study on the effect of quarantine on the diet and exercise of Lithuanians and the association between health behaviors and changes in body weight by Kriaucioniene et al. [30] showed a decrease in the consumption of carbonated or sugary drinks, fast foods, and sweets and an increase in the consumption of homemade sweets and fried foods. This was combined with a decrease in physical activity, resulting in an increase in body weight. Huber et al. [31] in a cross-sectional study from Bavarian universities showed that an increase in food consumption, mainly bread and sweets, combined with a lower level of physical activity led to a reduction in weight maintenance during the pandemic. Poelman et al. [32] analyzed consumer behavior in the Netherlands where they demonstrated that consumers kept their eating behavior or food purchases during COVID-19 lockdown, thus keeping their eating habits; however, in people with overweight and obesity, the lockdown had a negative effect on healthy food choices. A Polish study showed that health and weight control were more important during the pandemic compared to the period before it [33].

Although the short-term effects of lockdown practices differ between countries, women seem to be most affected [34]. Jaeger et al. [35], in a propositions to relevant authorities, proposed the need for educational programs to increase physical activity and to teach basic principles of healthy eating and the construction of a healthy food "plate" in case of a possible future lockdown.

**Proposition 1.** *The present data on health motives indicate that consumers decreased physical activity during the COVID-19 pandemic, with parallel increase in consumption of unhealthy foods which had negative effects on their health. A minority of consumers, though, preferred to choose homemade cooked meals for better results.*

### 3.2. Food Consumption and Convenience

The term 'convenience' is associated with 'convenience foods'—that is, those foods prepared and made available to shoppers designed for easy and quick consumption. Such foods include frozen or chilled foods, ready meals, confectionery, snacks and beverages, processed meat and cheese, canned products, and ready-to-eat foods for sale [36]. The convenience factor has always influenced the choice of food, with the result that the consumption of ready-made food is the outcome of the strategy followed by households to cope with time pressure [37]. However, Botonaki et al. [38] in their study on whether or not to choose a ready meal, which included spouse's work status and socio-demographic characteristics of consumers as the control variables, showed that the convenience of cooked meals may be negatively assessed as their consumption is connected with emotions of guilt, regret, and neglect.

Since the beginning of the pandemic, lifestyle and eating habits have been greatly affected [39]. An increase in the use of convenience foods, such as instant and frozen foods, has been recorded worldwide [40]. According to the study by Ko et al. [41], there was a significant decrease in visits to markets, fast food restaurants, catering restaurants, buffet restaurants, and snack bars, while food deliveries and home-cooked meals increased significantly during the pandemic period. The study by Marty et al. showed that the importance of convenience, familiarity, and price decreased during the pandemic [13]. Liu and Chen reported that the young Chinese have normalized takeaway food consumption and developed their own ways of reducing food/food-related waste, which reflect young people's lifestyles [36].

**Proposition 2.** *The present data clearly indicate an increase in the purchase of takeaway food and ready-to-go meals during the pandemic to avoid visits to supermarket or elsewhere.*

### 3.3. Food Consumption and Sensory Appeal

Sensory appeal is the taste, smell, texture, and appearance of food [42]. It is crucial in directing consumers' selection for various foods. Groups of consumers, such as consumers attaching high importance to all determinants ("demanding consumers" with high significance for all determinants), consumers attaching low importance to all determinants ("indifferent consumers" with low significance for all determinants), "healthy eaters" (health as the most important determinant of food choices), and "hedonists" (convenience, sensory appeal, and price as the most important determinants) experienced specific changes in their food consumption during COVID-19 [43]. Moreover, the "healthy eaters" were identified as those who preferred mostly vegetables; the "hedonists" showed a preference for meat/fish, dairy, and snacks; the "demanding consumers" showed a preference for all food categories; and the "indifferent consumers" showed a low preference for all food categories [44]. Sensory appeal seemed to be unaffected by "coronavirus pandemic" of rural China households [45]. Mood and sensory appeal became less important in Polish citizens [33] and Croatian males [46], but more important in French [13] and British people (except sensory appeal which was unchanged) [47].

**Proposition 3.** *The present data show that sensory appeal motives have not been a priority for consumers during COVID-19 and beyond; therefore, their preferences have not significantly changed for this motive.*

### 3.4. Food Consumption and Nutritional Quality

The health effect of food prevention is undeniable. The combination of food and drink in a concentrated period, combining taste and consumption, is called a meal. Analyzing meals and identifying what foods and drinks are consumed allows nutritionists to understand how different combinations of foods and drinks, throughout the day, affect overall diet quality and health [48]. The nutrients in foods combined with their effects can be interactive because, when consuming foods, humans primarily select to mix foods in meals or snacks according to their own formulations. Dietary advice and other nutritional recommendations are given on a daily basis to consumers so that they can understand and follow them [49]. However, the quality, food safety, and nutritional value of foods vary widely around the world. Serious constraints on global production include contamination of the food chain and water by persistent pesticide residues, and reduction in nutrient content and flavors due to intensive production and/or low-cost food processing [50].

During the pandemic, consumers chose healthy, safe, and better quality food compared to their previous practices [51]. However, in some countries, such as Greece [52] and UK [29], studies showed a consumer preference for unhealthy products, such as snacks and pre-packaged 'over-processed' foods high in fats, sugars, and salt. Ruiz-Roso et al. [53] reported a diversification in dietary habits and altered consumption of processed foods, fruits, and vegetables for consumers in Italy, Spain, Chile, Colombia, and Brazil. They further demonstrated new purchasing habits, such as 'conscious shopping', 'bartering' for cheap items, and attention on 'basics' [53]. Alternatively, consumers preferred groceries as the food of choice and consistently anticipated spending most of their money on foods since they are one of the basic human needs [54]. Finally, studies by Ellison et al. [55] and Huang et al. [56] showed that consumers spent money on foods with a longer shelf life and easier access to the market. Rahman et al. found significant differences in food and nutrient consumption, with marked differences in 'fruits and vegetables', vitamin A, folic acid, calcium, iron, magnesium, phosphorus, and potassium, resulting in higher rates of inadequate nutrient intake for those consumers who frequently consumed take-out foods [40].

**Proposition 4.** *Current results show that consumers with a preference for nutritional quality of foods became more sensitive during the pandemic and beyond, spending more money and consuming more nutritional foods, such as grocery and fruits.*

### 3.5. Food Consumption and Ethical Concern

Nowadays, environmental aspects are of main concern for consumers, such as pollution, food production, environment, and food waste, which are ethical issues related to the impact of food consumption on the environment or society [57]. Climate activists, who are concerned about the deterioration of the planet from consumption, food choosers who are vegetarians and vegans, and conservation activists who have concerns about the preservation of existing goods via their reuse and repair are three of the five types of anti-consumers that have emerged following the pandemic with ethical concerns about the conservation of the planet [58].

Food waste can be approached from an ethical perspective. The awareness, understanding, and embracing of ethical attitudes related to food waste may lead to a consumer's behavior change. Crisis, such as the COVID-19 pandemic, has curbed food waste, which can have an impact on climate change and environmental pollution, according to a study by Caloran [59]. Young people seem to be sensitive to food waste effects on the planet, and how this generates an environmental impact in large cities [57]. In addition, this generation will try to change their attitudes to the requirements of environmental conservation and generate innovative solutions to ease the negative impact of an increased population on the planet.

Food consumption behavioral changes have altered the variety of foods [60]. The impact that food waste has on the environment has also been changed by the pandemic as reflected by the fluctuations and short-term alternatives in the consumption of foods [54]. Not only these changes have exacerbated food waste, such as overcooked foods, foods exceeding long-term storage in the freezer, and overbuying, but they have also favored a decrease in food waste, including less frequent shopping, more carefully planned meals, and consumption of the long-term stored food [57].

Above all, food shopping in the context of COVID-19 is now a more careful process, with close attention to one's need and money available. Health maintenance concerns as well as ethical concerns can lead to better behavior on food waste and environmental footprint [59].

**Proposition 5.** *Overall, the data show that food waste and environmental effects are two ethical parameters receiving increased attention from consumers during COVID-19 and beyond.*

### 3.6. Food Consumption and Weight Control

Older people and women have always been more concerned about controlling their weight and following diet and exercise programs [61]. With increased exercise and eating low-calorie, portion-controlled meals, including liquid meal replacements, they try to maintain weight loss [62].

Stress and boredom were two factors that led to overweight as consumers ate 'comfort foods' with sugar and consumed more energy/calories during the COVID-19 period [63]. This is a type of emotional state driven by affective (strong eating desire), behavioral (food seeking), cognitive (thoughts about food), and physiological (salivation) sensations. Fatty-sweet products and sweet-tasting beverages were consumed (including fruit juices) during snacking. Sweets, biscuits, cakes, soft drinks, and sugary foods led to an increase in energy intake and, thus, an increase in body weight during the pandemic [13]. Warning elements in body weight have been recorded during lockdown worldwide [64], probably due to physical activity reduction and increased consumption caused by isolation measures during the pandemic, which resulted in a higher incidence of overweight, obesity, and relevant comorbidities [65]. Only half of the adult population, with increased sweet consumption and less exercise, kept their body weight during the first six months of pandemic-related confinement in Belgium [27]. According to Kalligeros et al. [66], cardiometabolic disorders caused by weight and body fat gain following physical inactivity increased among patients with coronavirus disease. Furthermore, the studies by Wiklund et al. [67] and Lighter et al. [68] have shown that obesity is associated with more severe disease and COVID-19 outcomes. An unhealthy diet is known to lead to chronic inflammation and reduced defense against viruses [69]. In addition, unhealthy eating habits during the pandemic led to increased obesity and caused a chronic systemic inflammatory condition that, along with other chronic non-communicable diseases, such as dyslipidemia, hypertension, heart disease, diabetes, and lung disease, increased the risk of severe complications [70,71]. These studies showed an increase in body weight of women during the Coliform pandemic. Social support during COVID-19 was part of many obesity management programs and was connected with better dietary adherence, better weight management, and even a lower risk of mortality [72]. For both sexes, it may be necessary to improve and adapt weight management goals. Ultimately, the best way to obtain all the necessary nutrients is a balanced diet to ensure normal immune system function while reducing the risk of obesity [72].

**Proposition 6.** *The present data indicate that lockdown resulted in an increase in overall food consumption and consumption of junk food on many occasions, which led to unbalanced body weight and disorders.*

### 3.7. Food Consumption and Mood and Stress

According to Singh and Mood [73], overeating and obesity are the results of changing food choice and intake due to changing mood and emotional eating, where these psychological "pathways" influence not only food choice but also the quantity and frequency of meals. Individuals are unable to perceive their state of hunger and satiety and show preference for palatable 'comfort foods' as a means of relieving their negative emotions. Furthermore, sweets, chocolate, cakes, and biscuits are more frequently consumed under stressful conditions, especially high-fat and energy-dense foods are chosen by people during stressful life events [74–76]. Food consumption has also been considered as a strategy for coping with stressful situations [77]. Indeed, it has been observed that anxiety and depressive symptoms lead to poorer food choices [78]. Moreover, it appears that individuals who experience periods of stress over-consume foods that they would usually avoid and this consumption makes them feel better [79].

The COVID-19 pandemic has drastically influenced consumers' consumption and food choice behavior in relation to depression, stress, and anxiety [80]. The huge disruption in social interactions, contacts, and daily lives of consumers, increased unemployment, and business disruption have caused increased loneliness, fear of illness, financial stress, food insecurity, and insecurity about the future and livelihood [81]. Even families were affected and put under a lot of pressure when parents educated their children at home during lockdown and fed their children more often than usual. In addition, stress and negative emotions led to emotional eating, i.e., eating as a result of negative emotions without any real evidence of hunger [82]. Larger amounts of foods, such as sweets, fatty foods, and salty snacks, were reported to be consumed during the pandemic for emotional reasons. The negative impact on normal food consumption was fully mediated by emotional distress during the pandemic [63]. The role of emotional distress as a key mechanism to explain coping behaviors, such as comfort food consumption, which were adopted as a consequence of the economic, interpersonal, and health impact of the pandemic, was also revealed. In another study comparing behaviors among different sexes, women consumed larger amounts of high-sugar and high-calorie foods during COVID-19 for reasons of emotion, leading to greater weight gain compared to men [83].

**Proposition 7.** *The present data show an increase in food consumption during COVID and beyond due to a deterioration in mental health, such as depression, stress, and anxiety, which has continued globally to date.*

### 3.8. Food Consumption and Familiarity

Familiarity is the cognitive ability to apply knowledge acquired via experience to objects or stimuli [84]. Regarding everyday food choices, familiarity is important as it relates to the close relationship between a person's eating habits during childhood, adolescence, and adulthood. Moreover, familiarity is due to previous personal experiences and tends to be linked to tradition, as many consumers prefer to choose foods that are familiar to them [85]. Still, familiarity is significant among those who have a relatively strong focus on prevention, who tend to be in good health, responsible, and safety oriented, and who consider their food a factor to cope with their stress and bad mood. However, consumers are demanding healthier food and, to meet this demand, technological solutions (such as reduced-fat and functional foods) have been implemented, together with a return to naturalness and purity of food [86].

During COVID-19, familiarity helped consumers address anxiety and mood when choosing foods, sustaining a healthy diet through adherence to personal nutrition by selecting foods they know and trust [87]. The lockdown led consumers to become familiar with the internet and other technologies to order the foods that they knew and consumed, demonstrating that familiarity depends also on personal past experiences [20]. Familiarity, convenience, and price became more important in Croatian adults [46], but less so in French [13] and British adults (except price more) [47], and remained the same for Polish adolescents [33].

**Proposition 8.** *Overall, the present data prove that familiarity is a motive that has helped consumers cope with the pandemic as far as food choice is concerned and will also help them with online purchase, which has drastically increased to date.*

### 3.9. Food Consumption and Price

One of the most important determinants of consumption patterns and living standards is food prices. In particular, high prices can have a significant negative impact on nutritional status and health, especially among poor people [88]. Green et al. [89] showed that price changes in the global food market have a greater impact on low-income countries and the poorest households within these countries. In addition, interferences in the purchase and consumption of goods due to self-control problems or temporally inconsistent preferences of consumers, who derive direct satisfaction from food consumption itself, influence future health costs [90]. Low-income consumers have lower fruit and vegetable consumption and reduced intake of nutrients (e.g., calcium and vitamins) [91].

During the pandemic and periods of lockdowns, the global restriction on 'normal' economic production affected all aspects of life, including decisions regarding food purchase, leading to an unstable food chain [92]. The consequence of this situation was that prices increased, and many consumers were unable to buy enough essentials and foods. In addition, jobs were lost and consumers cooked more at home in order to reduce the cost of their daily meals [93]. The crisis revealed the compromises that households were willing to make in times of shortages [94]. What led many households to consume less and make more careful food choices was the increase in food prices combined with any loss of disposable household income [95]. The International Food Security Assessment model that estimates changes in food consumption and food gaps in developing countries uses gross domestic product (GDP) and food price changes as the main inputs for its predictions. The results show that the lockdowns led to a decrease in global GDP of 7.2 per cent, and an increase in grain prices of 9 per cent. These changes led to an increase in the number of food-insecure people in 2020, totaling 211 million (a 27.8 per cent increase) [95]. In the post-COVID-19 era, price promotion policies are a common practice worldwide in order to control the price increase; however, this results in food waste by encouraging over-purchase, according to half of the reported studies [96]. In contrast, the other half of the studies prove that consumers buying price-promoted foods show average or even lower levels of household food waste [96]. Low-income households, due to the pandemic, may not have the financial resources to engage in any stockpiling behavior compared to higher-income households. In addition, important price shocks negatively affect household consumption patterns of low-income groups [97].

**Proposition 9.** *The findings indicate that price remains a major food choice motive during COVID and beyond, with low-income groups being more affected by the foreseeable economic global recession. Therefore, it may be the most important selection criterion, among the 10 presented motives, for food choice in the new era.*

*3.10. Food Consumption and Shopping Frequency Behavior*

Consumers' low income leads them to shop less and at longer intervals, which affects the sustainability and shelf life of perishable foods, such as vegetables and fresh fruits [98,99]. There are also those consumers who either do not have access to a supermarket or grocery store [100], or do not have transport to make it easier for them to buy foods from the store they want themselves after comparing prices [101].

During the pandemic, consumers were forced to adapt their behaviors, including their food purchasing habits and their preferences, to the new routine. Schools were closed, homework was imposed, and except in certain specific occupational areas (e.g., working in hospital, in grocery stores), leaving home was only allowed under restricted conditions following the completion of special certificates [82]. Consumers' eating habits were significantly affected by perceived risk and precautions related to the COVID-19 virus, resulting in major changes in consumers' shopping behavior [102].

Children's eating behavior and feeding practices changed through changes in their appetite, enjoyment of food, responsiveness to food, and emotional overeating, as well as frequency of snacking between meals which was enhanced by parents who became more indulgent [103]. Furthermore, as demonstrated by Moynihan et al. [104] an increased intake in energy was connected with high levels of boredom. The COVID-19 pandemic altered the content of meals for a proportion of consumers [105] as well as the frequency of their consumption [82], leading to an increase in demand for food [106]. As a consequence, the food industry and food production chain have been adapted to the new situation and consumers' demands [106].

In addition, online shopping had become the first choice during the home restriction, and the demand for online food shopping increased significantly for both food and wine [107,108]. As the COVID-19 pandemic had completely disrupted food production and food supply chains due to unavailable labor, lack of transport, and closure of various food services, such as restaurants [109], it is inevitable that a major change has been observed in the way households buy, prepare, and consume food [110]. A significant shift to traditional foods has also been studied with similar results [111,112]. Consumers must learn how to use e-commerce, ICT technologies, and credit card payment in order to facilitate food shopping and avoid crowding. This also demands the presence of an online mechanism for protection of personal and transactional data to avoid online attacks [113].

**Proposition 10.** *Overall, it appears that, due to lockdown, shopping frequency decreased with a parallel increase in online purchase and delivery, a tendency which has continued to date.*

A main limitation of this scoping review is the short-term nature of the studies included (2020/2021/2022) and, therefore, there is limited literature available based upon which a discussion of the findings was presented. However, the review type chosen was viewed as the most appropriate for the current topic.

**Table 1.** One hundred and seven papers in this review divided by theme and sub-theme.

| Theme for Discussion on Food Consumption | Sub-Sections | Paper Reference Numbers |
|---|---|---|
| | Chronic food-related diseases | [15–18] |
| (1) Health | Health behavior | [20–26] |
| | Health attitudes | [7,19–21] |
| | Physical activity | [3,27–30] |

**Table 1.** *Cont.*

| Theme for Discussion on Food Consumption | Sub-Sections | Paper Reference Numbers |
|---|---|---|
| (2) Convenience | Ready meals | [31–33] |
| | Fast food | [34–36] |
| (3) Sensory Appeal | Taste, smell, texture, and appearance | [13,37–42] |
| (4) Nutritional Quality | Diet quality | [6,9–11,43–45] |
| | Better quality | [46–51] |
| (5) Ethical Concern | Environmental aspects | [52,53] |
| | Food waste | [54,55] |
| (6) Weight Control | Weight loss | [56–58] |
| | Obesity | [59–64] |
| | Balanced diet | [65–67] |
| (7) Mood and Stress | Emotional eating | [1,8,68–74] |
| | Depression and stress | [75–78] |
| (8) Familiarity | Cognitive ability | [2,4,79,80] |
| | Trust | [81,82] |
| (9) Price | Low-income consumers | [83–86] |
| | Food compromises | [87–90] |
| | Price-promoted foods | [91,92] |
| (10) Shopping Frequency | Food shopping behavior | [93–97] |
| | Food shopping frequency | [5,98–101] |
| | Online shopping | [102–105,108] |
| | Traditional foods | [106,107] |

## 4. Conclusions

FCMs, based on the reviewed data, have been affected by the COVID-19 pandemic in certain ways, which are affecting consumers' choice beyond the pandemic in the new economic era. Of the ten motives presented in this review, food price seems to be the most important motive for consumers during and post-COVID-19 periods and will be more significant if a global recession is under way. Decreased physical activity, as well as increased mental disorders related to stress and anxiety, had a negative effect on health, weight control, and mood and stress motives, along with increased food consumption, especially junk food. On the other hand, the lockdowns had a positive impact on other motives, such as convenience and familiarity, and a negative impact on shopping frequency motive, with increased online and takeaway purchase of foods. Food waste and its effects on the environment seem to be the parameters concerning motives such as ethics and nutritional status. Nutritional quality and sensory appeal are two consumer motives which have not been affected significantly by the pandemic.

Raising consumer awareness of the incentives for food choice is of paramount importance in the new post-COVID-19 era where the world is changing drastically. Motivations, such as sensory appeal, taste, and food presentation, can act as a one-way street for emotional eating in the new era since they remain as important as before the pandemic. Family members, feeling secure and having high feelings of self-esteem when preparing a pleasant dish, bring the family together and create a context of daily stability, where people know what to expect with familiar dishes and can assess whether hunger and nutritional needs will be met. In addition, price as an incentive for food choice becomes important due to uncertainty about work and economic future and a sense of impending precariousness experienced by affected consumers. Still, changes in food choice incentives have led to an increased awareness of food choices, with the aim of sustaining health through quality food, ensuring healthy eating behaviors and attitudes toward food waste, and meeting environmental footprint and ethical concerns. In addition, online shopping is a rising choice for consumers, a habit that has emerged due to home confinement and the demand for online shopping has increased significantly.

Finally, could the pandemic be assumed to be the threshold at which the usual rules behind food choices are suspended? The definite answer will be known in three to five years when the new worldwide economic and social condition will be clear and stable, and an adequate number of studies will be published by then. In this review, we present the studies that have reported to date, with the above conclusions derived from their results so far.

It would be more workable if consumers are encouraged to explore healthier food options, such as fresh fruits, vegetables, and whole foods. In addition, when purchasing foods, they should be informed about the foods and their beneficial properties (e.g., vegetables, fruits, and organic wine) and reflect more on the importance of certain foods to themselves and their families through their cultural identity. Online food shopping can surely contribute to a reduction in food waste thanks to the elimination of frenzied shopping routines at supermarkets or groceries and can open up space to new fields of study. On the other hand, defining a "new" index of consumer satisfaction can alter the sale strategies of retail managers and entrepreneurs.

The present review, which is based on the findings reported so far, offers 10 specific propositions for each one of the 10 main food choice motives examined, which can be used as a practical and theoretical basis for the development of a "new" FCM index that can be used by retail managers, food companies, and any other parties involved in the agri-food chain.

- Regarding the health motive, physical activity should be re-emphasized to return to normal conditions and consumers should be directed to healthy, rather than junk, foods after the pandemic.
- Regarding the convenience motive, emphasis should be given to the purchase of takeaway foods and ready-to-go meals since they are going to be more and more in use by consumers in the new era.
- Regarding the sensory appeal motive, no significant changes are predicted for consumers in the future.
- Regarding the nutritional quality motive, consumers choosing their foods in the future will place more emphasis on their nutritional indications.
- Regarding the ethical concern motive, consumers will consider food waste and environmental impacts more when choosing their foods in the future.
- Regarding the weight control motive, an emphasis should be given to a balanced body weight with proper food selection for a healthy life, which can result in less disorders, after the pandemic
- Regarding the mood and stress motive, a return to normal mental conditions, following the end of lockdowns, should decrease the unusual and dangerous increase in food consumption recorded during the pandemic.
- Regarding the familiarity motive, consumers are going to use it as a major criterion to purchase food online in the future, and, therefore, it should be considered more carefully in the future.
- Regarding the price motive, consumers are going to depend heavily on it for their selection and purchase of foods in the future, thus becoming their priority motive.
- Regarding the shopping frequency behavior motive, consumers will avoid shopping in person in the future and turn more and more to online purchase and delivery of foods.

Despite the abovementioned conclusions, more studies are needed in the years to come to ensure their validity since only studies from a three-year period are recorded so far (2020–2022).

Furthermore, studies with longer time periods beyond the pandemic should be performed to ensure the long-term validity of the conclusions.

Finally, studies on consumer segments, such as young adults, older people, and children, will be very important to verify these findings and their applications to food choice motives.

**Author Contributions:** Conceptualization and methodology, D.S. and Z.C.K.; writing—original draft preparation, Z.C.K.; supervision and editing, D.S. All authors have read and agreed to the published version of the manuscript.

**Funding:** This research received no external funding.

**Institutional Review Board Statement:** Not applicable.

**Informed Consent Statement:** Not applicable.

**Data Availability Statement:** Data is contained within the article.

**Conflicts of Interest:** The authors declare no conflict of interest.

## References


1. Radwan, H.; Al Kitbi, M.; Al Hilali, M.; Abbas, N.; Hamadeh, R.; Saif, E.R.; Naja, F. Diet and lifestyle changes during COVID-19 lockdown in the United Arab Emirates: Results of a cross-sectional study. *Nutrients* **2020**, *12*, 3314.
2. Haleem, A.; Javaid, M.; Vaishya, R. Effects of COVID-19 pandemic in daily life. *Curr. Med. Res. Pract.* **2020**, *10*, 78–79. [CrossRef] [PubMed]
3. Wunsch, K.; Kienberger, K.; Niessner, C. Changes in Physical Activity Patterns Due to the COVID-19 Pandemic: A Systematic Review and Meta-Analysis. *Int. J. Environ. Res. Public Health* **2022**, *19*, 2250. [CrossRef] [PubMed]
4. Stockwell, S.; Trott, M.; Tully, M.; Shin, J.; Barnett, Y.; Butler, L.; McDermott, D.; Schuch, F.; Smith, L. Changes in physical activity and sedentary behaviours from before to during the COVID-19 pandemic lockdown: A systematic review. *BMJ Open Sport Exerc. Med.* **2021**, *7*, e000960. [CrossRef]
5. Bennett, G.; Young, E.; Butler, I.; Coe, S. The Impact of Lockdown During the COVID-19 Outbreak on Dietary Habits in Various Population Groups: A Scoping Review. *Front. Nutr.* **2021**, *8*, 626432. [CrossRef]
6. Januszewska, R.; Pieniak, Z.; Verbeke, W. Food choice questionnaire revisited in four countries. Does it still measure the same? *Appetite* **2011**, *57*, 94–98. [CrossRef]
7. Salari, N.; Hosseinian-Far, A.; Jalali, R.; Vaisi-Raygani, A.; Rasoulpoor, S.; Mohammadi, M.; Rasoulpoor, S.; Khaledi-Paveh, B. Prevalence of stress, anxiety, depression among the general population during the COVID-19 pandemic: A systematic review and meta-analysis. *Glob. Health* **2020**, *16*, 57. [CrossRef]
8. Aoun, C.; Nassar, L.; Soumi, S.; El Osta, N.; Papazian, T.; Rabbaa Khabbaz, L. The Cognitive, Behavioral, and Emotional Aspects of Eating Habits and Association With Impulsivity, Chronotype, Anxiety, and Depression: A Cross-Sectional Study. *Front. Behav. Neurosci.* **2019**, *13*, 204. [CrossRef]
9. Deschasaux-Tanguy, M.; Druesne-Pecollo, N.; Esseddik, Y.; De Edelenyi, F.S.; Allès, B.; Andreeva, V.A.; Baudry, J.; Charreire, H.; Deschamps, V.; Egnell, M.; et al. Diet and physical activity during the coronavirus disease 2019 (COVID-19) lockdown (March-May 2020): Results from the French NutriNet-Santé cohort study. *Am. J. Clin. Nutr.* **2021**, *113*, 924–938. [CrossRef]
10. Lamarche, B.; Brassard, D.; Lapointe, A.; Laramée, C.; Kearney, M.; Côté, M.; Bélanger-Gravel, A.; Desroches, S.; Lemieux, S.; Plante, C. Changes in diet quality and food security among adults during the COVID-19-related early lockdown: Results from NutriQuébec. *Am. J. Clin. Nutr.* **2021**, *113*, 984–992. [CrossRef]
11. Alhusseini, N.; Alqahtani, A. COVID-19 pandemic's impact on eating habits in Saudi Arabia. *J. Public Health Res.* **2020**, *9*, 354–360. [CrossRef] [PubMed]
12. Kushwah, S.; Dhir, A.; Sagar, M.; Gupta, B. Determinants of organic food consumption. A systematic literature review on motives and barriers. *Appetite* **2019**, *143*, 104402. [CrossRef] [PubMed]
13. Marty, L.; de Lauzon-Guillain, B.; Labesse, M.; Nicklaus, S. Food choice motives and the nutritional quality of diet during the COVID-19 lockdown in France. *Appetite* **2021**, *157*, 105005. [CrossRef] [PubMed]
14. Arksey, H.; O'Malley, L. Scoping studies: Towards a methodological framework. *Int. J. Soc. Res. Methodol. Theory Pract.* **2005**, *8*, 19–32. [CrossRef]
15. Ehrich, K.; Freeman, G.K.; Richards, S.C.; Robinson, I.C.; Shepperd, S. How to do a scoping exercise: Continuity of care. *Res. Policy Plan.* **2002**, *20*, 25–29.
16. Tricco, A.C.; Lillie, E.; Zarin, W.; O'Brien, K.; Colquhoun, H.; Kastner, M.; Levac, D.; Ng, C.; Sharpe, J.P.; Wilson, K.; et al. A scoping review on the conduct and reporting of scoping reviews. *BMC Med. Res. Methodol.* **2016**, *16*, 15. [CrossRef]

17. Pham, M.T.; Rajić, A.; Greig, J.D.; Sargeant, J.M.; Papadopoulos, A.; Mcewen, S.A. A scoping review of scoping reviews: Advancing the approach and enhancing the consistency. *Res. Synth. Methods* **2014**, *5*, 371–385. [CrossRef]
18. Anderson, S.; Allen, P.; Peckham, S.; Goodwin, N. Asking the right questions: Scoping studies in the commissioning of research on the organisation and delivery of health services. *Health Res. Policy Syst.* **2008**, *6*, 7. [CrossRef]
19. Peters, M.D.J.; Marnie, C.; Tricco, A.C.; Pollock, D.; Munn, Z.; Alexander, L.; McInerney, P.; Godfrey, C.M.; Khalil, H. Updated methodological guidance for the conduct of scoping reviews. *JBI Evid. Synth.* **2020**, *18*, 2119–2126. [CrossRef]
20. Cavallo, C.; Sacchi, G.; Carfora, V. Resilience effects in food consumption behaviour at the time of COVID-19: Perspectives from Italy. *Heliyon* **2020**, *6*, e05676. [CrossRef]
21. Carlson, K.A.; Gould, B.W. The Role of Health Knowledge in Determining Dietary Fat Intake. *Rev. Agric. Econ.* **1994**, *16*, 373. [CrossRef]
22. Etilé, F. Food consumption and health. In *Oxford Handbook on the Economics of Food Consumption and Policy*; INRA-ALISS–UR1303; France and Paris School of Economics: Paris, France, 2010.
23. Kushi, L.H.; Doyle, C.; Mccullough, M.; Rock, C.L.; Demark-Wahnefried, W.; Bandera, E.V.; Gapstur, S.; Alpa, P.V.; Andrews, K.; Gansler, T. American Cancer Society Guidelines on Nutrition and Physical Activity for Cancer Prevention Reducing the Risk of Cancer with Healthy Food Choices and Physical Activity. *CA Cancer J. Clin.* **2012**, *62*, 30–67. [CrossRef] [PubMed]
24. Li, T.; Yu, L.; Yang, Z.; Shen, P.; Lin, H.; Shui, L.; Tang, M.; Jin, M.; Chen, K.; Wang, J. Associations of Diet Quality and Heavy Metals with Obesity in Adults: A Cross-Sectional Study from National Health and Nutrition Examination Survey (NHANES). *Nutrients* **2022**, *14*, 4038. [CrossRef] [PubMed]
25. Zajacova, A.; Jehn, A.; Stackhouse, M.; Denice, P.; Ramos, H. Changes in health behaviours during early COVID-19 and socio-demographic disparities: A cross-sectional analysis. *Can. J. Public Health* **2020**, *111*, 953–962. [CrossRef] [PubMed]
26. Urhan, M.; Elif, A.O. Nutritional and health behaviour predictors of the weight gain during the COVID-19 pandemic. *Eur. J. Nutr.* **2022**, *61*, 2993–3002. [CrossRef] [PubMed]
27. Drieskens, S.; Berger, N.; Vandevijvere, S.; Gisle, L.; Braekman, E.; Charafeddine, R.; De Ridder, K.; Demarest, S. Short-term impact of the COVID-19 confinement measures on health behaviours and weight gain among adults in Belgium. *Arch. Public Health* **2021**, *79*, 1–10. [CrossRef]
28. Martínez-de-Quel, Ó.; Suárez-Iglesias, D.; López-Flores, M.; Pérez, C.A. Physical activity, dietary habits and sleep quality before and during COVID-19 lockdown: A longitudinal study. *Appetite* **2021**, *158*, 105019. [CrossRef]
29. Robinson, E.; Boyland, E.; Chisholm, A.; Harrold, J.; Maloney, N.G.; Marty, L.; Mead, B.R.; Noonan, R.; Hardman, C.A. Obesity, eating behavior and physical activity during COVID-19 lockdown: A study of UK adults. *Appetite* **2021**, *156*, 104853. [CrossRef]
30. Kriaucioniene, V.; Bagdonaviciene, L.; Rodríguez-Pérez, C.; Petkeviciene, J. Associations between changes in health behaviours and body weight during the COVID-19 quarantine in lithuania: The lithuanian covidiet study. *Nutrients* **2020**, *12*, 3119. [CrossRef]
31. Huber, B.C.; Steffen, J.; Schlichtiger, J.; Brunner, S. Altered nutrition behavior during COVID-19 pandemic lockdown in young adults. *Eur. J. Nutr.* **2021**, *60*, 2593–2602. [CrossRef]
32. Poelman, M.P.; Gillebaart, M.; Schlinkert, C.; Dijkstra, S.C.; Derksen, E.; Mensink, F.; Hermans, R.C.J.; Aardening, P.; de Ridder, D.; de Vet, E. Eating behavior and food purchases during the COVID-19 lockdown: A cross-sectional study among adults in the Netherlands. *Appetite* **2021**, *157*, 105002. [CrossRef] [PubMed]
33. Głąbska, D.; Skolmowska, D.; Guzek, D. Food preferences and food choice determinants in a polish adolescents' COVID-19 experience (Place-19) study. *Nutrients* **2021**, *13*, 2491. [CrossRef]
34. Tolhurst, T.; Princehorn, E.; Loxton, D.; Mishra, G.; Mate, K.; Byles, J. Changes in the food and drink consumption patterns of Australian women during the COVID-19 pandemic. *Aust. New Zealand J. Public Health* **2022**, *46*, 704–709. [CrossRef] [PubMed]
35. Jaeger, S.R.; Vidal, L.; Ares, G.; Chheang, S.L.; Spinelli, S. Healthier eating: COVID-19 disruption as a catalyst for positive change. *Food Qual. Prefer.* **2021**, *92*, 104220. [CrossRef] [PubMed]
36. Liu, C.; Chen, J. Consuming takeaway food: Convenience, waste, and Chinese young people's urban lifestyle. *J. Consum. Cult.* **2021**, *21*, 848–866. [CrossRef]
37. Warde, A. Convenience food: Space and timing. *Br. Food J.* **1999**, *101*, 518–527. [CrossRef]
38. Botonaki, A.; Mattas, K. Revealing the values behind convenience food consumption. *Appetite* **2010**, *55*, 629–638. [CrossRef]
39. World Health Organization. Coronavirus (COVID-19) Dashboard. Available online: https://covid19.who.int (accessed on 20 December 2022).
40. Rahman, N.; Ishitsuka, K.; Piedvache, A.; Tanaka, H.; Murayama, N.; Morisaki, N. Convenience Food Options and Adequacy of Nutrient Intake among School Children during the COVID-19 Pandemic. *Nutrients* **2022**, *14*, 630. [CrossRef]
41. Ko, Y.H.; Son, J.H.; Kim, G.J. An exploratory study of changes in consumer dining out behavior before and during COVID-19. *J. Foodserv. Bus. Res.* **2022**, 1–19. [CrossRef]
42. Boesveldt, S.; Bobowski, N.; McCrickerd, K.; Maître, I.; Sulmont-Rossé, C.; Forde, C.G. The changing role of the senses in food choice and food intake across the lifespan. *Food Qual. Prefer.* **2018**, *68*, 80–89. [CrossRef]
43. Laaksonen, O.; Ma, X.; Pasanen, E.; Zhou, P.; Yang, B.; Linderborg, K.M. Sensory characteristics contributing to pleasantness of oat product concepts by finnish and Chinese consumers. *Foods* **2020**, *9*, 1234. [CrossRef] [PubMed]
44. Głabska, D.; Skolmowska, D.; Guzek, D. Choice Determinants of Secondary School Students. *Nutrients* **2020**, *12*, 2640. [PubMed]
45. Tian, X.; Zhou, Y.; Wang, H. The Impact of COVID-19 on Food Consumption and Dietary Quality of Rural Households in China. *Foods* **2022**, *11*, 510. [CrossRef] [PubMed]

46. Sorić, T.; Brodić, I.; Mertens, E.; Sagastume, D.; Dolanc, I.; Jonjić, A.; Delale, E.A.; Mavar, M.; Missoni, S.; Peñalvo, J.L.; et al. Evaluation of the food choice motives before and during the COVID-19 pandemic: A cross-sectional study of 1232 adults from croatia. *Nutrients* **2021**, *13*, 3165. [CrossRef] [PubMed]
47. Snuggs, S.; Mcgregor, S. Food & meal decision making in lockdown: How and who has COVID-19 affected? *Food Qual. Prefer.* **2021**, *89*, 104145.
48. Gorgulho, B.M.; Pot, G.K.; Sarti, F.M.; Marchioni, D.M. Indices for the assessment of nutritional quality of meals: A systematic review. *Br. J. Nutr.* **2016**, *115*, 2017–2024. [CrossRef]
49. Murakami, K. Nutritional quality of meals and snacks assessed by the Food Standards Agency nutrient profiling system in relation to overall diet quality, body mass index, and waist circumference in British adults. *Nutr. J.* **2017**, *16*, 57. [CrossRef]
50. Lairon, D. Nutritional quality and safety of organic food. A review. *Sustain. Agric.* **2010**, *30*, 33–41. [CrossRef]
51. Carducci, B.; Keats, E.C.; Ruel, M.; Haddad, L.; Osendarp, S.J.M.; Bhutta, Z.A. Food systems, diets and nutrition in the wake of COVID-19. *Nat. Food* **2021**, *2*, 68–70. [CrossRef]
52. Morres, I.D.; Galanis, E.; Hatzigeorgiadis, A.; Androutsos, O.; Theodorakis, Y. Physical activity, sedentariness, eating behaviour and well-being during a COVID-19 lockdown period in greek adolescents. *Nutrients* **2021**, *13*, 1449. [CrossRef]
53. Ruiz-Roso, M.B.; Padilha, P.d.C.; Mantilla-Escalante, D.C.; Ulloa, N.; Brun, P.; Acevedo-Correa, D.; Peres, W.A.F.; Martorell, M.; Aires, M.T.; Cardoso, L.d.O.; et al. COVID-19 Confinement and Changes of Adolescent's Dietary Trends in Italy, Spain, Chile, Colombia and Brazil. *Nutrients* **2020**, *12*, 1807. [CrossRef] [PubMed]
54. Grashuis, J.; Skevas, T.; Segovia, M.S. Grocery shopping preferences during the COVID-19 pandemic. *Sustainability* **2020**, *12*, 5369. [CrossRef]
55. Ellison, B.; McFadden, B.; Rickard, B.J.; Wilson, N.L.W. Examining Food Purchase Behavior and Food Values during the COVID-19 Pandemic. *Appl. Econ. Perspect. Policy* **2021**, *43*, 58–72. [CrossRef]
56. Huang, K.M.; Sant'Anna, A.C.; Etienne, X. How did COVID-19 impact US household foods? an analysis six-month in. *PLoS ONE* **2021**, *16*, e0256921. [CrossRef] [PubMed]
57. Burlea-Schiopoiu, A.; Ogarca, R.F.; Barbu, C.M.; Craciun, L.; Baloi, I.C.; Mihai, L.S. The impact of COVID-19 pandemic on food waste behaviour of young people. *J. Clean. Prod.* **2021**, *294*, 126333. [CrossRef]
58. Kotler, P. The Consumer in the Age of Coronavirus. *J. Creat. Value* **2020**, *6*, 12–15. [CrossRef]
59. Carolan, M. Practicing social change during COVID-19:Ethical food consumption and activism pre- and post-outbreak. *Appetite* **2021**, *163*, 105206. [CrossRef]
60. Roe, B.E.; Bender, K.; Qi, D. The Impact of COVID-19 on Consumer Food Waste. *Appl. Econ. Perspect. Policy* **2021**, *43*, 401–411. [CrossRef]
61. Glanz, K.; Basil, M.; Maibach, E.; Goldberg, J.; Snyder, D. Why Americans eat what they do: Taste, nutrition, cost, convenience, and weight control as influences on food consumption. *J. Am. Diet Assoc.* **1998**, *98*, 18–26. [CrossRef]
62. Svetkey, L.P.; Stevens, V.J.; Brantley, P.J.; Appel, L.J.; Hollis, J.F.; Loria, C.M.; Vollmer, W.M.; Gullion, C.M.; Funk, K.; Smith, P.; et al. Comparison of strategies for sustaining weight loss: The weight loss maintenance randomized controlled trial. *JAMA* **2008**, *299*, 1139–1148. [CrossRef]
63. Salazar-Fernández, C.; Palet, D.; Haeger, P.A.; Mella, F.R. The perceived impact of COVID-19 on comfort food consumption over time: The mediational role of emotional distress. *Nutrients* **2021**, *13*, 1910. [CrossRef] [PubMed]
64. Bakaloudi, D.R.; Barazzoni, R.; Bischoff, S.C.; Breda, J.; Wickramasinghe, K.; Chourdakis, M. Impact of the first COVID-19 lockdown on body weight: A combined systematic review and a meta-analysis. *Clin. Nutr.* **2022**, *41*, 3046–3054. [CrossRef] [PubMed]
65. Martinez-Ferran, M.; de la Guía-Galipienso, F.; Sanchis-Gomar, F.; Pareja-Galeano, H. Metabolic impacts of confinement during the COVID-19 pandemic due to modified diet and physical activity habits. *Nutrients* **2020**, *12*, 1549. [CrossRef] [PubMed]
66. Kalligeros, M.; Shehadeh, F.; Mylona, E.K.; Benitez, G.; Beckwith, C.G.; Chan, P.A.; Mylonakis, E. Association of Obesity with Disease Severity Among Patients with Coronavirus Disease 2019. *Obesity* **2020**, *28*, 1200–1204. [CrossRef]
67. Wiklund, P. The role of physical activity and exercise in obesity and weight management: Time for critical appraisal. *J. Sport Health Sci.* **2016**, *5*, 151–154. [CrossRef]
68. Lighter, J.; Phillips, M.; Hochman, S.; Sterling, S.; Johnson, D.; Francois, F.; Stachel, A. Obesity in Patients Younger Than 60 Years Is a Risk Factor for COVID-19 Hospital Admission. *Clin. Infect. Dis.* **2020**, *71*, 896–897. [CrossRef]
69. Rogero, M.M.; Calder, P.C. Obesity, inflammation, toll-like receptor 4 and fatty acids. *Nutrients* **2018**, *10*, 432. [CrossRef]
70. Kuk, J.L.; Christensen, R.A.G.; Kamran Samani, E.; Wharton, S. Predictors of Weight Loss and Weight Gain in Weight Management Patients during the COVID-19 Pandemic. *J. Obes.* **2021**, *2021*, 48811430. [CrossRef]
71. Smaira, F.I.; Mazzolani, B.C.; Esteves, G.P.; André, H.C.S.; Amarante, M.C.; Castanho, D.F.; Campos, K.J.d.; Benatti, F.B.; Pinto, A.J.; Roschel, H.; et al. Poor Eating Habits and Selected Determinants of Food Choice Were Associated With Ultraprocessed Food Consumption in Brazilian Women During the COVID-19 Pandemic. *Front. Nutr.* **2021**, *8*, 672372. [CrossRef]
72. Marentes-Castillo, M.; Castillo, I.; Tomás, I.; Zamarripa, J.; Alvarez, O. Understanding the antecedents of healthy and unhealthy weight control behaviours: Grit, motivation and self-control. *Public Health Nutr.* **2022**, *25*, 1483–1491. [CrossRef]
73. Singh, M. Mood, food and obesity. *Front. Psychol.* **2014**, *5*, 925. [CrossRef] [PubMed]
74. Christensen, L.; Pettijohn, L. Mood and carbohydrate cravings. *Appetite* **2001**, *36*, 137–145. [CrossRef] [PubMed]
75. Wurtman, R.J.; Wurtman, J.J. Carbohydrates and depression. *Sci. Am.* **1989**, *260*, 68–75. [CrossRef] [PubMed]

76. Mikolajczyk, R.T.; El Ansari, W.; Maxwell, A.E. Food consumption frequency and perceived stress and depressive symptoms among students in three European countries. *Nutr. J.* **2009**, *8*, 31. [CrossRef]

77. Jenkins, S.; Horner, S.D. Barriers that influence eating behaviors in adolescents. *J. Pediatr. Nurs.* **2005**, *20*, 258–267. [CrossRef]

78. Liu, C.; Xie, B.; Chou, C.P.; Koprowski, C.; Zhou, D.; Palmer, P.; Sun, P.; Guo, Q.; Duan, L.; Sun, X.; et al. Perceived stress, depression and food consumption frequency in the college students of China seven cities. *Physiol. Behav.* **2007**, *92*, 748–754. [CrossRef]

79. Zellner, D.A.; Loaiza, S.; Gonzalez, Z.; Pita, J.; Morales, J.; Pecora, D.; Wolf, A. Food selection changes under stress. *Physiol. Behav.* **2006**, *87*, 789–793. [CrossRef]

80. Leeds, J.; Keith, R.; Woloshynowych, M. Food and Mood: Exploring the determinants of food choices and the effects of food consumption on mood among women in Inner London. *World Nutr.* **2020**, *11*, 68–96. [CrossRef]

81. Ricci, F.; Izzicupo, P.; Moscucci, F.; Sciomer, S.; Maffei, S.; Di Baldassarre, A.; Mattioli, A.V.; Gallina, S. Recommendations for Physical Inactivity and Sedentary Behavior During the Coronavirus Disease (COVID-19) Pandemic. *Front. Public Health* **2020**, *8*, 199. [CrossRef]

82. Philippe, K.; Chabanet, C.; Issanchou, S.; Monnery-Patris, S. Child eating behaviors, parental feeding practices and food shopping motivations during the COVID-19 lockdown in France: (How) did they change? *Appetite* **2021**, *161*, 105132. [CrossRef]

83. Ravichandran, S.; Bhatt, R.R.; Pandit, B.; Osadchiy, V.; Alaverdyan, A.; Vora, P.; Stains, J.; Naliboff, B.; Mayer, E.A.; Gupta, A. Alterations in reward network functional connectivity are associated with increased food addiction in obese individuals. *Sci. Rep.* **2021**, *11*, 3386. [CrossRef] [PubMed]

84. Reder, L.M.; Ritter, F.E. What determines initial feeling of knowing? Familiarity with question terms, not with the answer. *J. Exp. Psychol. Learn. Mem. Cogn.* **1992**, *18*, 435–451. [CrossRef]

85. Salvy, S.J.; Vartanian, L.R.; Coelho, J.S.; Jarrin, D.; Pliner, P.P. The role of familiarity on modeling of eating and food consumption in children. *Appetite* **2008**, *50*, 514–518. [CrossRef] [PubMed]

86. Verneau, F.; Caracciolo, F.; Coppola, A.; Lombardi, P. Consumer fears and familiarity of processed food. The value of information provided by the FTNS. *Appetite* **2014**, *73*, 140–146. [CrossRef]

87. Mertens, E.; Sagastume, D.; Sorić, T.; Brodić, I.; Dolanc, I.; Jonjić, A.; Delale, E.A.; Mavar, M.; Missoni, S.; Čoklo, M.; et al. Food Choice Motives and COVID-19 in Belgium. *Foods* **2022**, *11*, 842. [CrossRef]

88. Compton, J.; Wiggins, S.; Keats, S. Impact of the global food crisis on the poor: What is the evidence. *Overseas Dev. Inst.* **2010**, *44*, 1–99. Available online: https://cdn.odi.org/media/documents/6371 (accessed on 16 December 2022).

89. Green, R.; Cornelsen, L.; Dangour, A.D.; Honorary, R.T.; Shankar, B.; Mazzocchi, M.; Smith, R.D. The effect of rising food prices on food consumption:systematic review with meta-regression. *BMJ* **2013**, *347*, f3703. [CrossRef]

90. Schroeter, C.; Lusk, J.; Tyner, W. Determining the impact of food price and income changes on body weight. *J. Health Econ.* **2008**, *27*, 45–68. [CrossRef]

91. Rose, D.; Richards, R. Food store access and household fruit and vegetable use among participants in the US Food Stamp Program. *Public Health Nutr.* **2004**, *7*, 1081–1088. [CrossRef]

92. Laborde, D.; Martin, W.; Swinnen, J.; Rob, V. COVID-19 risks to global food security. Economic fallout and food supply chain disruptions require attention from policymakers. *Science* **2020**, *369*, 500–502. [CrossRef]

93. Sarda, B.; Delamaire, C.; Serry, A.J.; Ducrot, P. Changes in home cooking and culinary practices among the French population during the COVID-19 lockdown. *Appetite* **2022**, *168*, 105743. [CrossRef] [PubMed]

94. Henchion, M.; McCarthy, S.N.; McCarthy, M. A time of transition: Changes in Irish food behaviour and potential implications due to the COVID-19 pandemic. *Ir. J. Agric. Food Res.* **2021**, *60*, 1–12. [CrossRef]

95. Beckman, J.; Baquedano, F.; Countryman, A. The impacts of COVID-19 on GDP, food prices, and food security. *Q. Open* **2021**, *1*, qoab005. [CrossRef]

96. Tsalis, G.; Jensen, B.B.; Wakeman, S.W.; Aschemann-Witzel, J. Promoting food for the trash bin? A review of the literature on retail price promotions and household-level food waste. *Sustainability* **2021**, *13*, 4018. [CrossRef]

97. Combes, J.-L.; Meyimdjui, C. *Food Price Shocks and Household Consumption in Developing Countries: The Role of Fiscal Policy*; IMF Working Papers; International Monetary Fund: Washington, DC, USA, 2021; Volume 2021. [CrossRef]

98. Ma, X.; Liese, A.D.; Hibbert, J.; Bell, B.A.; Wilcox, S.; Sharpe, P.A. The Association between Food Security and Store-Specific and Overall Food Shopping Behaviors. *J. Acad. Nutr. Diet.* **2017**, *117*, 1931–1940. [CrossRef]

99. Shannon, N.Z.; Odoms-Young, A.; Dallas, C.; Hardy, E.; Watkins, A.; Hoskins-Wroten, J.; Hollandc, L. "You Have to Hunt for the Fruits, the Vegetables": Environmental Barriers and Adaptive Strategies to Acquire Food in a Low-Income African American Neighborhood. *Health Educ. Behavior.* **2011**, *38*, 282–292. [CrossRef]

100. Hirsch, J.A.; Hillier, A. Exploring the role of the food environment on food shopping patterns in philadelphia, PA, USA: A semiquantitative comparison of two matched neighborhood groups. *Int. J. Environ. Res. Public Health* **2013**, *10*, 295–313. [CrossRef]

101. Webber, C.B.; Sobal, J.; Dollahite, J.S. Shopping for fruits and vegetables. Food and retail qualities of importance to low-income households at the grocery store. *Appetite* **2010**, *54*, 297–303. [CrossRef]

102. Marinković, V.; Lazarević, J. Eating habits and consumer food shopping behaviour during COVID-19 virus pandemic: Insights from Serbia. *Br. Food J.* **2021**, *123*, 3970–3987. [CrossRef]

103. Aristovnik, A.; Keržič, D.; Ravšelj, D.; Tomaževič, N.; Umek, L. Impacts of the COVID-19 pandemic on life of higher education students: A global perspective. *Sustainability* **2020**, *12*, 8438. [CrossRef]

104. Moynihan, A.B.; van Tilburg, W.A.P.; Igou, E.R.; Wisman, A.; Donnelly, A.E.; Mulcaire, J.B. Eaten up by boredom: Consuming food to escape awareness of the bored self. *Front. Psychol.* **2015**, *6*, 369. [CrossRef] [PubMed]
105. Eftimov, T.; Popovski, G.; Petković, M.; Seljak, B.K.; Kocev, D. COVID-19 pandemic changes the food consumption patterns. *Trends Food Sci. Technol.* **2020**, *104*, 268–272. [CrossRef] [PubMed]
106. Aday, S.; Aday, M.S. Impact of COVID-19 on the food supply chain. *Food Qual. Saf.* **2020**, *4*, 167–180. [CrossRef]
107. Alaimo, L.S.; Fiore, M.; Galati, A. Measuring consumers' level of satisfaction for online food shopping during COVID-19 in Italy using POSETs. *Socio-Econ. Plan. Sci.* **2021**, *82*, 101064. [CrossRef]
108. Skalkos, D.; Roumeliotis, N.; Kosma, I.S.; Yiakoumettis, C.; Karantonis, H.C. The Impact of COVID-19 on Consumers' Motives in Purchasing and Consuming Quality Greek Wine. *Sustainability* **2022**, *14*, 7769. [CrossRef]
109. Garnett, P.; Doherty, B.; Heron, T. Vulnerability of the United Kingdom's food supply chains exposed by COVID-19. *Nat. Food* **2020**, *1*, 315–318. [CrossRef]
110. Rizou, M.; Galanakis, I.M.; Aldawoud, T.M.S.; Galanakis, C.M. Safety of foods, food supply chain and environment within the COVID-19 pandemic. *Trends Food Sci. Technol.* **2020**, *102*, 293–299. [CrossRef]
111. Skalkos, D.; Kosma, I.S.; Chasioti, E.; Skendi, A.; Papageorgiou, M.; Guiné, R.P.F. Consumers' Attitude and Perception toward Traditional Foods of Northwest Greece during the COVID-19 Pandemic. *Appl. Sci.* **2021**, *11*, 4080. [CrossRef]
112. Skalkos, D.; Kosma, I.S.; Vasiliou, A.; Guine, R.P.F. Consumers' trust in greek traditional foods in the post COVID-19 era. *Sustainability* **2021**, *13*, 9975. [CrossRef]
113. Tran, L.T.T. Managing the effectiveness of e-commerce platforms in a pandemic. *J. Retail. Consum. Serv.* **2021**, *58*, 102287. [CrossRef]

MDPI

St. Alban-Anlage 66

4052 Basel

Switzerland

www.mdpi.com

*Sustainability* Editorial Office

E-mail: sustainability@mdpi.com

www.mdpi.com/journal/sustainability

www.ingramcontent.com/pod-product-compliance
Lightning Source LLC
LaVergne TN
LVHW071307170726
843515LV00006BA/1507